Bioethics Yearbook

VOLUME 2

REGIONAL DEVELOPMENTS IN BIOETHICS: 1989-1991

Bioethics Yearbook

VOLUME 2

The titles published in this series are listed at the end of this volume.

Bioethics Yearbook

VOLUME 2
REGIONAL DEVELOPMENTS IN BIOETHICS: 1989-1991

THE CENTER FOR ETHICS, MEDICINE AND PUBLIC ISSUES
Baylor College of Medicine
The Institute of Religion
Rice University
Houston, Texas, U.S.A.

Edited by

B. Andrew Lustig, *Senior Editor*
Baruch A. Brody, *Director*
H. Tristram Engelhardt Jr.
Laurence B. McCullough

KLUWER ACADEMIC PUBLISHERS
DORDRECHT / BOSTON / LONDON

ISBN 0-7923-1893-5

ISSN 0926-261X

Published by Kluwer Academic Publishers,
P.O. Box 17, 3300 AA Dordrecht, The Netherlands.

Kluwer Academic Publishers incorporates
the publishing programmes of
D. Reidel, Martinus Nijhoff, Dr W. Junk and MTP Press.

Sold and distributed in the U.S.A. and Canada
by Kluwer Academic Publishers,
101 Philip Drive, Norwell, MA 02061, U.S.A.

In all other countries, sold and distributed
by Kluwer Academic Publishers Group,
P.O. Box 322, 3300 AH Dordrecht, The Netherlands.

Printed on acid-free paper

Printed in the Netherlands

TABLE OF CONTENTS

INTRODUCTION

The Center for Ethics, Medicine, and Public Issues, in conjunction with Kluwer Academic Publishers, is pleased to offer Volume Two in our Bioethics Yearbook series. As we noted in Volume One, the Yearbook series alternates between a biennial volume tracing recent theological discussion on topics in bioethics and a biennial volume tracing recent regional discussions in bioethics. Volume Two offers our initial survey of regional developments from 1989-1991, and provides for the first time a comprehensive single-volume summary of recent international and regional developments on specific topics in bioethics.

In order to establish a uniformity to the discussions within this volume, as well as among volumes in the series, we asked our authors to order their remarks according to the following list of topics: new reproductive technologies, abortion, maternal-fetal conflicts, care of severely disabled newborns, consent to treatment and experimentation, confidentiality, equitable access to health care, ethical concerns raised by cost-containment measures, decisions to withhold or withdraw life-sustaining treatment, active euthanasia, the definition of death, organ donation and transplantation, and, as a final broad category, "other issues." Authors cover only those topics on the list which have been recently discussed in a particular nation or region.

Volume Two brings together internationally respected commentators to report on recent developments in bioethics in the following sixteen nations, regions, or regional organizations: the United States; Canada; Latin America; the Council of Europe and the European Community; the United Kingdom and Ireland; France; the Netherlands; Germany, Austria, and Switzerland; Eastern Europe; Spain, Portugal, and Italy; Scandinavia; India; Southeast Asia; China; Japan; and Australia and New Zealand. We hope to expand our coverage to other regions in future volumes.

In providing expert summary and analysis, our authors draw upon a number of resources, both formal and informal, in their surveys of recent regional developments. First, as appropriate, they discuss statutes, legislative proposals, and regulatory changes that directly influence, or have implications for, areas of bioethical concern. Recent statutes and regulations have been proposed or passed on virtually the entire list of topics our authors have been asked to address, though obviously not for every nation or region during the past two years. However, because Volume Two provides our first regional coverage, we left it to the scholarly discretion of our authors to include whatever past legislative and regulatory information they deem necessary to place more recent developments in historical context.

In considering the broad range of legislation and regulations that our authors survey, certain characteristic tendencies and approaches suggest

B. Andrew Lustig (Sr. Ed.), Bioethics Yearbook: Volume 2, vii-xi.

themselves as the bases for comparison and contrast between and among nations and regions. For example, Jose Mainetti and his colleagues distinguish Latin American bioethics from its northern counterpart by its characteristic emphasis upon beneficence rather than autonomy as a first principle. As a result, relations between physician and patient remain strongly paternalistic in their orientation, though Mainetti interprets that paternalism more kindly than would his colleagues in the United States. Reidar Lie and Jens Erik Paulsen, in discussing developments in Scandinavia, point out the marked differences in the legal responses of Sweden and Denmark to the AIDS epidemic. Sweden responds to HIV infection as a straightforward case of venereal disease to be covered under the provisions of its Infectious Disease Act, which call for registration of affected individuals, contact tracing, and possible quarantine. All of these measures characterize a Swedish response according to a stringent public health model. By contrast, Denmark emphasizes the confidentiality of affected individuals, and does not categorize HIV/AIDS as a "generally dangerous" disease. In this instance, one is struck by the differences between countries within a general geographical region and left to ponder the reasons that might help to explain those different perspectives and approaches. Lisa Parker and her colleagues briefly discuss recent efforts by several states in the United States, most notably Utah and Pennsylvania, to impose restrictions on abortion. Interestingly, John Williams, in his essay on Canadian developments, comments on Nova Scotia's recent efforts to set certain modest restrictions on abortion at the provincial level. These initiatives in both countries have been taken at the state or provincial rather than federal level, and again prompt the interested reader to consider the reasons for such similar tactical choices.

In either sort of case, of apparent similarities or seeming differences, the interested reader may wish to draw larger comparisons or distinctions national or regional temperament, culture, politics, or history. At the same time, however, these examples, and countless others in Volume Two, prompt an immediate word of caution. A yearbook, if successful, achieves the important but limited purpose its name implies. This volume is offered primarily as a summary and general analysis of recent developments on particular topics. For scholars and others interested in seeing how more fundamental themes and attitudes may play themselves out in national and regional debates on particular issues, the Yearbook may serve to exemplify those tendencies in concrete fashion. But it should not be misconceived as an effort to analyze basic themes or trends in political science, legal philosophy, or the sociology of institutions.

Second, our authors, as appropriate, discuss relevant case law and court judgments that shape, either decisively or suggestively, recent legal interpretations of particular issues or areas in bioethics. For example, Parker and her colleagues provide concise summaries of, and incisive commentary on, notable U.S. court cases in a number of areas, including abortion, maternal-fetal conflict, and withholding/withdrawing treatment. Williams carefully draws out the implications of several recent court judgments in Canada on the same issues. David Greaves and his colleagues analyze important cases in the United Kingdom and Ireland involving maternal-fetal conflicts, care of severely disabled newborns, consent to treatment and experimentation, and other issues. And readers should note the distinctive legal logic at work in the assessment by a Chinese court of the harm posed by a case of active euthanasia. While none of our authors claims to provide exhaustive coverage of case law and court judgments, careful attention to the nuances of particular arguments may again suggest characteristic similarities and differences in jurisprudential approaches and legal philosophy perhaps reflective of larger matters of diverse cultures and histories.

Third, our authors, as appropriate, discuss the formal statements of governmentally appointed commissions, advisory bodies, and representative professional groups, as well as less formal statements recommendations, of other organizations. The usefulness of these various documents and statements to the analyses of our authors cannot be overstated. Consider only the following representative examples of these rich resources. Parker and her colleagues report on the recommendations of the Centers of Disease Control and the American College of Obstetricians and Gynecologists on prenatal HIV counseling and testing programs, on the 1990 report of the Office of Technology Assessment on "Medical Monitoring and Screening in the Workplace," and on the Code of Ethics accepted by the National Society of Genetic Counselors in 1991. Williams analyzes a number of recent policy announcements on the new reproductive technologies presented to the Royal Commission by the Canadian Bar Association and the Canadian Medical Association. Anne Fagot-Largeault discusses two recent major reports of French commissions appointed to make recommendations on the full menu of bioethical issues. Maurice de Wachter and his co-authors review major policy recommendations by the Netherlands National Council for Public Health on topics from cost-containment to treatment of patients in a persistent vegetative state. Kazumasa Hoshino reports on the recent activities of the Japanese Commission for the Study of Brain Death and Organ Transplantation and its recommendations to the prime minister. Francesc Abel discusses the documents of the recently appointed *Comitato Nazionale per la Bioetica* in Italy on the new reproductive technologies and issues of consent to experimentation. Jos Welie and Henk Ten Have detail the numerous statements and reports of the Council of Europe. Ishmar Verma discusses the consensus statements drawn up at a number of Indian workshops on several topics, including the definition of death and the legitimacy of physician-assisted suicide. Ren Zong Qiu details the recent Code of Professional Ethics drawn up by the Chinese Ministry of Health.

As these examples suggest, the activities of many governmental groups and professional advisory bodies, although varied, tend to converge upon a number of especially important issues. If one peruses the index of documents discussed in Volume Two, certain topics are more often the focus of legislation and official concern than others: withholding and withdrawing treatment, access to health care, consent to treatment and experimentation, and concerns about AIDS and HIV testing. Such commonality should not be exaggerated, for the discussion of topics is wide-ranging. But that commonality, when in evidence, is also not surprising. It suggests that key issues and concerns in bioethics may be fairly common to modern culture and society, for all the distinctiveness of a particular national or regional response to them. Issues of informed consent, after all, implicate more fundamental matters of respect for persons and the rights of individuals in the contexts of therapy and research. Issues of access to medical care concretize deeper questions about the nature and scope of a society's welfare obligations to its citizens. And choices of appropriate policies on HIV testing involve fundamental social judgments about how to balance the privacy of individuals and the commonweal, while also posing important questions about the nature and extent of the obligations of medical professionals.

On the one hand, it is important to acknowledge the practical force of these common concerns, if only as a preliminary to broader musings on the adequacy of ethical or cultural relativism as a theme in recent academic discussion. On the other hand, one may be equally struck by the differences that persist between nations and regions, differences reflective of characteristic attitudes or ethos, despite the presence of modernity. For example, still prevalent in Japan and dominant in China are approaches to patient confidentiality and informed consent which are clearly at odds with approaches familiar to Western readers. Such differences, in turn, may reflect characteristically different perspectives on the normative priorities at work in clinical medicine, and perhaps on the fundamental values that shape the relationship of the individual to society.

Finally, in their essay on "Bioethics in a Supranational Context," Welie and Ten Have report, in fascinating detail, on the resolutions and recommendations of the Council of Europe and on the research and technical development activities of the European Community. Especially interesting is their conclusion about the problems of coordination and authority that frustrate the implementation of policies recommended by both bodies, especially those of the Council of Europe. The voice of the European Community (EC) may be marginally more significant, because the EC exerts greater authority on its member states. Still, in the judgment of Welie and ten Have, the EC's conclusions on bioethics also lack decisive influence on national deliberations at the present time. The Welie – Ten Have discussion exemplifies a fascinating aspect of regional similarities and differences, *viz.*, the wide variations in the professional authority and legal bindingness of particular reports and recommendations, and the reasons for that variety. Many of the essays in Volume Two, confirm the not always obvious truth that politics remains a staple of international dialogue in bioethics as in every other sphere of policy and scholarship.

In addition to providing timely summaries of recent developments, our authors also offer rich and useful bibliographical references to a wide array of documents, many of which would be difficult for readers to learn about, given the lack of centralized international collection of such documents. In addition, our authors have often included translations of relevant passages on particular topics from documents and reports. Although these translations are certainly trustworthy, they should not be considered definitive for legal purposes. As always, we invite our readers to contact us with their own advice about ways to improve upon and to expand the range of regional resources upon which our authors may draw in future volumes.

We trust that the scholarly excellence of our contributors will be evident to our readers. Those familiar with the editing process appreciate that the state of an editor's mental health is directly proportional to the cooperation of authors in meeting deadlines. In light of the number of authors we have assembled for Volume Two and of the literally global sweep of their coverage, it is a pleasure to acknowledge their general cooperation in helping the staff at the Center to complete this project within a reasonable period of time. Finally, we owe special thanks to Delores D. Smith for her tireless commitment to the daily chores involved in preparing this volume for publication.

B. Andrew Lustig, Senior Editor
Baruch A. Brody, Editor, Director of the Center for Ethics
H. Tristram Engelhardt, Jr., Editor
Laurence B. McCullough, Editor

LISA S. PARKER, ATHENA BELDECOS, LISSA WETTICK, MICHAEL MANOLAKIS, AND ROBERT ARNOLD

BIOETHICS IN THE UNITED STATES: 1989-1991

I. INTRODUCTION

This report provides a summary and analysis of recent developments in bioethics in the United States on numerous topics. Its discussion has been organized according to eight broad areas. Section II reports on recent cases implicating several issues: the right to refuse medical treatment, the meaning of medical futility, and the legitimacy of physician-assisted suicide. Section III discusses recent cases involving abortion, including *Webster* and *Rust V. Sullivan*, as well as recent public policy decisions to ban the importation of the abortifacient pill RU-486. Section IV analyzes issues posed by the new reproductive technologies, including recent cases involving surrogate motherhood and the disposition of frozen embryos. Section V summarizes recent cases of maternal-fetal conflicts that raise questions about the legitimacy of court-ordered medical treatment, the appropriateness of prosecution for fetal abuse, and the status of fetal protection policies in the workplace. Section VI analyzes several ethical and legal issues raised by AIDS, including the propriety of notifying third parties, appropriate policies on HIV testing, and the challenges to health insurance posed by the AIDS epidemic. Section VII discusses topics in genetics, including issues of property rights posed by genetic engineering, and questions of appropriate guidelines in the management of genetic information. Section VIII traces recent proposals for health care financing reform, including the Oregon Basic Health Services Act, the "Health Access America" plan of the American Medical Association, the recommendations of the Pepper Commission, and the proposal offered by Physicians for a National Health Program. Finally, Section IX discusses recent discussion concerning the regulation of professional conduct, especially regarding relations between physicians and the pharmaceutical industry, and problems of sexual misconduct between physicians and their patients.

II. THE RIGHT TO REFUSE MEDICAL TREATMENT

The earliest history of bioethics in the United States is the history of patients asserting their right to refuse treatment, especially life-sustaining treatment. It is largely a history of adversarial relationships among family members or

B. Andrew Lustig (Sr. Ed.), Bioethics Yearbook: Volume 2, 1-50.

between patients and health care workers or hospital administrators. Thus, courtrooms have been the setting for much of the history of patients' refusal of treatment, and the past two years have added two landmark cases to this history.

A. *The Right of Competent Patients to Refuse Treatment*

Court decisions in recent years show an increasing trend toward allowing the rights of competent patients to refuse medical treatment to outweigh "countervailing state interests" ([119], Secs. 4.12, 4.13). These state interests include the preservation of life, the prevention of suicide, the protection of third parties and, protecting the ethical integrity of the medical profession. ([119], Secs. 4.12, 4.13, 4.15, 4.16). The Georgia Supreme Court held that a competent patient had the right to refuse medical treatment and to be free from pain in exercising this right. In *State v. Mcaffe*, the court maintained that a quadriplegic incapable of spontaneous respiration had the right to turn off his ventilator and have sedatives administered to alleviate the pain [165]. The court acknowledged that the state's interest in preservation of life did not take precedence over the individual's right to refuse medical treatment. In a similar Nevada case, *McKay v. Bergstedt*, the Nevada Supreme Court also held that a competent, quadriplegic patient had the right to disconnect his ventilator [115]. The Nevada court held that the state's interest in the preservation of the non-terminally ill patient's life is greater than in the case of a terminally ill patient, but acknowledged that the interest decreases in light of a patient's diminished quality of life ([119], Sec. 4.12). The court additionally stipulated that the patient be fully informed about the various "care options available" [115].

Although earlier decisions had required parents of minor children to be treated against their will; this justification has not overridden competent patients' choices. *Wons v. Public Health Trust* [184] and *Forsmire v. Nicoleau* [64] both involve competent mothers refusing medical treatment. In the first case, the Florida Supreme Court upheld the right of a competent woman to refuse life-saving blood transfusions despite having two children. The court found that her constitutional rights outweigh the state's interest in protecting her children ([119], Sec. 4.15). Noting that the rights of a competent patient to refuse medical treatment without regarding his status as a parent, the New York State Court of Appeals (which is that state's highest court) in *Forsmire* overturned the New York Supreme Court order authorizing a mother's blood transfusions, following a cesarean section ([119], Sec. 4.15).

B. *The Right To Die and Surrogate Decision-Making*

In the past two years, the right of incompetent patients to refuse life-sustaining treatment has been addressed in judicial decisions and both Congressional and state legislation. Prior to 1989, there had been no federal appellate court cases dealing with the right to die. The United States Supreme Court issued its first ruling on the right to die in its landmark, if somewhat anti-climactic, June 1990 decision in *Cruzan v. Director, Missouri Department of Health* [49].

When the parents of Nancy Cruzan, who had been in a persistent vegetative state for seven years, requested that artificial feeding and hydration be withdrawn, the hospital's administrator demanded that her legal guardians obtain a court order authorizing the discontinuance of tube feeding. Although the state trial court issued the order, the Missouri State Supreme Court reversed that decision. The court held that in the absence of proof by clear and convincing evidence that the patient had authorized such termination of treatment prior to losing decisionmaking capacity, artificial nutrition and hydration could not be withdrawn.

The U.S. Supreme Court decision in *Cruzan* was somewhat anticlimactic, because despite much speculation in the popular press about the impact of the decision, few suspected that the Court would find Missouri's "clear and convincing evidence" rule unconstitutional. Most anticlimactic of all, however, is the actual scope and content of the Court's decision. The *Cruzan* decision only affirms the Missouri Supreme Court decision; states are constitutionally entitled to require a substantial proof that substituted judgment of surrogate decision makers reflects the views of the patient ([119], Sec. 3.2;[162]). *Cruzan* does not, however, require other states to adopt Missouri's "clear and convincing evidence" rule. Indeed, only New York has done so. Interestingly, the Supreme Court grounded its *Cruzan* decision in the fourteenth amendment's constitutional guarantee of liberty, not in the right of privacy, which formed the basis for many state court decisions on the right to refuse treatment ([119], Sec. 3.2). Another area addressed by the Supreme Court decision is the status of artificial hydration and nutrition. The decision, though still somewhat unclear, implies that the majority of the Court believes that forgoing artificial nutrition and hydration is analogous to refusing other types of medical treatment [116].

Public and professional awareness of the *Cruzan* decision has prompted doctors and their patients to address more thoroughly advance directives and surrogate decisionmaking. To date, however, only three to 7 percent of patients have actually prepared advance directives ([107];[110];[162]). To encourage patients to follow through with the preparation of advance directives, Congress passed the Patient Self-Determination Act, which became effective in December 1991. It requires hospitals, nursing homes, hospices, and Medicare/Medicaid health care providers to provide to patients upon admission written information stating their rights under state law to accept or refuse treatment and to prepare advance directives ([119], Sec 10.14;[107]). Health care providers are additionally required to document in the patient's record whether or not the person has executed advance directives, to educate staff regarding these issues, and to maintain and implement advance directives policies concerning all adult patients ([107];[119], Sec. 10.14;[132]). Health care providers are forbidden from discriminating against an individual based on whether or not he has executed an advance directive. States are required to develop a written description of the law of the state concerning advance directives, and health care providers are charged with ensuring compliance with state law. The U.S. Department of Health and Human Services is required to develop a national educational campaign concerning advance directives. (The Act is not to be construed as requiring the implementation of an advance directive by a health care provider if state law provides for conscientious objection and if the provider so objects).

An increasing number of states have enacted statutes extending their durable power of attorney statutes to enable people to designate a surrogate decision maker who has the legal authority to make medical treatment decisions in the event that they lose decisionmaking capacity. The living will and the health care durable power of attorney are the primary written forms of advance directives. In a living will, the patient states which medical interventions he would accept and which he would reject if they were to be offered to him in the future and he were unable to express his preference at the time. People might, however, wish to accept some interventions in some circumstances and reject the same interventions in others. Moreover, it is difficult to foresee the very large numbers of interventions which one might be offered. Therefore, the health care durable power of attorney is often the favored advance directive instrument. The durable power of attorney permits one to designate a surrogate decision-maker to make decisions in light of changing circumstances and to determine medical interventions on one's behalf should one become incompetent.

Florida is one of the states which has enacted a durable power of attorney statute, and *In re Guardianship of Browning*, the Supreme Court of Florida held that the incompetent patient's guardian is authorized to "exercise the patient's right to forego artificially provided sustenance, where the patient is not in a permanent vegetative state, and is suffering from an incurable but not terminal condition" [90]. The court reasoned that the patient had expressed her wishes regarding life-sustaining treatment, through both written and oral statements, when she was competent, and that her guardian was authorized to enact these desires.

Massachusetts and, more recently, New York passed what are regarded as model proxy laws. Both laws allow a competent adult to select another person to make medical decisions in the event that he becomes incompetent. Adults may select whomever they desire and are not limited to selecting family members ([63];[21]). Other states have similar statutes except that the appointed health care decision maker is not specifically permitted to make decisions about withholding or withdrawing life support (whether or not they will be able to do so has yet to be adjudicated). In addition, some states have enacted surrogate decisionmaking statutes which enable the appointment of a surrogate decision maker in the event that the patient has failed to do so prior to becoming incapacitated ([119], Sec. 8.17). In general, however, the legislation stresses that in order for either advance directives to be truly effective, these issues must be addressed prior to a patient becoming incompetent.

C. *Religion & Medical Decision-Making*

People adhering to certain religious tenets often refuse potentially life-saving treatment for themselves and their family members. Although this right is usually upheld for competent adults, when this medical decision-making involves a minor, the courts must weigh the parents' freedom of religious expression against their child's best interest. Recent court rulings additionally considered the severity of the disease, the child's prognosis for survival, and the child's age or maturity in reaching their decisions.

In two cases involving parental objections by Jehovah's Witnesses to blood transfusions for their respective children, each court ruled against them. The Superior Court of Pennsylvania held *In re Cabrera* that (1) the state's interest in protecting the child overrode parents' religious objections, and (2) a one year limitation on therapy was permissible [88]. The child had sickle cell anemia and suffered a stroke due to this medical condition. There was a 70 percent chance that the child would experience another stroke without the transfusion. Though the child was not in imminent danger of losing her life, recurring strokes could very well lead to physical or mental impairment or both. Given that blood transfusions were the best possible treatment for her condition and that implementing the transfusions vastly improved her prognosis, the court held that state intervention is warranted even when the health risk is not immediately life-threatening. The court additionally ruled that the case be monitored on a yearly basis to ensure that any new alternative therapies for sickle cell anemia, which may be as effective and less offensive to the parents, be considered [88].

In a similar case in Massachusetts, *In re McCauley*, the Supreme Court of Massachusetts noted that there are three interests at stake: (1) the natural rights of parents; (2) the interests of the child; and (3) the interests of the state [91]. The case involved an eight year-old child diagnosed with leukemia whose parents are Jehovah's Witnesses and as such objected to the necessary blood transfusion for their child's treatment. There was a high risk of the child's death without the necessary transfusions and chemotherapy and a high probability of remission with the proper treatment. A Superior Court judge issued a temporary order authorizing the blood transfusion and this decision was upheld by the Massachusetts Supreme Court. The Supreme Court held that though the parent-child relationship must be protected from unwarranted state interference, the "State acting as parens patriae, may protect the well-being of children" [91]. This state's interest in protection of the child, as well as the child's best interest, are cited as being more important than the parent's religious rights to refuse the medical treatment.

The children in the prior two cases had reasonable prognoses if each received the court ordered treatment. Conversely in the case *Newmark v. Teresa Williams/D.C.P.* the child suffered from a lethal, advanced form of pediatric cancer, and even with aggressive treatment, his prognosis was exceedingly poor [127]. The Delaware Division of Child Protective Services petitioned Family Court for temporary custody of the child because his parents, who are Christian Scientists, rejected the medical treatment proposed for their son in favor of "spiritual aid and prayer" [127]. The Family Court granted the petition, but the Delaware Supreme Court reversed the order and returned the child to his parents. The Supreme Court determined that the State did not provide clear and convincing evidence of abuse or neglect and noted spiritual treatment exemptions within the Delaware Code. The exemptions specifically state that "no child who in good faith is under treatment solely by spiritual means through prayer in accordance with the tenets and practices of a recognized church or religious denomination by a duly accredited practitioner thereof shall for that reason alone be considered a neglected child for purposes of this chapter" [127]. Moreover, in determining the child's best interest the Supreme Court noted that the course of chemotherapy was highly invasive with many potentially permanent side-effects and that if he survived this course of treatment, he had at best a 40 percent chance of survival. Given these exacerbating circumstances, the Delaware Supreme Court was loathe to permit state intervention.

The Illinois Supreme Court, in the case *In re E.G.*, permitted a 17-year-old Jehovah's Witness, dying of leukemia, to refuse the recommended blood transfusions finding that she possessed the "requisite degree of maturity" [89]. Initially, when both the child and her mother refused to accept the transfusion, the State filed a neglect petition in Juvenile Court. Doctors testified that without the transfusions she would probably die in a month, but that with the recommended treatment course of both transfusions and chemotherapy, there was an 80 percent chance of remission. The trial court found that E.G. was medically neglected and appointed a guardian to allow the transfusions. On appeal, the court held that a mature minor, like an adult, is permitted to refuse medical treatment. This right for adults is established by Illinois common law and statutes ([119], Preface). The State Supreme Court upheld this decision stating that "if the evidence is clear and convincing that the minor is mature enough to appreciate the consequences of her actions, and that the minor is mature enough to exercise the judgment of an adult, than the mature minor doctrine affords her the common law right to consent to or refuse medical treatment" [89]. The Supreme Court additionally noted the importance of satisfying "third party interests," and noted that a parent's or guardian's opposition "would weigh heavily against the minor's right to refuse" [89].

D. *Futility and Physician-Assisted Suicide*

In the past two years, controversial cases have caused the public to re-evaluate the ethical issues surrounding the death of terminally ill patients and the role of the physician in the death of a patient. In the case of Helga Wanglie, physicians assumed the role of advocating the cessation of what they regarded as her futile treatment. In other cases, physicians have, with the patients' permission, assisted in their suicide. Dr. Kevorkian's use of his "suicide machine" for a woman suffering from Alzheimer's disease and Dr. Quill's admission of providing a terminally ill patient with the necessary information to end her life are two of a handful of widely publicized cases of physician-assisted suicide. The image of physicians as "angels of mercy", and growing patient demands for control over their own deaths, spurred the movement in Washington state to propose Initiative 119, which included the legalization of physician-assisted suicide.

The Wanglie case markedly contrasts with the more common right-to-die cases in which the family advocates the cessation of treatment and is opposed by the physicians. When family members wish to continue treatment, if the physicians disagree, they generally either reach an agreement or simply acquiesce to the family's wishes, perhaps to avoid the unsavory publicity of advocating a patient's death [24]. The Wanglie case involved an 87-year old woman who had been in a persistent vegetative state for nearly eight months when, against her families wishes, the Hennepin County Medical Center in Minneapolis sought a court order to appoint an independent guardian who could authorize her removal from the ventilator [92]. The physicians claimed that the respirator did not benefit the patient and served no medical purpose. The family asked that her husband be appointed as guardian, and claimed that Mrs. Wanglie, who had never prepared a living will, was devoutly religious and firmly believed that only God is allowed to end life ([83];[92]).

This case raised larger issues, including how to define when medical care becomes futile, and how to weigh the right of the family to determine the course of treatment for an incompetent patient against the right of doctors to not pursue what they consider to be unwarranted or unethical treatment [163]. This case was also unique because the patient's insurance covered her medical expenses; in the absence of this particular economic pressure, the case was regarded as a "pure ethics case" [24]. The judge appointed her husband as her guardian stating that "he is in the best position to investigate and act upon Helga Wanglie's conscientious, religious, and moral beliefs" [101]. This decision affirmed the family's right to make medical decisions for an incapacitated patient. The increasing number of "futility" cases which have appeared in the lay and professional literature make it likely, however, that this issue will again come before the courts.

The debate over euthanasia escalated as two doctors admitted to helping their respective patients commit suicide [18]. In the first case, Dr. Jack

Kevorkian connected Janice Adkins, who was suffering from Alzheimer's disease to his "suicide machine", and she pushed the button to release the drugs which ended her life. He chose to do this in the state of Michigan because it has no statute specifically forbidding assisting a suicide. The heated controversy over this case can be partially attributed to Dr. Kevorkian's somewhat nebulous role in Janet Adkins' death. Although he did not actually push the button himself, he did invent and advertise the use of his machine. Moreover, there was some question concerning his degree of care in ascertaining that Mrs. Adkins made a competent decision to end her life. Dr. Kevorkian was acquitted of first degree murder, but had to stand trial, and in a civil case had to defend his right to use his "suicide machine". Although the judge issued a permanent injunction forbidding his use of the device, Kevorkian assisted in two additional suicides.

Soon after the Kevorkian controversy, Dr. Timothy Quill published an article in the *New England Journal of Medicine* admitting to aiding his patient's suicide in New York state. His patient was dying of cancer and requested information from Quill about ending her life. He referred her to the Hemlock Society, prescribed barbiturates for her, and gave her information about a lethal dose [53]. Though assisted suicide is illegal in New York, the Board for Professional Medical Conduct did not bring charges against him citing "that the longstanding relationship between patient and doctor in Dr. Quill's case and the fact that he himself did not directly participate in any taking of life set this case apart" [170]. The trend toward being more open about topics relating to euthanasia is indicated by the fact that the suicide manual *Final Exit* climbed to the top of the New York Times best sellers list [86].

Even more conclusive evidence of this trend is provided by the introduction of Initiative 119 in Washington state [2]. The support stems from public interest in maintaining control over treatment decisions and not being kept alive by sophisticated medical technology [86]. This proposal extends the autonomy in treating decisionmaking to include legalization of physician-assisted suicide for competent patients. In order to qualify, however, the patient must suffer from an "incurable or irreversible condition which, in the written opinion of two physicians having examined the patient and exercising reasonable medical judgment, will result in death within six months" [2]. Though Initiative 119 had a surprisingly large amount of public support, it was turned down by the Washington voters in November, 1991. The debate over the ethics of euthanasia is likely to re-emerge as advocates are already planning similar initiatives in both Oregon and California.

III. ABORTION

Over the last several years, there has been growing intensity in the moral and political debate over abortion in the United States. The Supreme Court's decision in *Webster v. Reproductive Health Services* on July 3, 1989 marked a new era in the abortion debate as it appears more possible that the Court might overrule *Roe v. Wade*, the 1973 landmark case which afforded women the right to make decisions about abortion free of governmental interference ([181];[155]).

A. *Roe v. Wade*

In the 1973 *Roe* decision, the Supreme Court decided in a 7 to 2 vote that the constitutional "right to privacy...is broad enough to encompass a woman's decision whether or not to terminate her pregnancy", and therefore, the state could not intervene unless it could demonstrate a "compelling state interest" [155]. According to *Roe*, a state could intervene to protect maternal health in cases when abortion would be more dangerous than childbirth and to protect fetal life after viability. *Roe* found abortion to be increasingly objectionable as a fetus developed, with a landmark at about 24 weeks when a fetus is deemed capable of surviving outside the womb [12]. *Roe* divided pregnancy into trimesters. In the first, women were guaranteed the right to abortion, protected by the constitutional right to privacy. After the first trimester, the state was allowed to regulate abortion only on behalf of maternal protection. In the final trimester, when the fetus becomes viable, the state could regulate abortion to protect fetal life and even proscribe abortion so long as it did not threaten maternal health or life [12].

B. *Webster v. Reproductive Health Services*

Roe appears especially vulnerable to review in a Supreme court which has become increasingly receptive to anti-abortion arguments with its new majority of more conservative justices. In a brief filed by the Justice Department, the Court was explicitly requested to use *Webster* to overturn the 1973 *Roe* decision [33]. Though *Webster* did not overturn *Roe* outright, it did indicate that the Court may allow states far greater latitude to place restrictions on a woman's right to have an abortion. It also significantly weakened *Roe*'s trimester approach to abortion regulation and signaled the Court's future willingness to overrule or attenuate the protections afforded by the right of privacy in *Roe*.

The 5-4 *Webster* decision upheld a Missouri statute that prohibited the performance of abortions both by public employees and in state-financed institutions. This decision accords with previous court decisions allowing states and the federal government to restrict Medicaid funding for abortion [22]. The Court also upheld a provision requiring physicians to determine fetal viability prior to performing an abortion of a fetus believed to be at least 20 weeks old. Most significantly, the Court let stand the preamble to the statute which stated that "the life of each human being begins at conception" and that "unborn children have protectable interests in life, health and well-being" ([22], p. 858;[181]). *Webster* upheld that unborn children have "all the rights, privileges, and immunities available to other persons, citizens, and residents of this state" ([29], p. 404; [181]). The Court thus supported an opinion which could extend state interest to the whole period of pregnancy.

The Missouri statute defines conception as "the fertilization of the ovum of a female by the sperm of a male", even though implantation does not occur until at least six days after fertilization [122]. It therefore implies regulation not only of abortions in the first two trimesters, but also of forms of contraception such as the IUD (intra-uterine device) and the morning-after pill ([14], p. 181). Furthermore, the logic of the Missouri statute precludes research on human embryos. (As of 1989, 11 states had bans on human embryo research [13]).

In the *Webster* decision, the five majority justices, writing three separate opinions, indicated that they believe the trimester approach of *Roe* is no longer tenable. Chief Justice Rehnquist, writing for the plurality, argued that the concepts of viability and trimesters should be abandoned because they cannot be found in the text of the Constitution ([178];[17], p. 28; [181]). He concluded that the state's interest in protecting human life exists throughout pregnancy, not from viability onward [181]. The Court left the constitutionality of the preamble unaddressed concluding that it imposed no substantive restrictions on abortion and merely expressed a value judgment favoring childbirth over abortion ([106], p. 160).

In a minority opinion, Justice Blackmun defended *Roe*'s viability rule and argued that "it reflects the biological facts and truths of fetal development" and that the state's interest in potential human life becomes compelling only when the fetus loses its dependence on the uterine environment ([20], p. 28; [181]). As one *Amicus* brief in *Webster* argued, this benchmark for viability determination is not arbitrary, but is based on a real anatomical threshold ([138];[181]). Before 23-24 weeks of gestation, fetal lungs are not sufficiently developed to permit normal or even assisted breathing ([138], p.185).

The plurality opinion was relatively silent with respect to the constitutional right to privacy, recognized in *Roe*, and previously in *Griswold v. Connecticut*, as grounding the unenumerable right of women to make procreative decisions [73]. The plurality in *Webster* disregarded the ninth amendment, which recognizes that individuals' rights are not limited to those enumerated in the Constitution, and departed from *Roe* in not viewing a woman's right to make procreative decisions as such an unenumerated right.

Moreover, *Roe* recognized that abortion is essentially a joint medical decision arrived at through doctor-patient consultation, free of state intervention and, therefore, protected the privacy of the doctor-patient relationship. As some *Amici* briefs in *Webster* argued, this right is drawn from the Due Process Clause of the fourteenth amendment ([14];[138]). The statutory "gag rule" upheld by *Webster* not only prevents doctors from giving and obtaining informed consent, but also is viewed by many as a violation of the first amendment right of free speech, which goes beyond the legitimate regulation of speech by public employees subsidized by the state [97]. Supporters of the appellees in *Webster* concluded that the doctrine of informed consent properly places the abortion decision with the woman and her physician [14].

C. *Post-Webster Legislative Actions*

The Supreme Court decisions since *Webster* ensure a prolonged debate over abortion in the United States as they invite state legislatures to revise their abortion statutes. Because the Court is so divided concerning abortion, the decisions provide legislators and judges with relatively little guidance about which restrictions will be deemed constitutional [29].

After *Webster*, the state legislatures of Florida, Alabama, Illinois, Indiana Maryland, Michigan, and Idaho were unsuccessful in attempts to impose more restrictive abortion statutes [29]. In 1990, the Connecticut legislature passed legislation ensuring a woman the right to an abortion [29]. However, in January 1991, Utah signed into law the country's most stringent anti-abortion law which permits abortions only in order to prevent the birth of a child with grave defects or in cases of rape or incest where pregnancy threatens the woman's health [109]. Pennsylvania enacted an abortion law requiring a 24-hour waiting period for women seeking abortions, spousal notification and a requirement that the physician inform these women about fetal development and alternatives to abortion [142]. The Pennsylvania law also prohibited abortions for the purpose of sex selection and abortions performed after the 24th week, except to save the mother's life [142]. The law was enjoined by the courts and declared unconstitutional; however, a Federal court of appeals upheld most of the state's restrictive abortion law in October 1991, but eliminated the spousal notification provision [28]. Abortion rights activists, represented by the American Civil Liberties Union and the Planned Parenthood Foundation of America, asked the Supreme Court to rule on the Pennsylvania statute as a means of intensifying the political debate on abortion before the 1992 presidential election [28].

D. *Parental Notification and Judicial Bypass*

The Supreme Court, in June 1990, handed down its first post-*Webster* abortion decisions, *Ohio v. Akron Center for Reproductive Health* and *Hodgson v. Minnesota* ([81];[133]). In both cases, the Court ruled that parental notification for minors seeking abortions is constitutional. These decisions are expected to deter young women from seeking safe and timely abortions. They also reflect the Court's increasing willingness to restrict the abortion rights of minors on the ground of protecting parental rights.

In *Ohio v. Akron Center for Reproductive Health*, the court upheld an Ohio statute, which called for a minor to notify one parent when seeking an abortion and provided a judicial bypass alternative if the minor demonstrates by clear and convincing evidence that she has the maturity to make such a decision or that parental notification is not in her best interest. The Court found that the statute's bypass procedure was sufficient to prevent "an undue, or otherwise unconstitutional, burden on a minor seeking an abortion" [133]. It remains to be seen whether the judicial bypass provides a real alternative for young women.

In *Hodgson v. Minnesota* the Supreme Court upheld a Minnesota statute that requires a minor to notify both parents before obtaining an abortion because the law provides the alternative of a judicial bypass for pregnant girls who do not want to inform their parents of their decision. The Court also upheld the provision that requires a minor to wait 48 hours after parental notification before obtaining an abortion [81]. Although the Court ultimately upheld the statute on the basis of its judicial bypass procedure, before it reached that conclusion it ruled, by a 5-4 vote, that without the judicial bypass provision the statute would have been unconstitutional. The Court found that the judicial bypass procedure saves the notification and delay requirements from constitutional infirmity by providing an alternative for minors who would be harmed by these requirements. In addition, Justice O'Connor argued that the two-parent notification requirement was unreasonable in light of the fact that only half of the teenagers in Minnesota lived with both parents ([29], p. 405). The court's presumption is that the protection of parental rights and promotion of familial communication will safeguard the best interests of minors, while the judicial bypass provision safeguards minors' interests in circumstances where this presumption is rebutted by the facts.

The *Hodgson* and *Akron Center* decisions confirm the Supreme Court's intention, as first expressed in *Webster*, to allow the states far greater leeway in placing limits on a woman's right to an abortion.

E. *The "Gag Rule": Rust v. Sullivan*

Following *Webster*, government officials have restricted what federally funded physicians and family planning clinics may say to their patients about abortion. In 1988, the Department of Health and Human Services (HHS) adopted revised regulations that forbade family planning programs funded under Title X from counseling patients about abortion and from referring patients for abortion ([61], in [22]). These regulations, otherwise known as the "gag rule", affect the 4000 family planning clinics and the approximate 4 million poor women served by them [20]. In Section 1008 of Title X, Congress had authorized aid to family-planning clinics specifying that the money was not to be used for abortions [144]. The 1988 revised regulations Title X to prohibit all medical discussion of abortion in federally funded family planning clinics and to require counselors at these clinics to say that abortion is not an approved method of birth control. The "gag rule" has been criticized in the lay and professional literature for interfering with the doctor-patient relationship by restricting the topics of medical discussions ([21];[144]). Furthermore, it threatens the constitutional right of free speech and suggests the possibility of censorship in other institutions that rely on federal funding such as education, research, and the arts.

Federal appellate courts were divided over the constitutionality of these regulations ([44];[128];[129]) until the Supreme Court decided in May 1991, in the 5-4 decision of *Rust v. Sullivan* that the regulations were constitutionally acceptable [158]. *Rust* will have broad ramifications for the provision of family planning and contraceptive services for the poor, young, and members of minority groups, who rely most heavily on federally funded family planning organizations. Chief Justice Rehnquist, writing for the majority, found the regulations to be within the guidelines of a statute prohibiting the use of Title X funds in programs in which abortion is a method of family planning [158]. He said that the government had simply chosen not to subsidize speech about abortion and that the government "may validly choose to fund childbirth over abortion" [158]. In a dissenting opinion, Justice Blackmun said that the regulations violated the free-speech rights of clinic employees and amounted to the coercion of impoverished women to continue unwanted pregnancies by withholding information and by erecting obstacles to referrals for prenatal care [158].

The AMA and the American College of Obstetricians and Gynecologists (ACOG) argued that the regulations infringed on the doctor-patient relationship and would forbid physicians from counseling about abortion even when continuing the pregnancy is medically contraindicated, *e.g.*, in cases where severe fetal abnormalities are detected through prenatal diagnosis. In June 1991, in response to *Rust*, the AMA issued a policy statement condemning government restrictions on medical advice about abortion or about any other matter [31].

It is possible that the regulations upheld in *Rust* will set a precedent for Congress to prescribe the content of the doctor-patient dialogue in other medical contexts. The indigent, who are recipients of Medicaid, will especially suffer from these restrictions on free speech by being denied opportunities to learn of their legal options. State control of the doctor-patient dialogue may prevent physicians from performing their ethical and legal obligations to their patients and ultimately hold the uninformed patient as a desirable result ([22], p. 860).

The most recent victory for anti-abortion forces came in November 1991 with President Bush's veto of legislation that would have once again allowed doctors and counselors in federally financed family planning clinics to discuss abortion with pregnant women [40]. This veto galvanized the Bush Administration's alliance to the pro-life movement. Despite intense Democratic efforts, the House, voting 276 to 156, fell short of the two-thirds majority needed to override the veto.

F. *The Abortion Pill: RU-486*

The United States government has taken measures to suppress the controversial abortion-inducing drug, RU-486. RU-486 is a progesterone-blocking drug that can be used to interrupt pregnancy in the first seven weeks. RU-486 inhibits the activity of progesterone, a hormone that is necessary to begin or sustain pregnancy [164]. In conjunction with prostaglandin, RU-486 achieves a 96 percent rate of success in the termination of pregnancies. (Studies have indicated that RU-486 may also function as an anti-cancer agent ([67], p. 11). Although RU-486 has been commercially available in France since 1988, neither RU-486 nor the prostaglandin essential for its efficacy has been approved by the FDA for distribution in the United States. There is some evidence that the FDA is inclined to impede approval for political, rather than medical reasons [70].

In 1988, the FDA issued an "import alert" specifically banning the importation of RU-486. An updated alert in June 1989 concluded that the restrictions placed on RU-486 were to protect the user from its risks and that therefore importation of RU-486 was prohibited, despite an FDA policy generally allowing the importation of drugs for the individual importer's personal use [70]. In April 1990, the FDA adopted a policy which called for the automatic detainment of all abortifacient drug shipments [70]. Lack of corporate sponsorship is another obstacle to the dissemination of the abortion pill; fearing the response of the anti-abortion movement, no American companies have yet applied for FDA approval to market this pill [3].

Despite governmental resistance to the abortion pill, in June 1990 the AMA adopted a resolution introduced by the California Medical Association to endorse testing and possible use of RU-486 ([3];[70]). In November 1990 the FDA retracted the June 1989 order barring RU-486 from importation into the United States and stated that it would not block research on RU-486 as a possible weapon against some cancers [79].

G. *Status of the Fetus*

The current controversy over abortion in the United States has increasingly focused on the legal and moral status of the fetus. Among the issues that have arisen in lower courts are the status of the fetus under vehicular homicide and insurance statutes, embryo research and the custody of embryos, illegal imprisonment of a fetus, forced cesarean sections, wrongful birth, life and death actions, fetal protection policies in the workplace, and fetal abuse.

Largely responsible for this increase of interest in the fetus are recent advances in reproductive technologies. Prenatal diagnosis of genetic abnormalities, improved management of prematurely born infants, and the manipulation of eggs, sperm and embryos to treat infertility all contribute to the notion that a fetus is a patient with interests of its own.

IV. REPRODUCTIVE TECHNOLOGIES

Modern reproductive technologies such as oocyte retrieval, embryo transfer, cryopreservation of embryos, in vitro fertilization (IVF), gamete intra-fallopian transfer (GIFT), and surrogate mothering have made it possible for many infertile couples to realize previously inconceivable procreative goals. Since the first IVF birth in 1978, for example, IVF has rapidly developed as a treatment for infertility and accounts for over 5000 births in the United States alone. The scope of these technologies is forever broadening as evidenced by the recently reported success in extending reproductive potential to post-menopausal women using donated eggs [160]. These "new" means of reproduction, however, have precipitated an array of debates regarding the concepts of motherhood and family. Furthermore, the legal status of embryos remains a contested issue especially with respect to determining dispositional authority over cryopreserved embryos.

A. *Gestational Surrogacy*

The ability to separate motherhood into its components of genetics (egg production) and gestation enables two different women to make biological contributions to the creation of a new life [113]. The problem of gestational surrogates, those who are genetically unrelated to the fetuses they carry and deliver, poses the question of whether a genetic or gestational relationship ought to determine maternal parentage and legal rights. This question was recently addressed by a California court [99]. In *Johnson v. Calvert*, the judge denied the request of the surrogate mother, Anna Johnson, for parental rights to the son she bore for Mark and Crispina Calvert. The Calverts hired Anna Johnson, a young, single black nurse, to gestate an embryo composed of their egg and sperm. Ms. Johnson was to receive a fee of $10,000 upon giving the resultant child to the Calverts. Near the end of her pregnancy, Ms. Johnson sought to retain custody of the child. The judge awarded custody to the Calverts on the basis that Johnson and the child were "genetic hereditary strangers" ([15], p. 36; [99]). He supported this opinion by noting that the surrogacy contract was valid and that giving the child to the Calverts was in its best interests.

It has been noted in the literature that this opinion contradicts the current legal presumption favoring the gestational mother ([15];[113]). This current presumption holds that because of the greater biological and psychological investment, the woman who gives birth should *prima facie* be considered the child's legal mother, who has the right and responsibility to raise the child and may agree to adoption only after the birth of the child [15]. One commentator cautions that determining parenthood exclusively on the basis of genes serves to marginalize and demean pregnancy and childbirth [15]. Such a genetically-based policy questions the legitimacy of the gestational mother's bonding to her infant, while upholding the genetic contributors to the child as the primary "owners" of their genetic products.

There is also concern that surrogate motherhood is demeaning to women and may be used by financially secure people to exploit needy women ([37];[67]). Following the 1988 "Baby M" case, which held that commercial surrogacy contracts are completely void and possibly criminal, the legal status of surrogate arrangements remains unclear [96]. Legislation in different states has been enacted both facilitating and inhibiting such arrangements. State court decisions prior to the *Johnson* case found surrogacy contracts to be unenforceable in the event of a custody dispute. To clarify these matters, the American Bar Association (ABA) has recently approved model legislation for the legalization of surrogate motherhood and the enforceability of surrogacy contracts [1].

B. *Disposition of Frozen Embryos*

An important adjunct to IVF is cryopreservation of embryos. It enhances the success of IVF by allowing greater flexibility in the number and timing of embryo implantations [152]. Conflicts have recently arisen regarding the disposition of frozen embryos. These conflicts point to the contentious legal status of pre-implantation embryos, or "pre-embryos". They are genetically unique living human entities that can potentially develop into live offspring, but are nonetheless at a very rudimentary stage of development. A Tennessee court of appeals was recently forced to reconcile the interests of prenatal life with the right of procreative choice within the context of a property dispute. The custody of seven frozen embryos produced during an IVF procedure was contested in the divorce trial, *Davis v. Davis* [52]. The woman claimed that the embryos were pre-born children and bid for her right to be impregnated with the frozen embryos. The estranged husband invoked the right to control his own reproduction and wanted the embryos to remain indefinitely frozen. The central question was whether the disputed embryos were property that should be disposed of as other commonly held goods in a divorce case or whether they were human life with rights of their own.

The Tennessee court rejected the husband's claims and granted custody of the seven embryos to the woman for the purpose of implantation. The trial judge ruled that the embryos were in fact "human beings existing as embryos" and "children in vitro" whose best interests required that they be made available for implantation [52]. This conclusion was unprecedented and had no discernible basis in common law precedents nor in Tennessee law, which recognizes a separate legal interest in prenatal human life only at viability ([151];[153]). Extracorporeal embryos have no legal standing in their own right [152].

This decision was widely criticized and was subsequently overturned in an appellate court decision granting joint custody of the seven frozen embryos to the divorced couple and giving each parent an equal voice over their disposition ([51];[151]). The appellate court recognized that the protection of pre-embryos, by giving them a chance to implant and come to term, was not a sufficiently compelling justification to override the husband's constitutional right not to procreate [151]. The potential risks of financial liability and the psychosocial burdens of unwanted parenthood contribute to the argument against subordinating one person's procreative liberties to another's [153]. It is recommended that couples specify the disposition of their frozen embryos at the time of cryopreservation to minimize the chances of litigation in case of dispute, divorce, death or unavailability [151].

The need for dispositional guidelines with respect to extracorporeal embryos also arose in the first case of a custody dispute between a couple and an IVF program. In *York v. Jones*, an infertile couple had one frozen embryo remaining after three unsuccessful attempts at an IVF pregnancy at a leading program in Norfolk, Virginia [185]. After the couple moved to California, they sought to transport their embryo to a Los Angeles IVF program for thawing and implantation. The Norfolk program refused to release the frozen embryo [152]. The court found that the IVF program had no right to retain the embryo against the couple's joint wishes or to make dispositions unless the couple had specifically ceded that right to them. The judge found the cryopreservation agreement between the couple and the IVF program created a bailor-bailee relationship in which the bailee was obligated to return the "good" to the bailor when the purpose of the bailment had been terminated ([153], p. 462). Therefore, the couple had a *prima facie* claim to have the embryo released to them. This case secures the primary dispositional authority or ownership with the gamete providers. This decision affords prospective parents leeway in procreative decisionmaking through the creation, storage, thawing and transfer of embryos, yet allows them to avoid IVF reproduction when their needs or circumstances change.

V. MATERNAL-FETAL CONFLICTS

The pregnant woman and her fetus have a unique relationship. One cannot be treated without affecting the other. Furthermore, the pregnant woman's behavior has a profound influence on the growth and development of her fetus. On occasion, a woman may reject medical recommendations or behave in ways that are potentially detrimental to her fetus. There is a growing national debate over judicial efforts to balance the rights of pregnant woman with fetal interests. A great deal turns on the moral status accorded a fetus. Within the current framework of procreative law, as established by *Roe v. Wade* and *Webster*, the state's interest in promoting the fetus' well-being becomes compelling at the point of viability, unless the pregnant woman's life or health is at stake ([155];[181]). These laws, however, are the subject of a great deal of legal and moral controversy.

As the fetus is increasingly the object of medical intervention and as mounting evidence implicates prenatal exposure to drugs and toxic chemicals in fetal impairment, the courts are increasingly being asked to balance a woman's constitutionally protected right to make reproductive choices against the state's interest in the viable fetus ([41];[130]).

A. *Court-Ordered Medical Treatment*

Preserving the right of a competent pregnant patient to refuse medical treatment may sometimes come at the risk of preventable fetal harm. Earlier court decisions revealed a great deal of ambivalence about letting competent women make decisions which could harm their fetuses. In fact, some courts have found that the state's interest in preserving the life of a viable fetus is sufficiently compelling to override the mother's right to refuse treatment ([41];[125];[131]).

In a landmark decision heralded by patients' rights activists, an appellate court in the District of Columbia ruled that a pregnant woman may not be forced against her will to undergo cesarean delivery to save her fetus [87]. The suit involved a court-ordered cesarean delivery of a 26 week old fetus from Angela Carder, who was terminally ill with cancer. Both her doctors and family objected to the effort to deliver her baby before her imminent death in 1987. The hospital, fearing legal liability, sought a court order to perform the cesarean.

The decision, *In re A.C.*, upheld the pregnant woman's right of bodily integrity, finding that when the rights of fetuses conflict with the rights of pregnant women, the latter must take precedence. Furthermore, the opinion held that where a woman pregnant with a viable fetus cannot give an informed consent to a proposed surgical intervention to save an unborn child, a court must use the substituted judgment standard of surrogate decisionmaking, thereby reflecting the woman's own beliefs and values [87].

In re A.C., though only binding in the District of Columbia, will influence how other courts balance maternal rights with fetal interests when the pregnant woman refuses proffered medical care. Whether the absolute right to make decisions relevant to fetal welfare will be extended to cases of narcotic or alcohol abuse is much less certain.

In a related development, George Washington University Medical Center has established a policy to ensure that ethically difficult decisions concerning how to treat severely ill pregnant women and their fetuses be made by the woman, her family, and her doctor, thus avoiding recourse to the courts. This recently instituted policy, respecting the autonomous decisions of pregnant patients, even when counter to medical recommendations, is the most comprehensive of its kind and may influence hospital policies across the country [69].

B. *Harmful Behavior by Pregnant Women: Drug and Alcohol Abuse*

The over 350,000 infants exposed to illicit drugs in utero each year raise a crucial question: Should pregnant, drug-abusing pregnant women be considered a criminal or a public health problem? [130] Judicial systems across the country are increasingly acting to protect fetuses from drug-abusing pregnant women [5]. Since 1987, about 60 criminal cases have been set in motion in 19 states and the District of Columbia against women for abusing drugs during pregnancy [182]. Women have been prosecuted under the statutes of child abuse, manslaughter and, most recently, criminal delivery of drugs to a minor ([41];[100];[111]). In 1989 a Florida appellate court upheld the first conviction of a woman charged with delivering cocaine to her newborn baby through the umbilical cord [97]. Similar charges have been filed and dismissed in other jurisdictions causing controversy regarding how drug abuse in pregnant women should be handled. In the decision of *Johnson v. the State of Florida*, the state court of appeals approved the prosecution's strategy of charging mothers, whose babies are born with traces of cocaine in their blood, under laws designed to punish drug dealers who deliver controlled substances to their children [100].

Few would argue that it is morally justifiable to knowingly cause a fetus significant harm through drug abuse. However, if one believes that it is morally wrong to harm a fetus it is still unclear that criminalization is the proper response. Questions of the voluntariness of drug abuse lead some to argue abuse should be treated as a disease rather than a crime. The problem of efficacy also may discourage criminal sanctions. It is widely held that criminal sanctions will both fail to deter pregnant women from substance abuse and also exacerbate the harm done to fetal health by deterring them from seeking needed prenatal care and medical help for chemical dependency ([41];[114];[130]). If the goal is to help the largest number of fetuses, a treatment-based program may be more successful. In 1990, the AMA determined criminal sanctions or civil liability for harmful behavior by the pregnant woman toward her fetus are inappropriate and instead emphasized the primacy of informed consent and rehabilitative care for such women [41].

Finally, issues of equity arise. Most women who have been prosecuted for exposing fetuses to harmful substances have been poor members of racial minorities who are already more likely to lack access to prenatal care and drug treatment programs [41]. A study conducted in 1990, for instance, found that the majority of drug treatment programs in New York City do not accept pregnant women and that 87 percent refused to admit pregnant Medicaid patients who used crack [38].

Despite the widespread consumption of alcohol during pregnancy and the seriousness of its effects on the fetus, its legal and social acceptance make it difficult to prevent its use ([41];[130]). Efforts have therefore concentrated on changing the public's acceptance of drinking alcohol during pregnancy. In

February 1990, New York City instituted a requirement that signs be posted in public places serving alcohol to warn pregnant women of the dangers to the fetus of alcohol consumption. The National Organization for Women criticized this measure saying that it both unfairly singles out women and suggests that fetal rights exceed those of pregnant women [159].

The most draconian legislative response to pregnant drug abusers was passed by the state of Illinois, which extended its definition of child neglect or abuse to include any newborn "whose blood or urine contains any amount of a controlled substance..." ([85];[111], p. 16). This statute allows the state to remove a child from its drug-abusing mother without proving either the addiction of the child or harmful effects of the drug on the child ([85];[111], p. 16).

C. *AIDS and Reproductive Decision-Making*

The prevalence of HIV infection in childbearing women has made the perinatal transmission of the AIDS virus an urgent ethical and public health problem. It is estimated that 1.5 per 1000 women giving birth to infants in 1989 were infected with HIV in the United States [75]. Assuming a conservative perinatal transmission rate of 30 percent, approximately 1,800 newborns acquired HIV infection during 12 months, mid-1988 to mid-1990 [75]. Given that some women know they are infected with HIV and choose to carry through with their pregnancies, a tension arises between the pregnant woman's right to make reproductive decisions and the public objective of preventing the transmission of a lethal disease to offspring.

Many state department of health materials recommend unequivocally that HIV-infected women should not become pregnant [25]. Prenatal HIV counseling and testing programs have been recommended by a variety of agencies (*e.g.*, Centers for Disease Control, American College of Obstetricians and Gynecologists) and have been instituted in many prenatal centers [121]. Interestingly, some in the Federal government have come close to suggesting abortion may be an acceptable option in this scenario. How this tension is to be resolved is an open question. Legal and policy precedents have yet to be firmly established regarding this issue, but it will inevitably be a contentious debate regarding maternal rights and 'preventable' fetal suffering.

D. *Fetal Protection Policies in the Workplace*

Much controversy has surrounded fetal protection policies and the exclusion of women from hazards in the workplace ([94];[95];[96]). Such policies have historically enabled employers to restrict job opportunities for women based on a concern for fetal health. The Supreme Court in the recent *Johnson Controls* case, decided that federal law prohibits employers from excluding fertile women from certain job categories on the basis that it might endanger a fetus [94]. The case originated as a class-action suit, involving several unions and individual employees. They alleged that the company's fetal protection policy, which barred all fertile women from working in high lead exposure positions, was sexually discriminatory and thus violated Title VII of the Civil Rights Act of 1964, which was passed by Congress to prohibit discrimination of the basis of sex, race, color, religion, or national origin ([80];[176]). The appellate court found the company's policy to be justified under Title VII as a business necessity and a *bona fide* occupational qualification (BFOQ) [95].

In March 1991, the Supreme Court unanimously reversed this decision on two grounds [94]. First, it held that the company did not seek to protect all unconceived children, but discriminated against female employees under the guise of protecting their children. Second, the court found that under the BFOQ any limitations involving pregnancy must relate to a woman's ability to perform the job. This ruling stands as a clear victory for advocates of equal employment opportunities for women; the court found it illegal to recast sexual discrimination in the name of fetal protection ([16], p.742).

As a response to the appellate court's *Johnson Controls* decision the federal Equal Employment Opportunity Commission (EEOC) issued an internal policy guideline in January 1990 stating that employers should endeavor to shield all persons from hazardous substances and to similarly protect the offspring of males and other third parties [80]. This policy statement, predating the Supreme Court's ruling, is in accordance with its decision to reduce gender-based discrimination in the workplace.

VI. AIDS

Just as the epidemic of the Acquired Immunodeficiency Syndrome (AIDS) has severely taxed public health care budgets and medical resources and research efforts, it has also tested the meaning and limits of ethical, legal, and medical concepts, such as confidentiality, the duty to warn, and the notions of "reportable diseases" and invasive procedures. Perhaps most strikingly in recent years, the AIDS epidemic has prompted public concern that infected health care workers may present a threat to patients' health.

A. *Classification of AIDS and HIV and Notification of Third Parties*

Considerable controversy continues to surround the classification of AIDS and HIV infection. Public perception of AIDS as a disease of homosexuals, and subsequently of drug addicts, led gay rights activists and their supporters to argue that HIV-related diseases should be treated as exceptions to the usual public health policies developed to control other sexually transmitted and communicable diseases. Their efforts and concern about public backlash against the groups most affected by the disease – homosexual men and black and Hispanic IV drug users – led to what is termed "HIV exceptionalism", differential policies for controlling the spread of HIV [26]. Considerable controversy has surrounded this HIV exceptionalism.

In 1990 the House of Delegates of the American Medical Association (AMA) called for HIV infection to be classified as a sexually transmitted disease. As of December 1990, 11 states had classified AIDS and HIV as sexually transmitted or venereal diseases, and 22 had classified them as communicable diseases, infectious diseases, or both ([26], p. 1503). The three hardest hit by the AIDS epidemic – three states – California, New Jersey, and New York – have not.

In 1988, four New York clinical societies sued to force the state's health commissioner to classify AIDS and HIV infection as sexually transmitted and communicable diseases and to abandon the state's policy of anonymous and voluntary testing. In 1991 the New York Court of Appeals rejected the suit and upheld the New York state policy which did not require physicians to report the names of HIV-infected patients to state health authorities for contact tracing. Opponents of these measures contend that reporting and contact tracing are coercive. Indeed, although contact tracing depends upon the cooperation of the index patient and protects that patient's anonymity, at a social level, some coercive elements do infect the process: since 1988, for instance, the Centers for Disease Control (CDC) has made development of a contact notification program a condition for states to receive HIV-prevention funds ([26], p. 1502).

Although clinical AIDS has been a reportable condition since 1983 and despite the fact that in 1990 the Centers for Disease Control and the AMA House of Delegates endorsed the reporting of the names of those infected with HIV, only a few states have required reporting of HIV infection. Although both the AMA and the Association of State and Territorial Health Officials endorsed legislative provisions which would permit disclosure to partners of patients placed at risk for HIV infection, as of 1990 only two states had imposed on physicians a legal duty to notify spouses that they were at risk for HIV. Approximately a dozen states granted physicians a privilege to warn sexual and needle-sharing partners of HIV-infected patients and thereby freed physicians in those states from liability whether or not they chose to warn ([26], p. 1502;[93]). The policy of the Ad Hoc Committee on AIDS Policy, approved by the American Psychiatric Association (APA) in 1987 states that "if a patient refuses to agree to change his or her behavior or to notify the person(s) at risk or the physician has good reason to believe that the patient has failed to or is unable to comply with this agreement, it is ethically permissible for the physician to notify an identifiable person who the physician believes is in danger of contracting the virus" [10]. The policy urges the psychiatrist to make clear at the beginning of treatment, particularly before inquiring about a patient's HIV status, the limits of confidentiality.

Several legal cases and legislative actions are representative of the conflicts that have arisen surrounding AIDS testing. Illinois, for example, implemented a controversial mandatory premarital HIV testing from January 1, 1988 until the measure was repealed in September 1989 [116]. It was naively hoped that the test would allow couples to know each other's HIV status before entering into marriage or sexual relationships. During the period of mandatory testing, Illinois experienced an abrupt decrease in the number of marriages within the state, while many couples from Illinois either married in bordering states or lived together without benefit of marriage.

In court cases, judges have yet to formulate a consistent way to balance confidentiality with other interests. In response to a sheriff's request to release to prison officials the results of an inmate's AIDS test, an Alabama court held that the results of a test for AIDS could not be released without the permission of a health official [160]. The sheriff's argument that release of the test results was necessary to protect other prisoners was rejected, and the individual's right to privacy was upheld, albeit in a somewhat limited degree, because state authorized health officials were empowered to divulge test results when they deemed it necessary.

The Wisconsin Court of Appeals set aside the privacy interest of an HIV-infected plaintiff in a malpractice suit and upheld the court's discretion to admit or suppress HIV test results whose disclosure is not prohibited by statute [56]. One anticipated result of this decision is that HIV-infected plaintiffs may receive smaller damage awards predicated on their anticipated reduced quality of life and life expectancy. HIV-infected convicted criminals, on the other hand, according to a New York Supreme Court decision, may not have their jail sentences reduced because the sentence constitutes a *de facto* life sentence because of their shortened life expectancy [141].

B. *Health Care Workers: Testing and the Duty to Warn*

The plight of dental patient Kimberly Bergalis, who contracted HIV during treatment by her Florida dentist, sparked heated medical and political debate over the responsibilities of HIV-infected health care workers and the appropriateness of mandatory HIV testing of health care workers. In January 1991, Bergalis announced that she had reached a $1 million settlement in her claim against her dentist, who is likely responsible for the infection of at least four other patients ([8];[36]). Despite the fact that health care professional-patient transmission is extraordinarily unlikely, this case sparked national controversy. Later in the year, in televised congressional hearings, Bergalis called on Congress to enact legislation mandating testing of health care workers [35].

The CDC did not recommend mandatory testing of health care workers (HCWs), but did state that HIV or hepatitis B-infected health care workers "should not perform exposure-prone procedures unless they have sought counsel from an expert review panel and been advised under what circumstances, if any, they may continue to perform these procedures. Such circumstances would include notifying prospective patients of the HCWs, seropositivity before they undergo exposure-prone invasive procedures" ([35], p. 775). The CDC charged medical, surgical, and dental societies and institutions with the task of identifying these exposure-prone procedures, and stated that infected health care workers who perform invasive procedures not identified as exposure-prone should not be restricted in their practice. Health care workers who perform exposure-prone procedures should know their HIV and hepatitis B status, and those who are infected and thus are forced to modify their patient-care activities should be provided with counseling and job retraining whenever possible.

The AMA, the American Academy of Orthopedic Surgeons, and the American College of Obstetricians and Gynecologists have agreed with the CDC conclusions that some procedures present too high a risk of transmission of HIV and should not be performed by HIV-infected health care workers ([4];[43];[55]). Disagreements within and outside of these societies about whether HIV-infected physicians should perform various procedures do not center on how much risk it is reasonable to expose patients to, but instead focus on the empirical question of how much risk patients actually face from HIV-infected physicians ([137], p. 1135).

In July 1991, responding to public outcry over the Bergalis case, the Senate adopted a proposal mandating prison terms of at least 10 years and fines of up to $10,000 for health care workers who knew that they had AIDS but failed to inform patients on whom they performed invasive procedures [172]. Departing from CDC recommendations, the Senate, in a 99-0 vote, voted on a proposal directing states to require health professionals who perform invasive procedures to be tested for the AIDS virus. States which failed to comply would risk losing federal health care funds. The proposal would bar infected professionals from performing invasive procedures unless they received permission from a panel of experts and informed their patients of their infection. Opponents of the measure pointed out that only five of the 182,000 cases of AIDS reported since 1981 could be traced to a health professional. The Senate proposal would also apply to professionals infected with the virus causing hepatitis B for which the risk of transmission is significantly greater than for HIV ([35], p. 774;[71], p. 664). The House of Representatives had not acted on the proposal.

Mandatory testing of health care workers has been criticized on several grounds. Mandatory testing invades privacy. The time lag between infection and seroconversion may prevent infected professionals from being identified for weeks or months and would require almost constant testing of health care workers [137]. Such testing would be exceedingly costly and would severely strain already inadequate health care funds. Moreover, if HIV testing of health care workers were mandated, they might demand the testing of their patients with the attendant loss of privacy. In addition, negative results might lead to a false sense of security and prompt them not to use the universal precautions which serve to protect both patients and professionals. Finally, as a matter of fairness, policies should not be implemented which single out, for testing and avoidance, the risk of HIV infection; professionals and patients generally tolerate other risks, including other risks of practicing medicine, iatrogenic risks, and the life-threatening risks of daily living.

A Louisiana case tested the mandatory testing policy of one hospital district. The court in *Leckelt v. Board of Commissioners of Hospital District No. 1* permitted public hospital officials to require HIV testing of health care workers whom they suspect were exposed to HIV only if the employees' duties create a potential risk of infection to patients or other employees [108]. The plaintiff, a licensed practical nurse, charged that he was discriminated against when he was discharged for failure to supply the results of his HIV test to hospital officials. The scope of the decision and of the policy which it upheld was limited; only those health care workers whose duties presented a risk of transmission could be required to undergo testing.

In April of 1991, a New Jersey court upheld a hospital's decision to restrict the surgical privileges of an HIV-infected physician, William Behringer, to protect patients ([60];[137]). The hospital instituted a policy requiring HIV-infected physicians to inform their patients of their infection and to obtain written informed consent prior to performing invasive procedures. Dr. Behringer's estate sued the hospital alleging that this policy violated state anti-discrimination provisions which prohibit discrimination against handicapped people unless their handicap "reasonably precludes the performance of the particular employment" [126]. The court determined that under state law persons suffering from AIDS are handicapped, but that the hospital's restriction of Dr. Behringer was reasonable because of the risks he presented to patients in performing invasive procedures: first, the risk of infection, and second, the anxiety which would be experienced by patients if they were exposed to Dr. Behringer's blood, even if infection did not occur.

The New Jersey court did, however, award Behringer's estate damages for the hospital's failure to take reasonable precautions to protect the physician's confidentiality. The Behringer decision is technically binding only in Mercer County, New Jersey; nevertheless, it is the first case to decide the issue of practice restrictions for HIV-infected physicians.

The court did uphold the hospital's informed consent requirement. The informed consent requirement was included to guard against physicians' potential conflict of interest in determining which procedures present a risk of transmission and which, therefore, an HIV-infected physician would not be permitted to perform [137]. The requirement that HIV-infected physicians inform patients of their HIV status allows patients to serve as the final arbiters of whether or not to be treated invasively by an HIV-infected physician. The *Behringer* decision imposes a standard of "zero-tolerance" risk, a standard which is likely to be overridden in other court decisions in light of both the CDC's recommendations and the Americans With Disabilities Act which should protect physicians from discrimination unless they present a significant risk of harm to patients ([137], p. 1135).

C. *AIDS: Challenges to Health Insurance*

The AIDS epidemic has shown in stark terms the cracks in the American health care system. The problems of the uninsured and the underinsured, as well as problems of individual and employer-based insurance which excludes particular diseases and pre-existing conditions, are brought into sharp focus by the difficulty of financing the provision health care to AIDS patients.

The encouraging findings that drugs – expensive drugs – can delay the onset and slow the progress of HIV-related diseases and the increasing incidence of HIV among people who do not have health insurance coverage or who lose their coverage exacerbate the problem of financing health care for HIV-infected patients. By the end of 1989, 42 states had appropriated nearly $100 million for HIV-related patient care and support services in addition to their Medicaid contributions ([65], p. 224). Some states are purchasing private health insurance for HIV-infected patients, and at least 12 states have used regulatory authority to organize treatment for those with AIDS, most often through their Medicaid programs. In addition, whereas the states' financial concerns in the 1980s focused on how to fund health care for people with AIDs, in the 1990s states are also concerned with funding early detection and preventive drug treatment, as well as providing health and social services to patients whose life expectancy is unknown ([65], p. 225-226).

Many federal actions have addressed these concerns ([65], p. 228-229). Most employment-based insurance does not cover long-term care for chronic illnesses, which presents a problem which is not uniquely faced by AIDS patients. The Consolidated Budget Reconciliation Act of 1986 required employers to permit former employees to continue their health insurance coverage at a capped premium and did extend insurance to many AIDS patients who were in the work force. A diagnosis of AIDS creates presumptive eligibility for disability benefits under the Supplemental Security Income entitlement and has enabled many patients to qualify for Medicaid more quickly, but the Health Care Financing Administration has not recommended that Congress eliminate the two year waiting period before those qualifying for Social Security Disability Insurance become eligible for Medicare.

The expense of azidothymidine (AZT), a drug which delays the onset of HIV-related diseases, presents specific financing challenges to states. By mid-1989, all states had received federal subsidies to purchase AZT ([65], p. 228). Missouri attempted to limit Medicaid coverage of AZT treatments to those cases of AIDS patients who met certain diagnostic criteria based on Food and Drug Administration (FDA) approved indications for the drug. In *Weaver v. Reagen* the court held that Missouri Medicaid may not deny coverage for AZT treatments when the patients' physicians have certified that AZT is medically necessary treatment, that the state's attempt to label AZT experimental was unreasonable in light of widespread recognition that AZT was at that time the only known treatment for patients with AIDS, and that the FDA did not issue its approved indications to interfere with medical practice, but to regulate manufacturers' marketing practices [180].

The challenges presented by HIV-related diseases to private health care insurance, to duties of confidentiality and of notifying third parties, and to the public's perception of health care and health care providers are not unique. These same issues, among others, are raised in the contexts of genetic testing and screening and concerns about genetic discrimination.

VII. GENETICS

In sharp contrast to developments in human genetics itself, the number of government and professional organization responses to ethical concerns raised by these numerous rapid advances are thus far relatively few. Recent HIV legislation and public policy statements, especially regarding HIV+ asymptomatic patients, may suggest future governmental responses to similar concerns raised by genetics.

Perhaps the most significant response by the federal government, to both recent genetic developments and long-term possibilities, is the U.S. Congress appropriation of funds to support efforts to map the human genome. As part of this international effort, expected to span the next 15 years, the U.S. Congress appropriated $43,500,000 to the Department of Energy (DOE) and $87,760,000 to the National Institute of Health (NIH) for 1989-90 and has proposed appropriation of $45,000,000 to the DOE and $108,029,000 to the NIH in 1991 [84].

In recognition of the myriad ethical issues raised by the possibility of detecting genetic predisposition to disease, prescribing preventive measures, and developing genetic therapies, approximately 3 percent of the NIH's budgeted sums are to be used to fund research on the social, ethical, legal, and social implications of the human genome project.

This federal commitment to bioethical research is itself significant for two reasons. First, these appropriations mark the first time that federal funds have been earmarked for bioethical research. Second, it is significant that developments in genetics have occasioned the first mandate for a *prospective* review of anticipated ethical concerns. The field of bioethics has developed primarily by reacting to, rather than anticipating, the ethical conflicts and quandaries resulting from new technological capabilities, evolving medical goals, and clashing religious, ethical, and cultural values. With memories of twentieth-century eugenic policies still fresh, ([143];[148]) the popular sentiment that human genetic research should proceed, but proceed with caution [82], explains the call for prospective consideration of ethical concerns raised by genetic engineering, screening, and counseling. Forty-two human geneticists formed an international group, the Human Genome Organization (HUGO), "to provide a forum for the discussion of ethical, social, commercial, and legal considerations relating to the genome project" [120].

A. *Genetic Engineering: Property Rights*

Moore v. Regents of the University of California (1990) raised the legal issues most frequently associated with genetic engineering: ownership of individuals' genetic information and body tissues and fluids, as well as concerns about the patentability of newly created living organisms [124]. In 1976, upon the recommendation of his physician, Moore had his spleen removed as treatment for hairy-cell leukemia. From Moore's spleen cells, his physician and other researchers developed a cell line, which they patented in 1984 and from which they anticipated substantial commercial profit. Moore sued his physician for economic damages resulting from his physician's failure to disclose his financial interest in producing a cell line from Moore's spleen, and he sued the University and two corporations for a share of their profits from the cell line. Moore's suit against his physician was successful, he collected damages, and the decision was upheld by the California Supreme Court [19].

The 1990 California Supreme Court five-to-two majority opinion in *Moore* expands the disclosure requirements of informed consent to include full financial disclosure. Physicians are required to disclose their interests which are "extraneous to the patient's health" – presumably, research and commercial interests – which might prejudice the physician's judgment. Determining just which interests of physicians might affect their judgment and who should determine this question was left for future analysis. Moore was therefore entitled to sue his physician for economic damages for breach of the fiduciary relationship because of his nondisclosure. Dissenting Justice Broussard went even farther, arguing that Moore's case differed from typical failure of informed consent cases, because the case involved economic damages; therefore, he argued, Moore should only have to show that the physician's failure to disclose caused him compensable damage, not that he would have refused to consent to the surgery if he had been informed.

The majority in *Moore's* opinion did not grant Moore a property right in his excised cells and thereby blocked his right to sue for conversion (the unauthorized assumption of another's goods). The court argued that Moore's interests could be protected by his right to sue his physician, that the question of ownership requires a legislative solution, and that countervailing social interests – namely, society's interest in the development of biotechnology – outweighed Moore's property claim. The majority's failure to grant Moore any interest in his cells and to demand that this interest be taken into account as a cost of the biotechnology business is quite surprising, given the California court's previous history of defending individuals against business interests in product liability cases since the 1940s [59]. The irony of the decision is that "everyone *except* the patient can own the patient's removed cells" and at the time of excision,

Moore was denied the right to do with his own tissue what the defendants did ([19], p. 37).

Dissenting Justice Broussard argued that the majority's decision to deny Moore a property right in his excised cells cannot consistently rest on the general claim that there is no legal title to excised body tissues, because the court would certainly uphold the University of California's legal title against, for example, another laboratory which stole the cell line ([19], p. 37). Dissenting Justice Mosk argued that in comparing Moore's case to that of organ donation, the majority's opinion confused scientific use of body tissues with commercial exploitation [50]. First, he notes that Moore's informed consent to his "donation" was allegedly not obtained. Second, the proximity and anticipation of commercial gain – namely, the initial researchers' own profit – distinguishes the case both from those of organ donation to needy recipients who, with their health restored go on to profit financially and from those cases in which commercial ventures ultimately, but not directly, result from scientific research undertaken on donated body parts.

The questions of ownership of one's excised body tissues and of one's pecuniary interest in resulting commercial gain remain in need of legislative remedy. To decide that persons have no property rights in their excised tissues and therefore cannot reap financial benefit from the use of those tissues, while others can, seems both unfair and illogical. Demanding that patients be compensated for all excised tissues, of which few will prove to be valuable, may place undue financial burden on researchers and thus be contrary to social interests. Prohibiting profit from the sale of human tissue may eliminate substantial incentive for medical research and thus also be contrary to the social good. Therefore, some scheme establishing and granting to patients some fixed percentage of gross sales (*e.g.*, 1 percent) might constitute a legislative solution ([19], p. 39).

B. *Management of Genetic Information*

Government guidelines concerning the management of genetic information extend protections already guaranteed in other contexts. The proposed Human Genome Privacy Act and Senate Resolution No. 75 (1991), as well as the Americans with Disabilities Act (ADA) address concerns about the misuse of individuals' genetic information. Two areas of potential misuse focus on DNA databanking and on genetic discrimination in the workplace.

1. *Genetics and Forensics: DNA Profiling*

DNA identification tests have tremendous potential forensic value. So-called DNA fingerprinting will permit forensic scientists to move from mere exclusion (as in traditional paternity cases) to positive identification of individuals. It is possible to identify the donor of a body tissue or fluid to the exclusion of all other people except an identical twin.

In recognition of this potential, legislation has been passed in four states that permits blood or saliva samples to be taken from those convicted of sex crimes as a condition for their parole [32]. Of concern, according to the American Civil Liberties Union, is the strategy of initially focusing on sex criminals toward whom there is little public sympathy and subsequently expanding the practice of DNA banking to include other classes of convicted criminals and social undesirables ([54], p. 387). In the absence of a government-established uniform method of DNA analysis, laboratories preserve the actual body fluid for future testing. Officials in California reportedly freeze 200 blood and saliva samples each month ([54], p. 388). Extraneous information, such as the individual's HIV or genetic disease status, is thus stored along with the DNA fingerprint. The legislation requiring the collection and banking of blood and saliva samples does not adequately limit access to the DNA test results, typically stating that such information will be released to any law enforcement agency upon request [42]. Moreover, although the legislation applies only to convicted criminals whose civil rights are circumscribed, the statutes do not recognize that the paroled prisoners' sentences will expire and that they will regain their rights as citizens.

2. *Genetics in the Workplace*

In 1989 the Congress' Office of Technology Assessment conducted a survey in support of its report on the legal, ethical, and social aspects of medical screening in the workplace [174]. This survey considered genetic monitoring and screening in 1500 U.S. companies, the 50 largest utilities, and the 33 largest unions. Only 5 percent of the companies reported having a policy regarding hiring persons with increased genetic susceptibility to disease ([173], pp. 11, 35). Of these, 5 percent reported that their policies prohibited hiring employees with an increased susceptibility, 13 percent reported that their policies did not, and 22 percent did not answer the question. Although only 1 percent of the corporate health officers reported that their company had a formal policy regarding genetic screening or monitoring, just over 50 percent of the company health officers and personnel officers reported that their companies would approve of a preemployment medical examination to identify individuals with genetic susceptibility to workplace exposures. Forty-eight percent of the personnel officers felt that their companies would consider it unacceptable to use genetic screening or monitoring to exclude high risk individuals from the work place. Concerns about cost-effectiveness, tests' reliability and legality, and liability were cited as reasons for employers not adopting genetic screening and monitoring ([173], p. 44).

C. *The Americans with Disabilities Act*

The ADA, signed into law in July 1990, was not drafted to address concerns about genetic discrimination [11]. The ADA extends the scope of the Rehabilitation Act of 1973 [177] and prohibits discrimination on the basis of a person's disability in the contexts of employment, public services and accommodations, and telecommunications. One of its major impetuses was the call for more comprehensive protection against discrimination in light of the HIV epidemic ([140], p. 331). Forthcoming regulations interpreting the Act may clarify its applicability to genetic testing and genetic predisposition to disability.

The Act prohibits employers of more than 14 employees, excluding the federal government, from discriminating on the basis of a person's disability unless that disability interferes with his ability to perform his job [11]. Moreover, if the disabled person can meet the requirements of a job then reasonable accommodations are made, and the employer is required to make those accommodations, *e.g.*, installing wheelchair ramps. Only if making such accommodations would impose undue hardship on the employer or would fundamentally alter the nature of the job, is the employer entitled to deny employment on the basis of the person's disability.

At issue in using the ADA to protect employees against genetic discrimination is whether those not currently disabled, but who have a genetic risk of disabling disease, are protected by its provisions. Importantly, the ADA does protect those who are perceived as disabled, as well as those who are in fact disabled. Furthermore, in combination with existing statues which prohibit employers from denying employment to those whose future ability to work will be impaired by ill health, the ADA permits use of only those medical tests which provide information about the person's current ability to perform a job ([27];[139];[183]). Thus, genetic tests to determine a person's genetic disposition to develop a disease might not be permitted. Thus, for example, asymptomatic HIV-infected patients are believed to be protected under the ADA.

In addition, the ADA advised that medical tests may not be performed until after a job offer has been extended, and the results of medical tests must be kept separate from records used for making employment decisions, such as promotion and salary determinations. The fact that an employee's illness may impose higher health care costs on the business should not be used as grounds for refusing employment, therefore eliminating one incentive for using medical tests, including genetic screening. Self-insuring employers may, however, exclude certain diseases from coverage.

In commenting on the ADA, the AMA's Ethics and Health Policy Counsel urges that routine functional testing – testing of a worker's actual capacity to function in a particular job – is in all cases superior to untargeted genetic testing. Testing targeted to determine functional capacity would best protect the individual's interest in employment, employers' interests in an economically efficient work force, and public safety interests (in cases where particular diseases may affect an employee's ability to perform a safety sensitive job, *e.g.*, an airline pilot at risk for developing Huntington's chorea). Especially because genetic conditions are often characterized by incomplete penetrance, variable expression, and variation in time of onset, reliance on genetic screening could result in persons being denied employment although their genetic condition does not (at least currently) affect job performance.

Outside the employment arena, the ADA prohibits discrimination on the basis of disability by all public entities whether or not they receive federal funding, and the Act expands previous notions of "public accommodations" to include, for example, professional offices, hospitals, schools, and commercial establishments, as well as places of lodging and food service ([140], fns. 44,48,52,54).

D. *The Human Genome Privacy Act and Insurance Concerns*

The Human Genome Privacy Act was introduced (but has not been enacted) in the House of Representatives as a means of protecting individuals' privacy of genetic information from misuse by government agencies, their contractors, and government grantees [84]. It shares with the Privacy Act of 1974 the underlying principle that information collected for one purpose cannot be used for a different purpose without the individual's consent [175]. The act would also guarantee individuals access to their genetic records maintained by governmental agencies. The scope of the Act is obviously limited; it would not provide redress for misuse of genetic information by private and commercial agencies not receiving federal funds. In particular, the Act would not address concerns about the use of genetic testing by insurance companies to determine insurability. In response to these concerns, a Senate Resolution was introduced requesting placement of a two-year moratorium on the denial of individuals insurance coverage due to genetic abnormalities and urging that, during that period, the Department of Health study the use of genetic information to deny insurance [163]. (Resolutions are non-binding recommendations.)

E. *Guidelines for Clinicians*

In 1991, the membership of the National Society of Genetic Counselors (NSGC) accepted its Code of Ethics, effective January 1992. Reflecting some of the ethical concerns addressed in the 1983 President's Commission for the Study of Ethical Problems in Medicine and Biomedical and Behavioral Research in its report, "Screening and Counseling for Genetic Conditions", [158] the two-page NSGC code advises genetic counselors to keep information offered by clients confidential, to avoid exploitation of clients for the counselor's advantage, and to serve all clients equally. Perhaps more than the codes of the American Medical Association or the American Psychiatric Association, the NSGC Code stresses respecting clients' own values and views the professional counselor's role as enabling "their clients to make informed independent decisions, free of coercion ...". It also urges counselors to "prevent discrimination on the basis of race, sex, sexual orientation, age, religion, genetic status, or socioeconomic status".

The ethical concerns of clinical geneticists and genetic counselors are not novel. Responsibilities to third parties, duty to protect patients' confidentiality, or concern about their patients' ability to understand the probability-ridden information disclosed to them so that they can make informed decisions – all are already prominent issues in the bioethics literature. Even data-banking, genetic discrimination, and the use of medical information for non-medical purposes are not exactly new; the practice of storing information and the possibility for its misuse predate developments in genetics.

Increased use of genetic screening and diagnosis will, however, severely strain the American institutions which currently foot the health care bill, especially individual and employer-based insurance. Some suggest that the ability to obtain a genetic profile of every individual's predisposition to disease undermines the very notion of aggregating people into actuarial classes; each person may be viewed as an actuarial class of one. Less dramatic commentators suggest that developments in genetics, like the epidemic of HIV infection, will at least highlight the current failure of the American health care system to provide equitable access to its resources.

VIII. FINANCING HEALTH CARE

Financing America's health care bill is one of the country's primary ethical, economic, and political challenges. Expenditures on health care are estimated to equal approximately 13 percent of the gross national product and growing twice to three times as fast as other consumer services and goods [58]. In 1988, 13 percent of 244 million Americans are insured by Medicare, 6 percent by Medicaid, 57 percent by employer health plans, 9 percent by individual insurance or other sources. Most disconcerting, given the amount spent on health care is that 15 percent of Americans are uninsured. Of the 37 million uninsured, 26 million were employed, while approximately 3 million – some of whom were employed – were considered "medically uninsurable" by private insurance companies because of preexisting health conditions ([66];[171]). Progress in the detection of genetic predisposition to disease will exacerbate this problem. Further, a 1985 estimate, projected from 1977 survey data, suggested that approximately 56 million were underinsured [66].

Ensuring that Americans have equitable access to basic health care is certain to be a major issue during the 1992 Presidential campaign. Most agree that current system is "broken", but there is no consensus about how to fix it. Many states, task forces, and professional organizations have already paid considerable attention to methods of providing health care to the uninsured, the underinsured, and the uninsurable. Two initiatives stand out among the efforts of the past two years. First, in Oregon, legislators have been struggling with a plan to rank according to importance the health care services offered to Medicaid recipients. Second, an initiative of a very different sort was undertaken by the editors of *The Journal of the American Medical Association*, who asked their readers to propose major options to address the problems of providing adequate health insurance and health care for all Americans.

A. *Oregon Basic Health Services Act*

The goal of the 1989 Oregon Basic Health Services Act is to expand Medicaid coverage to include all Oregonians living in poverty by covering only sufficiently important health services ([76];[136]). The Act charged the Oregon Health Services Commission (OHSC) to develop "a priority list of health services, ranging from the most to the least important for the entire population to be served" [136]. The OHSC initially conducted a cost-effectiveness study of over 1600 health services and issued a draft list of services ranked according to their importance. "Importance" was measured as a function of the service's expected outcome (*e.g.*, in terms of alleviation of suffering or prolongation of life), its cost, and the number of patients who stand to benefit from it. Office visits (usually for non-life-threatening or even self-limiting conditions such as thumb sucking and low back pain) ranked high on the list, while more complex services for life-threatening conditions (*e.g.*, appendectomies) ranked relatively low. Negative public reaction to this ranking, as well as the counterintuitive results themselves, led the OHSC to abandon cost-effectiveness as a basis for ranking medical interventions in its final priority list ([76];[77]).

Instead the OHSC developed 17 categories of services, describing either a type of service (*e.g.*, maternity care) or an expected outcome (*e.g.*, treatment of life-threatening illness where treatment restores life-expectancy and returns to previous health). These categories were then ranked by Commissioners according to three criteria: value to the individual, value to society, and whether the category seemed "necessary". The OHSC then assigned individual treatments to what it deemed to be the most appropriate category. Ultimately, some services were rearranged in the final list to produce a much more intuitively plausible priority list than the earlier draft.

Which treatments are funded from the list will depend on how much money Oregon is willing to spend on health care for Medicaid recipients. Independent actuaries are to provide an estimate of the cost of the services on the final list [167]. The Oregon legislature will, in light of the amount of available funds, draw a line somewhere on the list to separate the services covered under Oregon Medicaid from those lower on the priority list.

Behind the Oregon initiative is a rough notion of equity: if a medical service is to be made available to anyone in Oregon, it will have to be made available to all Oregonians living below the poverty level. If the state cannot afford to provide a particular service to all those living at the poverty level, then none shall have it provided through Medicaid funds. This equity notion is, of course, coupled with a utilitarian notion of providing the most important services to the greatest number of poverty-stricken Oregonians; thus, the services were ranked according to their importance. Therefore, most organ transplants, for example, will not be paid for with public funds, but the fund will be used to provide maternity care for approximately 1500 women [105]. It therefore constitutes an attempt to "change the debate from who is covered to what is covered" ([103];[104]).

Also of importance is the Oregon Act's provisions for active solicitation of public involvement to build a consensus on the values to be used in making health care resource allocation decisions [105]. Indeed the Oregon experiment, as it has often been called, has been hailed as grass roots attempt to define what constitutes a package of basic health care services. It has also been criticized as showing a "naive faith in scientism" and in the possibility of depoliticizing resource allocation [104].

From a political perspective the Oregon plan will likely prove difficult to implement. Coby Howard, a seven year old leukemia patient in Oregon, was denied funding for a bone marrow transplant and died $30,000 short of the $100,000 necessary to obtain the procedure [105]. He thus became the focus of state and national attention and sympathy. Cases like Howard's will test the resolve of state politicians who fear the loss of their constituents' support. Oregon has also been criticized for developing a rationing system which solely affects the poor. Opponents have claimed that a) the system will have a disproportionately negative impact on poor women and children; b) that "those who are worst off... will have their benefits shaved to pay for expanded coverage to those only slightly better off than themselves"; and c) "while the poor get less, Oregon's providers and taxpayers are benefitted" [57]. Oregon has also been criticized for its failure to appropriate adequate total funds for its plan; indeed, Oregon ranks 39th in per capita spending for Medicaid [165].

Oregon's bold move will be closely scrutinized in the next few years as other states attempt to develop plans to address the needs of their indigent, uninsured, and uninsurable citizens while at the same time controlling run away health care costs.

B. *JAMA Editors' Call for Solutions*

If the Oregon Basic Health Services Act is the most noteworthy among recent state health care initiatives, the proposals received by the editors of *The Journal of the American Medical Association*, in response to their call for innovative solutions to the problem of providing health care for the uninsured and the underinsured, are the most noteworthy proposals to come from professional organizations and public policy task forces. These proposals, published in the May 15, 1991 issue of the *Journal*, are of four basic types: an all-government insurance system; a program of income-indexed tax credits for the purchase of private insurance; plans requiring employers to provide their employees with health insurance or pay a tax, with the government insuring the poor and nonworkers; and plans involving compulsory, employer-based private insurance, again with the government insuring the poor and nonworkers [34].

The American Medical Association (AMA) proposal – entitled "Health Access America" – contends that improvements in the health care system should build upon existing strengths of America's mixed private-public funding of health care through insurance companies rather than radically transform the current system of financing health care. The stated goals of their program include ensuring that affordable insurance coverage for appropriate care is available to all Americans, expanding continued access to health services for the elderly, delivering services at appropriate costs, and guaranteeing that patients are free to select their health care benefits and provider.

Among other points, the AMA proposal calls for Medicaid reform to ensure that all medically necessary services are provided to people below the poverty level, for the elimination of state by state differences in the provision of Medicaid benefits, and for employers to provide health insurance for all full-time employees and their families, and for tax incentives for employers. The plan also calls for the development of an actuarially sound, prefunded Medicare program to ensure the elderly's access to health care services and for the expansion of financing for long-term care for the elderly by "increasing private sector coverage, encouraged by tax incentives and an asset protection program, and to provide Medicaid coverage for those below the federal poverty level" [171]. The AMA estimated that the plan, once fully implemented, would cost the federal government approximately $21 billion per year (in 1990 dollars) [171].

The Pepper Commission is the blue ribbon, congressional bipartisan Commission on Comprehensive Health Care. Its recommendation also involves extending employment-based and public insurance coverage. However, it proposes replacing Medicaid with a new federal program which would cover all those not covered through the workplace, as well as workers whose employers find public coverage more affordable than private insurance. Reform of the private insurance industry, tax credits for small employers, and the opportunity to purchase public coverage are proposed as means of guaranteeing affordable coverage for employers [154].

The Physicians for a National Health Program (PNHP) plan proposes to replace the current mixed system with a Canadian style unitary payer system. PNHP proposes to cover all Americans under a publicly administered and tax-financed national health program, which would replace all private insurers, Medicaid, and Medicare. The proposal contends that "a unitary program could initially pay for expanded care out of administrative savings without adding new costs to the overall health care budget and would establish effective mechanisms for long-term cost control" [74]. New taxes would replace employer-employee insurance premiums and would join payroll taxes and existing government revenue sources in funding the program.

The diversity of these proposals, submitted to JAMA in response to its editors' plea for focused debate, but developed in recent years in response to public and professional organizations' desire for reform of the American health care system, is striking. These proposed solutions raise many important philosophical issues ranging from what constitutes justice, to how to define "medically necessary". The economic and political complexity of the problem and the diversity of proposed solutions guarantee that plans to reform and finance American health care will influence, if not dominate, the political agenda for the next few years.

IX. REGULATION OF PROFESSIONAL CONDUCT

As a largely self-regulating profession which is responsible for promoting the public's health and warranting the public's trust, the medical profession has become increasingly concerned with the ethical conduct of its members. In particular, conflicts of interest, sexual misconduct, and malpractice have received attention in the lay and professional literature and have recently been addressed in professional organization guidelines.

A. *Physicians and the Pharmaceutical Industry*

The practice of gift giving by companies in the pharmaceutical and medical equipment industries to physicians – both clinicians and researchers – has come under increasing scrutiny since the early 1980s when various scandals about the influence of drug companies came to light [135]. Ethical questions about this practice have been raised in the editorial columns of medical and pharmacy journals, on the floor of the House of Delegates of the American Medical Association, and in the United States Senate [145].

In 1989, T.B. Graboys pointed out that the management of silent ischemia is complicated not only by clinical factors, but also by nonclinical economic factors [72]. The enormous market in treating silent ischemia has led the pharmaceutical industry to encourage physicians to treat patients despite the lack of scientific evidence for this therapy. In a survey of faculty at seven midwestern teaching hospitals and of housestaff at two of them, 25 percent of the faculty and 32 percent of the housestaff reported having altered their prescribing practices at least once based on their contact with pharmaceutical representatives [112].

Although drugs and medical device industry share the common goal of alleviating patients' pain and suffering, there is an inherent problem with the relationship. To sell its drugs and devices, the industry cannot deal directly with patients who are the consumers; instead, companies must deal with the prescriber or drug purchaser, which makes these parties the target of direct marketing practices may not best serve patient interests [30]. In contrast to most marketing practices, in the case of pharmaceuticals, the party receiving the benefit of the sales campaign is not the one paying for the products.

The most thorough ethical analysis of the practice of gift giving and its ethical repercussions was discussed by Chren *et al.* in 1989 [39]. They note three effects of gift giving and acceptance. First, in medicine, the gift's cost is ultimately passed on to patients without their explicit knowledge in the cost of the medical supply. Second, physicians' acceptance of gifts contributes to the developing perception that the medical profession no longer serves the best interest of patients. Third, the acceptance of a gift establishes a giver-recipient relationship, which entails vague but real obligations. Chren *et al.* point out that "accepting a gift ... triggers an obligatory response from the recipient that involves a complex web of concerns or sentiments that is directed toward the giver" ([39], p. 3449). Because physicians do not respond to gifts by giving gifts in return, the obligatory response may take the form of prescribing the detailed product over a less expensive and equally effective therapeutic agent. Thus, note Chren *et al.*, obligations that result from gifts may threaten the physician-patient relationship in which the physician's role is that of the patient's agent or trustee whose first consideration in all clinical decisions is the patient's interest. The analysis by Chren *et al.* concluded with a call for professional societies to establish guidelines which would acknowledge the ethical dangers of obligations created by gifts from industry.

In 1990, the American College of Physicians (ACP) responded with a position paper addressing the influence of industrial gifts on clinical judgment, conduct in practice-based clinical trials, and continuing medical education [6]. The paper again called upon professional societies to develop guidelines to discourage excessive gift giving and urged individual practitioners to ask: "Would you be willing to have these arrangements generally known?" This attempt to address the ethics of physician encounters with industry, despite its good intentions, came under immediate criticism for being too vague and lacking in enforcement mechanisms [147].

In December 1990, the Council on Ethical and Judicial Affairs of the American Medical Association issued its opinion on gifts to physicians from industry [46]. They cited the same ethical concerns discussed in earlier articles, including the influence on physician practices, the appearance of impropriety, and the cost of gifts. In contrast to the ACP position paper, the guidelines specifically outline behavior which physicians should observe to avoid the acceptance of inappropriate gifts. Gifts should be of minimal value, relate to the physician's work, and primarily benefit patients. Cash payments should not be accepted. Subsidy from industry for continuing medical education should be made to the conference's sponsor, not directly to physicians. Subsidies for travel and time compensation should not be accepted by individual physicians. In response to concerns about industry influence in medical schools and residency programs, the guidelines provide that medical schools and residency training programs may accept scholarship funds to permit residents and students to attend conferences, as long as the selection of recipients is made by the academic institution [68]. Finally, physicians should especially avoid gifts tied to prescribing practices.

Although these guidelines resolve many of the problems related to industrial gift giving, two problems persist. First, the guidelines lack efficient enforcement mechanisms. When the Collagen Corporation sent 55 dermatologists who purchased the largest volume of injectable collagen during a nine-month period on an eight-day South Pacific educational cruise, the AMA could only write to Collagen Corporation to request their names and write to them informing them that they are violating the AMA's code of ethics [146]. Although the Pharmaceutical Manufacturers Association (PMA) agreed that its members would abide by the AMA's guidelines, non-PMA members, like the Collagen Corporation, are outside the PMA's authority.

Second, these guidelines do not apply to pharmacists who are playing an ever increasing role in drug product selection in many institutions via their membership on institutional pharmacy and therapeutics (P&T) committees. The main function of the P&T committee, whose membership usually consists of representatives from medicine, pharmacy, nursing, and administration, is the establishment and maintenance of a drug formulary. For economic reasons hospitals are increasingly limiting the number of drugs on their formulary. Because of this pharmacists have found themselves not only being called upon to recommend drug products in the selection process, but also being the target of the industry's marketing efforts [186]. Examination of the relationship between pharmacists and the industry leading to the development of standards and ethical guidelines for pharmacists is now necessary [123].

In addition, in 1989 Arnold Relman called into question ties between industry and medical research [149]. He cited a reduction in scientific objectivity and an erosion of public trust as his primary concerns. The Council on Ethical and Judicial Affairs and the Council on Scientific Affairs recommended in 1990 that guidelines be developed to help curtail conflicts of interest in the medical center-industry research relationship. They recommend that investigators not buy or sell company stock when they are involved with that company, that their remuneration be commensurate with their research efforts, that they disclose any company ties, and that review committees be formed at medical centers to review disclosures about financial associations.

B. *Physicians' Sexual Misconduct*

The relationship between physicians and the pharmaceutical and medical device industry is not the only relationship in recent years to prompt the formulation of guidelines for conduct. Sexual relationships between physicians and their patients or between medical supervisors and their trainees have recently come under close scrutiny. Studies on supervisor-trainee sexual relationships documented that many trainees surveyed had experienced some form of sexual harassment, and approximately 5 percent of psychiatric residents surveyed had experienced some form of sexual involvement with their educators [150]. Studies of doctor-patient sexual relationships focused primarily on psychiatrists and revealed that between 5 percent and 10 percent had sexual contact with their patients ([47];[120]). The impact of these studies prompted guidelines from both the AMA and the American Psychiatric Association (APA). Many states already had criminal or civil sanctions against such behavior, and most educational institutions have policies prohibiting sexual harassment.

In 1989, the Council on Ethical and Judicial Affairs of the AMA issued its *Current Opinions*, which outlined a code of ethics for physicians, "Principles of Medical Ethics of the AMA" [134]. In the same year, the APA also released a report elaborating the seven tenets set forth in the AMA report and defining more explicit codes of conduct for psychiatrists [9].

One of the guidelines in the AMA report states that "a physician shall deal honestly with patients and colleagues and strive to expose those physicians deficient in character or competence, or who engage in fraud or deception" ([9], p.4). In interpreting this principle, the APA report states that "sexual activity with a patient is unethical" and that sexual involvement with former patients is almost always unethical because it "exploits emotions deriving from treatment" [9]. While there is consensus on the former position, sex with former patients is the subject of more controversy.

At its 1990 interim meeting, the AMA House of Delegates adopted a report which noted that physicians may intentionally exploit patients either by presenting the sexual contact as part of the therapeutic relationship or by taking advantage of an unconscious or otherwise incapacitated patient, or an error in judgment may lead the physician to pursue a sexual relationship during the course of a therapeutic one [47]. Whether intentionally exploitative or not, sexual involvement jeopardizes the patient's best interest by allowing the physician's needs to eclipse the interests of the patient. Like the APA, the AMA House of Delegates did not condone the establishment of a sexual relationship following the termination of the therapeutic one. In doctor-patient sexual relationships, the physician may exploit his greater power and is the party most likely to emerge from the sexual relationship unscathed. The issue of power is highlighted by studies focusing on psychiatrists which revealed that 85 percent to 90 percent of their sexual contact with patients involved a male physician and a female patient, while studies concluded that the patients who submitted to the sexual relationship were "more likely to consider exploitative relations with an authority figure to be normal" [47]. Further, much anecdotal evidence shows that the consequences of a sexual relationship with one's physician is usually deleterious to female patients' health. The resulting symptom complex, which resembles Post-Traumatic Stress Disorder, is typically severe and prolonged [62].

The exploitative nature of such sexual involvement is mirrored in supervisor-trainee sexual relationships. The APA and the AMA's Council on Ethical and

Judicial Affairs both cite the disparity between the relative positions of each group in the hierarchy of medical education as exacerbating the possibility for exploitation and note that engaging in these sexual relations potentially jeopardizes the medical care of patients ([9];[150]). Results from a survey indicate that the trainee's perception of the coercive and exploitative nature of this relationship increases over time [150].

C. *National Practitioner Data Bank*

The creation by the AMA of the National Practitioner Data Bank (NPDB) may play a major role in curtailing sexual exploitation by physicians, as well as other forms of professional misconduct. The data bank was established as part of the Health Care Quality Improvement Act passed by Congress in 1989 [78]. The data bank is to serve as a "nationwide repository for information related to professional conduct and competence" [7]. Prior to the establishment of the data bank, physicians could lose their licenses in one state and go to another state to open a practice.

All Medical Examiners' Boards, hospitals and other health care facilities, and professional societies are required to report any licensure, membership, or clinical privilege actions taken against a physician to the NPDB [7]. In addition, any payment on behalf of a physician as the result of a malpractice claim or judgment must be reported to both the data bank and the state medical board. Hospitals are, in turn, required to obtain information from the NPDB about all physicians when they apply either to become a member of the staff or for clinical privileges, and hospitals must obtain updates from the data bank every two years.

The information contained in the NPDB is confidential and available only to authorized parties, and several provisions have been made to permit physicians to dispute the information reported about them. Physicians may periodically request a copy of their data bank file. If a physician wishes to dispute reported information, he must first attempt to do so with the reporting party. Ultimately the Secretary of Health and Human Services will, upon request, review the disputed information to determine its accuracy. Despite these and other safeguards, there is concern that the NPDB might be used to intentionally damage a physician's career, that information may not be accurate or complete, or that the information reports obtained from the NPDB will come to be viewed by institutions receiving the reports as the sole or primary source of relevant information about physicians' conduct and competence.

The APA report on sexual misconduct charged psychiatrists' APA district branch with the responsibility of investigating charges of unethical conduct [9]. In accordance with the reporting requirements of the NPDB, this process would, in theory, protect patients from psychiatrists who were proven to have engaged in unethical sexual behavior. Similar channels of investigation and reporting provide the same protection from physicians in other types of practice.

Both the APA's report and the report adopted by the AMA Council stress that it is a physician's ethical responsibility to report colleagues' sexual transgressions to the appropriate authorities ([9];[47];[120]). The AMA Council report additionally cites the importance of using medical education to clarify the ethical and emotional issues surrounding sexual transgression [47]. In addition, medical training programs are given the responsibility of protecting trainees by adopting and enforcing policies addressing sexual exploitation and coercion by medical supervisors [150]. These measures may enhance the effectiveness of the NPDB, whose effectiveness is compromised by the fact that victims of sexual exploitation may be emotionally unprepared to initiate and pursue the charge of sexual misconduct. Indeed, for the National Practitioner Data Bank to function effectively with respect to all types of misconduct, it must serve as an integral part of an active peer review process in which physicians monitor the competence and ethics of all of their colleagues professional conduct.

Department of Genetics
University of Pittsburgh
Pittsburgh, Pennsylvania

University of Tennessee
Knoxville, Tennessee
U.S.A.

BIBLIOGRAPHY

1. "ABA's Two Models for 'Baby M' Laws", *New York Times*, February 9, 1989, C13.
2. "Aid In Dying", Initiative Measure No. 119.
3. "AMA Supports Testing of French Abortion Pill", *New York Times*, June 29, 1990, A16.
4. American Academy of Orthopedic Surgeons: 1991, "HIV-Infected Orthopedic Surgeons", *American Academy of Orthopedic Surgery Bulletin*, 39 (Supplement to No. 2), 1-16.
5. American Civil Liberties Union: 1989, "Criminal Prosecutions for Fetal Abuse Increasing at an Alarming Rate", *Reproductive Rights Update*, 1 (September 1), 2.
6. American College of Physicians: 1990, "Physicians and the Pharmaceutical Industry", *Annals of Internal Medicine* 112, 624-626.
7. American Medical Association: *National Practitioner Data Bank*, Information For Physicians.
8. *American Medical News*, "CDC: Dentist Likely Infected 3 Patients", February 4, 1991.

9. American Psychiatric Association: 1989, *The Principles of Medical Ethics With Annotations Especially Applicable to Psychiatry*, Washington, D.C., American Psychiatric Association.
10. American Psychiatric Association Ad Hoc Committee on AIDS Policy: 1988, "AIDS Policy: Confidentiality and Disclosure", *American Journal of Psychiatry*, 145, 541.
11. Americans with Disabilities Act (ADA), P.L. 101-336, 104 Stat. 327 (1990)
12. Anderson, A.: 1989, "Abortion Ruling Divides the United States", *Nature*, 340 (July 13), 83.
13. Anderson, A.: 1989, "Science Ahead of the Law", *Nature*, 340, 492.
14. Annas, G.J. *et al.*: 1989, "Brief for Bioethicists for Privacy as *Amicus Curiae* Supporting Appellees", *American Journal of Law and Medicine*, 15, 169-183.
15. Annas, G.J.: 1991, "Crazy Making: Embryos and Gestational Mothers", *Hastings Center Report*, (Jan/Feb), 35-38.
16. Annas, G.J.: 1991, "Fetal Protection and Employment Discrimination: The *Johnson Controls Case*", *NEJM*, 325, 740-743.
17. Annas, G.J.: 1989, "Four-One-Four", *Hastings Center Report*, (September/October), 27-29.
18. Annas, G.J.: 1991, "Killing Machines", *Hastings Center Report*, (March/April), 33-35.
19. Annas, G.J.: 1990, "Outrageous Fortune: Selling Other People's Cells", *Hastings Center Report*, (Nov/Dec), 36-39.
20. Annas, G.J.: 1991, "Restricting Doctor-Patient Conversations in Federally Funded Clinics", *NEJM*, 325, 362-364.
21. Annas, G.J.: 1991, "The Health Care Proxy and the Living Will", *NEJM*, 324, 1210-1213.
22. Annas, G.J. *et al.*: 1990, "The Right of Privacy Protects the Doctor-Patient Relationship", *JAMA* 263, 858-861.
23. Arras, J.D.: 1990, "AIDS and Reproductive Decisions: Having Children in Fear and Trembling", *Milbank Quarterly*, 68, 353-382.
24. "As Family Protests, Hospital Seeks An End to Woman's Life Support," *New York Times*, January 10, 1991.
25. Bayer, R.: 1990, "AIDS and the Future of Reproductive Freedom", *Milbank Quarterly*, 68 (suppl.2), 179-204.
26. Bayer, R.: 1991, "Public Health Policy and the AIDS Epidemic: An End to HIV Exceptionalism?", *NEJM*, 324, 1500-1504.
27. *Bentivegna v. US Dept of Labor*, 694 F2d 619, 623 (9th Cir 1982).
28. Berke, R.L.: 1991, "Groups that Back Right to Abortion Ask Court to Act", *New York Times*, (November 8), 1.
29. Boland, R.: 1990, "Recent Developments in Abortion Law in Industrialized Countries", *Law, Medicine and Health Care*, 18, 404-418.

30. Bricker E.M.: 1989, "Industrial Marketing and Medical Ethics", *NEJM* 320, 1690-1692.
31. Burton, T.: 1991, "AMA Opposes Government Interference with Doctor's Counseling of Patients", *Wall Street Journal*, June 26, B3.
32. California Penal Code Section 290.2 (West 1988).
33. Campbell, C.S.: 1989, "Abortion: Searching for the Common Ground", *Hastings Center Report*, (July/August), 22-29.
34. "Caring for the Uninsured: Choices for Reform", Editorial: 1991, *JAMA* 265, 2564-2565.
35. Centers for Disease Control: 1991, "Recommendations for Preventing Transmission of Human Immunodeficiency Virus, Hepatitis B Virus to Patients During Exposure-Prone Invasive Procedures", *JAMA* 266, 771-776.
36. Centers for Disease Control: 1991, "Update: Transmission of HIV Infection During Invasive Dental Procedures – Florida", *Morbidity and Mortality Weekly Report*, 40, 377-381.
37. Charo, R.A: 1988, "Legislative Approaches to Surrogate Motherhood", *Law, Medicine and Health Care*, 16, 96-112.
38. Chavkin, W.: 1990, "Drug Addiction and Pregnancy: Policy Crossroads", *American Journal of Public Health*, 80, 483.
39. Chren M.M., Landefeld, C.S., Murray T.H.: 1989, "Doctors, Drug Companies, and Gifts", *JAMA* 262, 3448-3451.
40. Clymer, A.: 1991, "Bill to Let Clinics Discuss Abortion is Vetoed by Bush", *New York Times*, November 19, 1.
41. Cole, H.M.: 1990, "Legal Interventions During Pregnancy: Court-Ordered Medical Treatments and Legal Penalties for Potentially Harmful Behavior by Pregnant Women", *JAMA* 264, 2663-2670.
42. Colo. Rev. Stat. Section 17-2-201 (5)(g)(I) (1988).
43. Committee on Ethics, American College of Obstetricians and Gynecologists: 1990, "Human Immunodeficiency Virus Infection: Physicians' Responsibilities", *Obstetrics and Gynecology*, 75, 1043-1045.
44. *Commonwealth of Mass v. Secretary of Health and Human Services*, 899 F 2d 53 (1st Cir.1990).
45. Cotton, P.: 1991, "Preexisting Conditions 'Hold Americans Hostage' to Employers and Insurance", *JAMA* 265, 2451-2453.
46. Council on Ethical and Judicial Affairs of the American Medical Association: 1991, "Gifts to Physicians from Industry", *JAMA* 265, 501.
47. Council on Ethical and Judicial Affairs of the American Medical Association: 1991, "Sexual Misconduct in the Practice of Medicine", *JAMA* 266, 2741-2745.
48. Council on Scientific Affairs and Council on Ethical and Judicial Affairs: 1990, "Conflicts of Interest in Medical Center/Industry Research Relationships", *JAMA* 263, 2790-2793.

49. *Cruzan v. Director, Missouri Department of Health*, 110 S.Ct. 2841 (1990).
50. Curran, W.J.: 1991, "Scientific and Commercial Development of Human Cell Lines", *NEJM*, 324, 999.
51. *Davis v. Davis*, 1990 Tenn. App. LEXIS 642 (13 September 1990).
52. *Davis v. Davis v. King*. Fifth Jud. Ct.Tennessee, E-14496, September 21, 1989 (Young, J.)
53. "Dealing Death or Mercy", *New York Times*, March 17, 1991.
54. De Gorgey, A.: 1990, "The Advent of DNA Databanks", *American Journal of Law and Medicine*, 16, 387.
55. Dickey, N.W.: 1991, "Physicians Infected with HIV", *JAMA* 265, 2338.
56. *Doe by Doe v. Roe*, 151 Wis. 2d 366, 444 N.W.2d 437 (1989).
57. Dougherty, C.J.: 1991, "Setting Health Care Priorities", *Hastings Center Report*, (May/June), 1-9.
58. Enthoven, A.C., Kronick, R.: 1991, "Universal Health Insurance through Incentives Reform", *JAMA* 265, 2532-2536.
59. *Escola v.Coca-Cola Bottling Co.*, 24 Cal.2d 453, 150 P.2d 436, 1944.
60. *Estate of Behringer v. Medical Center*, No. l88-2550 (NJ Super Ct Law Div April 25, 1991).
61. Fed.Reg. Feb 2, 1988; 53:2921-46.
62. Feldman-Summers, S.: 1989, "Sexual Contact in Fiduciary Relationships", in G.O.Gabbard, *Sexual Exploitation in Professional Relationships*, Washington, D.C., American Psychiatric Press.
63. "Filling the Gap Where a Living Will Won't Do", *New York Times*, January 17, 1991.
64. *Forsmire v. Nicoleau*, 75 N.Y.2d 218;551 N.E.2d 77 (N.Y. 1990).
65. Fox, D.M.: 1990, "Financing Health Care for Persons with HIV Infection: Guidelines for State Action", *American Journal of Law and Medicine* 16, 223-247.
66. Friedman, E.: 1991, "The Uninsured: From Dilemma to Crisis", *JAMA* 265, 2491-2495.
67. Garcia, S.A.: 1990, "Reproductive Technology for Procreation, Experimentation and Profit", *Journal of Legal Medicine* 11, 1-57.
68. Gelbart H.: 1990, Letter to the Editor, *JAMA* 263, 2177.
69. "George Washington University Medical Center Sets Policy", *New York Times*, November 29, 1990.
70. Gervais, K. and Miles, S: 1990, "RU-486: New Issues in the American Abortion Debate", The University of Minnesota's Center for Biomedical Ethics. Biomedical Ethics Reading Packet. Minneapolis UMHC.
71. Gostin, L.: 1991, "The HIV-Infected Health Care Professional: Public Policy, Discrimination, and Patient Safety", *Archives of Internal Medicine* 151, 663-665.
72. Graboys, T.B.: 1989, "Conflicts of Interest in the Management of

Silent Ischemia", *JAMA* 261, 2116-2117.

73. *Griswold v. Connecticut*, 381 U.S. 479 (1965).
74. Grumback, K.: 1991, "Liberal Benefits, Conservative Spending: The Physicians for a National Health Program Proposal", *JAMA* 265, 2549-2554.
75. Gwinn, M.: 1991, "Prevalence of HIV-Infection in Childbearing Women in the United States", *JAMA* 265, 1704-1708.
76. Hadorn, D.C.: 1991, "The Oregon Priority-Setting Exercise: Quality of Life and Public Policy", *Hastings Center Report*, (May/June), 11-16.
77. Hadorn, D.C.: 1991, "Setting Health Care Priorities in Oregon: Cost-effectiveness Meets the Rule of Rescue", *JAMA* 265, 2218-2225.
78. Health Care Quality Improvement Act of 1989, Title IV of Public Law 99-660.
79. Hilts, P.J.: 1990, "FDA says It Allows Study of Abortion Drug", *New York Times*, November 20, C9.
80. Hoadley, D.L.: 1991, "Fetal Protection Policies: Effective Tools for Gender Discrimination", *The Journal of Legal Medicine* 12, 85-104.
81. *Hodgson v. Minnesota*, 110 S.Ct.2926 (1990).
82. Holtzman,N.A.: 1989, *Proceed with Caution: Predicting Genetic Risks in the Recombinant DNA Era*, Johns Hopkins University Press, Baltimore, Maryland.
83. "Hospital Seeks to Override Family Objections, Stop Respirator", *American Medical News*, January 28, 1991.
84. Human Genome Privacy Act, H.R. 2045, Introduced April 24, 1991.
85. Ill.Rev.Stat. ch 37, 802-3 (c) (1989).
86. "In Matters of Life and Death, The Dying Take Control", *New York Times*, August 18, 1991.
87. *In re A.C.* 573 A.2d. 1235 (D.C. 1990).
88. *In re Cabrera* 552 A.2d 1114 (Pa. Super. Ct. 1989).
89. *In re e.g.* 549 N.E.2d 322 (Ill. 1989).
90. *In re Guardianship of Browning* 543 So. 2d 258 (Fla. Dist. Ct. App. (1989).
91. *In re McCauley* 1991 WL 3337 (Mass.)
92. *In re: The Conservatorship of Helga M. Wanglie*, 4th Judicial District Court (Dist. Ct. Probate Ct. Div.) PX-91-283. Minnesota, Hennepin County.
93. Intergovernmental Health Policy Project: 1989 Legislative Overview, Intergovernmental AIDS Reports, January 1990:3.
94. *International Union v. Johnson Controls*, 111 S.Ct. 1196 (1991).
95. *International Union v. Johnson Controls*, 886 F.2d 871 (7th Cir. 1989), *cert. granted*, 110 S. Ct. 1522 (1990).
96. *In the Matter of Baby M*, 537 A.2d 1227, 109 N.J. 396 (N.J.S.Ct. 1988).
97. Johnsen, D.J. and Wilder, M.J.: 1989, "*Webster* and Women's Equality", *American Journal of Law and Medicine*, 15:2-3, 178-183.

98. *Johnson Controls, Inc v. California Fair Employment & Housing Commission* 218 Cal. App. 3d 517, 2647 Cal. Rptr. 158 (1990).
99. *Johnson v. Calvert*, Cal. Super. Ct., Orange Co., Dept.11, No. X633190 (22 October 1990).
100. *Johnson v. State*, No. 89-1765 (Fla. Dist.Ct.App.,5th Dist.,1989).
101. "Judge Rejects Request by Doctors To Remove a Patient's Respirator", *New York Times*, July 2, 1991.
102. King County, Washington Ordinance 88-105.
103. Kitzhaber, J.: 1989, *The Oregon Basic Health Services Act*, Oregon, State Capital (mimeograph).
104. Klein, R.: 1991, "On the Oregon Trail: Rationing Health Care", *British Medical Journal*, January 5.
105. Klevit, H.D. *et al.*: 1991, "Prioritization of Health Care Services: A Progress Report by the Oregon Health Services Commission", *Archives of Internal Medicine* 151, 912-916.
106. Kolbert, K.: 1989, "The *Webster Amicus Curiae* Briefs: Perspectives o on the Abortion Controversy and the Role of the Supreme Court", *American Journal of Law and Medicine* 15:2-3, 153-168.
107. La Puma, J., Orentlicher, D. and Moss, R.: 1991, "Advance Directives on Admission", *JAMA* 266:3 (July 17), 402-405.
108. *Leckelt v. Board of Commissioners of Hospital District No. 1*, 714 F. Supp. 1377 (D.C. La. 1989).
109. Lewin, T.: 1991, "Strict Anti-Abortion Law Signed in Utah", *New York Times*, January 26, 10.
110. Lo, B. and Steinbrook, R.: 1991, "Beyond the Cruzan Case: The U.S. Supreme Court and Medical Practice", *Annals of Internal Medicine* 114:10 (May 15), 895-901.
111. Logli, P.A.: 1990, "Drugs in the Womb: The Newest Battlefield in the War on Drugs", *Criminal Justice Ethics* 9:1 (Winter/Spring), 23-29.
112. Lurie, N. *et al.*: 1990, "Pharmaceutical Representatives in Academic Medical Centers: Interaction with Faculty and Housestaff", *Journal of General Internal Medicine* 5, 240-243.
113. Macklin, R.: 1991, "Artificial Means of Reproduction and Our Understanding of the Family", *Hastings Center Report*, (Jan/Feb), 5-11.
114. Mariner, W.K. *et al.*: 1990, "Drugs, and the Perils of Prosecution", *Criminal Justice Ethics* 9:1 (Winter/Spring), 30-41.
115. *Mckay v. Bergstedt*, 801 P.2d 617 (Nev. 1990).
116. McKillip, J.: 1991, "The Effect of Mandatory Premarital HIV-Testing on Marriage: The Case of Illinois", *American Journal of Public Health* 81, 650-653.
117. McKusic, V.: 1989, "Mapping and Sequencing the Human Genome", *NEJM* 320, 910-915.
118. Meisel, A.: 1990, "Lessons from Nancy", *Journal of Clinical Ethics*,

1:3 (Fall), 245-250.

119. Meisel, A.: 1991, *The Right to Die*, Cumulative Supplement No. 2, John Wiley & Sons, New York.
120. Merz, B.: 1990, "House Just Says No To Sex Between Doctors, Patients", *American Medical News*, December 14.
121. Minkoff, H.L. and Moreno, J.D.: 1990, "Drug Prophylaxis for Human Immunodeficiency Virus-Infected Pregnant Women: Ethical Considerations", *American Journal of Obstetrics and Gynecology* 163, 1111-1114.
122. Mo.Rev.Stat. Section 188.205 (1988).
123. Mitchell, J.F.: 1991, "Pharmacists and Drug Company Gifts", *American Journal of Hospital Pharmacy* 48, 457.
124. *Moore v. Regents of the University of California*, 793 P2.d 479, 271 Cal. Rptr. 505 (1972).
125. Nelson, L.J. and Milliken, N.: 1988, "Compelled Medical Treatment of Pregnant Women: Life, Liberty and Law in Conflict", *JAMA* 259, 1060-1063.
126. New Jersey Stat Ann Section 10:5-4.1.
127. *Newmark vs. Williams/D.C.P.* WL 50644(Del. Supr. 1991)
128. *New York v. Bowen*, 889 F.2d 401 (2nd Cir. 1989).
129. *New York v. Sullivan*, 889 F. 2d 401 (2nd Cir.1989).
130. Nolan, K.: 1990, "Protecting Fetuses from Prenatal Hazards: Whose Crimes? What Punishment?", *Criminal Justice Ethics* 9:1 (Winter/Spring), 13-23.
131. Obade, C.C.: 1990, "Compelling Treatment of the Mother to Protect the Fetus: The Limits of Personal Privacy and Paternalism", *The Journal of Clinical Ethics* 1, 85-87.
132. Obade, C.C.: 1990, "The Patient Self-Determination Act: Right Church, Wrong Pew", *The Journal of Clinical Ethics* 1, 320-321.
133. *Ohio v. Akron Center for Reproductive Health*, 110 S.Ct.2972 (1990).
134. *Opinions and Reports of the Council on Ethical and Judicial Affairs of the American Medical Association*, 1989, Chicago, American Medical Association.
135. "Open Scandal", Editorial: 1983, *Lancet* 1, 219-220.
136. Oregon Senate Bill 27 (1989).
137. Orentlicher, D.: 1991, "HIV-Infected Surgeons: *Behringer v. Medical Center*", *JAMA* 266, 1134-1137.
138. Orentlicher, D.: 1989, "*Webster* and the Fundamental Right to Make Medical Decisions", *American Journal of Law and Medicine* 15, 184-188.
139. Orentlicher, D.: 1990, "Genetic Screening by Employers", *JAMA* 263, 1007.
140. Parmet, W.E.: 1990, "Discrimination and Disability: The Challenges of the ADA", *Law, Medicine, and Health Care* 18, 331.

141. *People v. Chrzanowski*, 147 A.D.2d 652, 538 N.Y.S.2d 55 (1989).
142. *Planned Parenthood of Pennsylvania v. Casey*, 744 F. Supp 1323.
143. Proctor, R.N.: 1988, *Racial Hygiene: Medicine Under the Nazis*, Harvard University Press, Cambridge.
144. Public Health Serv. Act, 42 U.S.C.(1988 &Supp. 1990).
145. Randall, T.: 1991, "Kennedy Hearings Say No More Free Lunch – or Much Else – from Drug Firms", *JAMA* 265, 440-442.
146. Randall, T., 1991: "Not All Drug Firms Subject to Gifts Guidelines, but for Physicians, Their Gifts are Still Taboo", *JAMA* 265, 2305.
147. Randall T.: 1990, "New Guidelines Expected in 1991 for Relationship of Continuing Education, Financial Support", *JAMA* 264, 1080.
148. Reilly, P.R.: 1991, *The Surgical Solution: A History of Involuntary Sterilization in the United States*, Johns Hopkins University Press, Baltimore, Maryland.
149. Relman, A.S.: 1989, "Economic Incentives in Clinical Investigation", *NEJM* 320, 933-934.
150. Report of the Council on Ethical and Judicial Affairs, Report B (A-89), Subject: Sexual Harassment and Exploitation Between Medical Supervisors and Trainees.
151. Robertson, J.A.: 1991, "Divorce and Disposition of Cryopreserved Pre-embryos", *Fertility and Sterility* 55:4, 681-683.
152. Robertson, J.A.: 1990, "In the Beginning: The Legal Status of Early Embryos", *Virginia Law Review* 76, 437-517.
153. Robertson, J.A.: 1989, "Resolving Disputes Over Frozen Embryos", *Hastings Center Report* (Nov./Dec), 7-12.
154. Rockefeller IV, J.D.: 1991, "A Call for Action: The Pepper Commission's Blueprint for Health Care Reform", *JAMA* 265, 2507-2510.
155. *Roe v. Wade*, 410 US 113 (1973).
156. Rosenbaum, R.: 1991, "To Hell with the Ethicists. I'm a Real Physician", *Medical Economics* (November 4), 80-104.
157. Rouse, F.: 1990, "Advance Directives: Where are We Heading after Cruzan", *Law, Medicine and Health Care* 18, 353-359.
158. *Rust v. Sullivan*, 111 S.Ct. 1759 (1991).
159. Sack, K.: 1991, "Unlikely Union in Albany: Feminists and Liquor Sellers", *New York Times*, April 5, B1.
160. Sauer, M.V. *et al.*: 1990, "A Preliminary Report on Oocyte Donation Extending Reproductive Potential to Women Over Forty", *NEJM* 323, 1157-60.
161. *Screening and Counseling for Genetic Conditions*: 1983, President's Commission for the Study of Ethical Problems in Medicine and Biomedical and Behavioral Research.
162. Shuster, E.: 1990, "Cruzan: It's Not Over, Nancy"?, *Journal of Clinical Ethics* 1, 237-241.

163. Senate Resolution No. 75, introduced June 10, 1991.
164. Silvestre, L. *et al.*: 1990, "Voluntary Interruption of Pregnancy with Mifespristone (RU 486) and a Prostoglandion Analogue", *NEJM* 322, 645.
165. Somerville, J.: 1991, "Economist Notes Flaws in Oregon Plan", *American Medical News* (March 11), 10.
166. "Spouse Says He'd Never Agree to Cut Life Support": 1991, *Washington Post*, May 30.
167. Stason, W.B.: 1991, "Oregon's Bold Medicaid Initiative", *JAMA* 265, 2237-2238.
168. *State vs Mcaffee* 259 Ga. 579, 385 S.E.2d 651 (1989) as reported in *American Journal of Law and Medicine* 16, 442-443.
169. *State Department of Public Health v. Wells*, Ala. Civ. App. (1989).
170. "State Won't Press Case on Doctor in Suicide": *New York Times*, August 17, 1991.
171. Todd, J.S., Seekins, S.V., Krichbaum, J.A., Harvey, L.K.: 1991, "Health Access American – Strengthening the US Health Care System", *JAMA* 265, 2503-2506.
172. Tolchin, M.: 1991, "Senate Adopts Tough Measures on Health Workers with AIDS", *New York Times*, July 18.
173. U.S. Congress, Office of Technology Assessment: 1991, "Medical Monitoring and Screening in the Workplace: Results of a Survey – Background Paper", OTA-BP-BA-67.
174. U.S. Congress, Office of Technology Assessment: 1990, "Medical Monitoring and Screening in the Workplace".
175. 5 U.S.C. Section 552a (1974).
176. 42 U.S.C. 20006-2(a)(1)(1982).
177. 29 U.S.C.A. Sec. 701 et. seq. (1990).
178. Virginia Code Ann. Section 53.1-23.1 (1989).
179. Wash. Rev. Code Section 43.43.754 (1989).
180. *Weaver v. Reagen* 886 F.2d 194 (8th Cir. 1989).
181. *Webster v. Reproductive Health Services*, 109 S.Ct. 3040 (1989).
182. Wilkerson, I.: 1991, "Court Backs Woman in Pregnancy Case", *New York Times*, April 2.
183. Wilson, B.L. and Wingo, K.L.: 1988, "AIDS in the Workplace: Handicap Discrimination Laws and Related Statutes", *Journal of Legal Medicine* 9, 573-585.
184. *Wons v. Public Health Trust*, 541 So. 2d 96 (Fla. 1989).
185. *York v. Jones*, 717 F. Supp. 423-5 (E.D.Va. 1989).
186. Zoloth, A.M.: 1991, "The Need for Ethical Guidelines for Relationships between Pharmacists and the Pharmaceutical Industry", *American Journal of Hospital Pharmacy* 48, 551-552.

JOHN R. WILLIAMS

BIOETHICS IN CANADA: 1989-1991[1]

I. INTRODUCTION

In 1991 Canadians discovered, much to their surprise, that their health care system was the subject of intense interest and debate beyond their borders. The United States, in particular, was coming to realize that its own system was disintegrating and that reform, if not radical restructuring, was urgently needed. In their search for alternate models of health care financing and delivery, many U.S. academics and policy makers turned their gaze northwards. Other groups, most notably the American Medical Association, recoiled at the prospect of a Canadian (or European) style health care system and launched a well-funded campaign to convince Americans to avoid this path.

What are the features of the Canadian health care system which are so intriguing to outsiders, and what does all this have to do with bioethics? To answer the latter question first, bioethics is a species of practical ethics. It is firmly rooted in reality, in this instance the real world of health, illness, disease, and the professions and institutions which deal with these aspects of human existence. Since health care differs from one country to another, bioethics, as the field of study and practice which identifies and attempts to resolve the moral issues related to health and illness, must also vary on a national basis. This is not to say that those interested in the ethical dimensions of health care have nothing to learn from their counterparts in other countries. However, it is important to avoid the opposite extreme of supposing that one nation's system can be imposed on another without regard for their social and cultural differences. Imperialism in bioethics is to be condemned no less strongly than political, economic, and cultural imperialism.

Canadians have never been imperialists. They are, however, very proud of their health care system. Indeed, it is one of the few features of the national character which finds favor among the vast majority of the citizens. Whereas other aspects of the national identity, bilingualism, multiculturalism, and federalism, are generating much divisive debate as the country attempts to find a *modus vivendi* among its English, French, and Native founding peoples and innumerable newer arrivals, there is near unanimous agreement that the funda-

[1] I am grateful to Carole Lucock and Judith Bedford-Jones, Department of Ethics and Legal Affairs, Canadian Medical Association, and Louis Chauvin, Center for Bioethics, Clinical Research Institute of Montreal, for their invaluable assistance in the preparation of this report.

B. Andrew Lustig (Sr. Ed.), Bioethics Yearbook: Volume 2, 51-81.

mental character of the health care system should not be tampered with. However, whether this system is suitable for other countries is not for Canadians to say, although they would probably be very pleased if this were to be the case.

To speak of a Canadian Health Care System, in the singular, is somewhat inaccurate. There are no less than 10 distinct provincial and 2 territorial systems. Under the Canadian Constitution, health is a provincial/territorial responsibility. Each province and territory controls the financing and operation of hospitals and other health care institutions, regulates the health care professions, and administers the health insurance program. Despite the multiplicity of jurisdictions, the systems share several basic features. The source of this agreement was a series of federal-provincial accords whereby the federal government provides a considerable part of the funds to the provinces and territories for health care, and in return the latter governments agree to incorporate the basic features of the program into their systems.

The four pillars of the national program are enshrined in the *Canada Health Act* (1984), which consolidated the *Hospital Insurance Act* (1957) and the *Medical Care Insurance Act* (1966) and added several amendments. These pillars are:

- universal accessibility: these services are available to all Canadians, regardless of income or other considerations;
- portability: a resident of one province or territory is covered while travelling in other parts of Canada and, at least partially, in other countries;
- public administration: either by an independent commission or by the health department.

At present the provincial and territorial systems are being subjected to intense financial pressures, as costs skyrocket and the federal government reduces its share of expenditures. The chapter on Canada in the next volume of this yearbook will undoubtedly contain considerably longer sections on equitable access to health care and allocation of resources as the financial squeeze results in new policy proposals, legislation, and, possibly, court challenges.

Despite, or perhaps because of, the financial constraints on health care in Canada, bioethics has flourished in the period under review. Five new research centers and groups were established in universities and hospitals, bringing the Canadian total to sixteen. A new funding program in applied ethics, including bioethics, was inaugurated by the Social Science and Humanities Research Council of Canada. The Canadian Medical Association established a Department of Ethics and Legal Affairs with three professional staff. The teaching of bioethics in health care professional programs and in general humanities and social science courses continued the progress underway for over a decade. Moreover, as will be evident from what follows, governments, para-governmental bodies, and professional associations have produced a considerable number of policy statements and legislative proposals on bioethical issues.

Although Canada is not a litigious society, there have been a number of court decisions during this period which have greatly influenced both legal and ethical thinking on certain health related topics.

In keeping with the scope of this volume, the present article deals with bioethics in Canada from mid-1989 to mid-1991. Earlier developments in this field are discussed elsewhere ([28];[31];[32]).

II. NEW REPRODUCTIVE TECHNOLOGIES

On October 25, 1989 the Prime Minister of Canada, Brian Mulroney, announced that the federal government was establishing a Royal Commission to inquire into and report on current and potential medical and scientific developments related to the new reproductive technologies. Specifically, the Commission is to examine and make recommendations on:

- the implications of new reproductive technologies for women's reproductive health and well-being;
- the causes, treatment and prevention of male and female infertility;
- reversals of sterilization procedures, artificial insemination, in vitro fertilization, embryo transfers, prenatal screening and diagnostic techniques, genetic manipulation and therapeutic interventions to correct genetic anomalies, sex selection techniques, embryo experimentation and fetal tissue transplants;
- social and legal arrangements, such as surrogate childbearing, judicial interventions during gestation and birth, and ownership of ova, sperm, embryos and fetal tissue;
- the status and rights of people using or contributing to reproductive services, such as access to procedures, rights to parenthood, informed consent, status of gamete donors and confidentiality, and the impact of these services on all concerned parties, particularly children;
- the economic ramifications of these technologies, such as the commercial marketing of ova, sperm, and embryos, the application of patent law, and the funding of research and procedures including infertility treatment.

When first constituted, the Commission consisted of seven members, headed by Dr. Patricia Baird, Professor of Medical Genetics at the University of British Columbia, and was given two years to complete its work. In August, 1990, two additional Commissioners were named, and the deadline was extended by one year, to October 1992.

A. *Policy Announcements*

During the period under review, numerous professional associations and other interest groups prepared and presented submissions to the Royal Commission. Among the most extensive documents were those of the Canadian Bar Association (CBA) [3] and the Canadian Medical Association (CMA) [7].

The CBA brief was based on the *Report of the Special Task Force Committee on Reproductive Technology of the British Columbia Branch, The Canadian Bar Association*, which was commissioned in May, 1988 and completed in June, 1989. Although this document dealt specifically with the medical and legal status of the reproductive technologies in British Columbia, the CBA agreed with the general principles behind the recommendations of that report, and when developing its submission to the Royal Commission, it needed only to apply these principles on a nation-wide basis.

The CBA document focuses on two aspects of the Royal Commission's mandate:

- social and legal arrangements, such as surrogate childbearing, judicial interventions during gestation and birth, and ownership of ova, sperm, embryos and fetal tissue;
- the status and rights of people using or contributing to reproductive services, such as access to procedures, rights to parenthood, informed consent, status of gamete donors and confidentiality, and the impact of these services on all concerned parties, particularly the children.

The first three sections of the submission provide a framework for analyzing the issues relating to regulation of reproductive technology. This framework consists of an overview of this technology in Canada, a discussion of the values and objectives of regulation, and an outline of constitutional and legal considerations. The following sections discuss specific legal issues: the status of the child, parentage and birth registration, artificial reproductive technology and the practice of medicine, eligibility to benefit from these technologies, medical records, surrogacy, research and experimentation on human genetic material, and judicial intervention in gestation and childbirth.

The report concludes with nine individual and three sets of recommendations. The latter deal with surrogacy arrangements, research and experimentation, and judicial interventions in gestation and child birth.

The published version of the Canadian Medical Association submission to the Royal Commission is more than 300 pages in length. Following introductory chapters on background considerations and ethical considerations, the document describes, analyzes and makes recommendations on each of the different reproductive technologies: artificial insemination, ovulation induction and enhancement, gamete intra-fallopian transfer and peritoneal ovum/sperm transfer, ovum donation/reception, in vitro fertilization, zygote intra-fallopian transfer and tubal embryo transfer, and surrogate and gestational motherhood. This section also considers the ethical issues raised by gene therapy and genetic manipulation.

The general conclusions reached by the CMA brief are as follows:

- First, the new reproductive technologies should be developed and made available only on the condition that the principle of autonomy and honor for persons is fully respected at all times, and that the value of respect for human life is always retained as a guiding theme.

- Second, it must be fully recognized that the ultimate result of the use of this technology has momentous consequences for individuals who cannot speak for themselves namely children and future generations.
- Third, the decision to develop and apply these technologies should be taken only after a full and conscientious examination of all other options that are reasonably available.
- Fourth, the decision to publicly fund the use of these technologies should be taken only after a considered balancing process that takes into account their appropriateness and the demands of competing health care claims.

In addition, the report contains 118 specific recommendations on the following topics: social considerations, professional considerations, education, health and infertility, evaluation, funding, access, gamete donorship, screening, records, standards and licensing, counselling, ownership and disposition, preservation of embryos and gametes, parentage (rights and obligations), surrogate motherhood, research and experimentation, embryo status and destruction, gene therapy and manipulation, cloning, and sex selection.

A lengthy appendix to the CMA report deals with the legal aspects of the new reproductive technologies. It consists of a review of relevant legislation and case law on the following topics: regulatory structure, access to reproductive technologies, record-keeping, rights of custody, control and disposition of gametes and embryos, compensation for harm, and parental status and family relationships. For each of these topics, there is a discussion of the issues which need to be addressed, a summary of current Canadian law, a review of recommendations and initiatives in Canadian and other jurisdictions, and conclusions.

Well before the Royal Commission was established, the two medical societies most involved with the reproductive technologies, the Canadian Fertility and Andrology Society and the Society of Obstetricians and Gynaecologists of Canada, undertook a study of the ethical issues associated with these technologies. In 1987 they established a joint committee for this purpose. Its report, *Ethical Considerations of New Reproductive Technologies*, was released in September 1990 [5].

The joint committee report begins with a discussion of informed choice, which it considers to be the first ethical issue in reproductive technologies. It then proceeds to a list of definitions, medical indications, and risk/benefit considerations for each of these technologies. This is followed by a statement on the status of human gametes and pre-embryos. The heart of the report is its analysis of 27 ethical issues. For each issue the report lists alternative positions, discusses the advantages of each position, and recommends one of them. For the most part, the use of the technologies is endorsed, including egg and embryo donation and surrogacy. The development and implementation of sex preselection techniques is recommended for use in the prevention of certain genetic disorders and in a controlled and carefully monitored research setting to determine the potential impact when applied for completion of the family.

An appendix to the report discusses additional considerations regarding surrogate pregnancy.

Another early contribution to the reproductive technologies debate was the Working Paper of the Law Reform Commission of Canada entitled *Biomedical Experimentation Involving Human Subjects* [21]. Among the various subjects of research mentioned in this document are human embryos and fetuses. The Commission recommends that the creation of embryos solely for purposes of research, the re-implantation of embryos that have been used in experiments, and certain types of experimentation on embryos such as cloning should be prohibited; experimentation on embryos should be prohibited after the fourteenth day of development; the freezing of embryos should be allowed, but it should not be prolonged for more that five years; and no experiments which carry more than a minimal risk should be performed on fetuses. The Commission would have these provisions incorporated into the Criminal Code, along with appropriate penalties for infractions.

III. ABORTION

Canada is one of the very few countries in the world without an abortion law. In 1988 the Supreme Court of Canada, in *R. v. Morgentaler*, declared that the section of the Criminal Code which dealt with abortion was unconstitutional and therefore no longer in effect. Needless to say, this decision did not end the battle over the legal status of abortion. Between 1989 and 1991, this battle was fought in the courts and in Parliament.

A. *Court Decisions*

During the summer of 1989, Canadian courts had to deal with three separate cases where men sought injunctions to prevent the women they had impregnated from having abortions.

Murphy v. Dodd [40]. In Toronto an injunction granted July 4th by Mr. Justice John O'Driscoll of the Ontario Supreme Court prevented Barbara Dodd from terminating her 15-week pregnancy and declared the fetus to be under the court's protection. No reasons were given for this decision. It was appealed before another Ontario Supreme Court Judge, W. Gibson Gray, who heard the case on July 10 and rendered his decision the following day. He set aside the injunction on the grounds of procedural irregularities in the first hearing.

Nimi v. Morris [40]. In Winnipeg on July 6th, a man invoked Judge O'Driscoll's decision in an attempt to prevent his partner from having an abortion. However, Justice Aubrey Hirschfield of the Court of Queen's Bench denied the request for an injunction, citing the Supreme Court of Canada's judgment in *R. v. Morgentaler* [43]. On March 30, 1990, the Manitoba Court of Appeal affirmed this decision.

Tremblay v. Daigle [48]. On July 7th in Quebec City, Jean-Guy Tremblay obtained a temporary injunction from a Quebec Superior Court Judge preventing Chantal Daigle, his former partner, from having an abortion. Ms. Daigle, who was 18 weeks pregnant at the time, reappeared before the Quebec Superior Court on July 17th, at which time another judge maintained the injunction. He stated that according to the Quebec Charter of Rights and sections of the Civil Code of Quebec, a fetus is a human being with legal rights and is entitled to protection under the law. Ms. Daigle then appealed this judgment to the Quebec Court of Appeal. On July 26th, in a 3-2 decision, this Court reaffirmed the injunction, with each of the judges giving separate reasons for their decision. A request was then made to the Supreme Court of Canada to allow an appeal. The request was granted, and on August 8th, the full Court heard the case and rendered their unanimous verdict, which was to quash the injunction. The reason given for this decision was that in neither the Charter nor the Code is the fetus included within the term "human being". In this respect the legal status of the fetus is the same in Quebec as elsewhere in Canada, where it must be born alive to enjoy rights.

B. *Legislation*

In order to fill the legal void on abortion, the Federal Government proposed three alternative laws to Parliament in July 1988, but none of them received a majority vote, and the issue was set aside until after the Fall 1988 federal election.

To guide Parliament in formulating a new abortion law, the Law Reform Commission of Canada published a document entitled *Crimes Against the Fetus* in January 1989 [20]. The Commission recommended that abortion be once again made a criminal offense for both the pregnant woman and the one performing the procedure. Exceptions would be allowed under certain conditions, and the document provided three different proposals to legalize abortion. The majority position was that the crime of destroying a fetus should not apply to acts done before the twenty-second week of pregnancy to protect the mother's physical or psychological health or after the twenty-second week to save her life or to protect her against serious physical injury, or because the fetus suffers from an incurable and terminal medical condition. The second position was identical to the first except that abortion would be allowed for any reason up to twelve weeks. The third position would permit abortion only to save the pregnant woman's life or to protect her against serious and substantial danger to her health.

When the federal government did finally introduce its new proposal for abortion legislation (Bill C-43, *An Act Respecting Abortion*) in the Fall of 1989, it did not adopt the Law Reform Commission's recommendation to make abortion easier to obtain in the earlier stages of pregnancy than in the later. Instead, abortion would be permitted at any stage as long as it would be performed by a doctor "who is of the opinion that, if the abortion were not induced, the health or life of the female person would be likely to be threatened." Health was defined as including physical, mental, and psychological aspects. All other abortions would be illegal and could be punished by a prison term of up to two years. On May 29, 1990 the House of Commons approved this bill by a vote of 140-131. It then went to the Senate, which launched an intensive study of the issue. On January 31, 1991 the bill finally came to a vote and the result was a tie, 43-43. Under Senate rules, a tie is considered as a defeat. The government then announced that it would not pursue this matter during its current term of office.

While the federal government was trying unsuccessfully to restrict abortion through the Criminal Code, at least one province was attempting to achieve this goal by other means. In March 1989, the government of Nova Scotia passed a regulation under existing legislation which forbade the performance of abortions outside accredited hospitals. This provision was evidently intended to prevent Dr. Henry Morgentaler from opening a free-standing abortion clinic in Halifax. The regulation was introduced as legislation in June and quickly adopted as part of the *Medical Services Act*.

Two court challenges of this act were soon forthcoming. In *Canadian Abortion Rights Action League (CARAL) Inc. v. Attorney General of Nova Scotia* [35], the plaintiff questioned the constitutional validity of the Act on the grounds that it was an infringement of the exclusive jurisdiction of the federal government in matters of criminal law and that it was contrary to the *Canadian Charter of Rights and Freedoms*. However, the Attorney General of Nova Scotia successfully challenged CARAL's standing in this matter, a decision which was upheld by the Appeals Division of the Nova Scotia Supreme Court on March 27, 1990. The second challenge to the Act (*R. v. Morgentaler*) turned out differently [43]. On October 19, 1990, the Nova Scotia Provincial Court dismissed charges against Dr. Henry Morgentaler for breaching the Act by performing abortions in his clinic; the legislation was said to be an intrusion into the legislative competence of the federal government and therefore of no force or effect.

C. *Policy Announcement*

The federal government's proposed legislation on abortion met with strong opposition from the Canadian Medical Association. In two briefs, one to the House of Commons Legislative Committee studying Bill C-43 (February 6, 1990) and the other to the Standing Senate Committee on Legal and Constitutional Affairs (January 17, 1991), the CMA reiterated its long-standing policy that "the decision to perform an induced abortion is a medical decision made confidentially between the patient and her physician, within the context of the physician-patient relationship, after a conscientious examination of all other options." The passage of Bill C-43, according to the CMA, would reduce access to abortion services for Canadian women and would lead to, and perhaps even encourage, harassment of those physicians willing to provide these services. It would also threaten the nature of the physician-patient relationship in that some women might be inclined to deceive their physicians about their reasons for seeking abortions if these reasons were not admissible under the law. The CMA's recommendation was that abortion should not be included in the Criminal Code and that Bill C-43 should be defeated.

IV. MATERNAL-FETAL CONFLICTS

Several different types of conflict between pregnant women and their fetuses have arisen in recent years. These include abortion, compulsory medical treatment of pregnant women, and natural childbirth.

A. *Abortion*

The Supreme Court of Canada, in its 1988 *Morgentaler* decision, declined to deal with the question as to whether a fetus is included in the word "everyone" in section 7 of the *Canadian Charter of Rights and Freedoms*, so as to have a right to "life, liberty and security of the person [42]." Nine months after this decision, however, the Supreme Court upheld a judgment of the Saskatchewan Court of Queen's Bench against Joe Borowski's challenge to the existing abortion law (*Borowski v. Canada [Attorney General]*) [33]. The lower court had ruled that there is no basis in law which justifies a conclusion that fetuses are legal persons. Since fetuses have no legal rights, there can be no maternal-fetal conflict in law. At the same time, both the Saskatchewan Court and the Supreme Court suggested that Parliament could enact laws which would extend to the unborn any or all legal rights possessed by living persons.

As is evident in the title of their document, the Law Reform Commission of Canada (LRCC) proposed in their working paper, *Crimes Against the Fetus*, to provide some such rights within the Criminal Code. This would require the addition of a new crime, fetal destruction or harm, so that "Everyone commits a crime who (a) purposely, recklessly or negligently causes destruction or serious harm to a fetus; or (b) being a pregnant woman, purposely causes destruction or serious harm to the fetus by any act or by failing to make reasonable provision for assistance in respect of her delivery" ([20], p. 64).

The LRCC document recognizes that there are situations where the prevention of harm to the fetus conflicts with the well-being of the pregnant woman. In some of these situations, the woman is given priority; in particular, the general crime of destroying a fetus should not apply to acts done to save the mother's life or to protect her against serious physical injury. In other situations, the interests of the fetus take priority, especially when the conflict is between the life of a third-trimester fetus and the autonomy of the pregnant woman who wants to terminate the pregnancy for reasons other than her physical health. The five commissioners could not agree on how other conflicts should be resolved. As mentioned above, their document offers three different proposals to deal with these situations: the majority position would permit abortions to protect the mother's physical or psychological health up to the twenty-second week of pregnancy; a second position would reduce this time limit to twelve weeks; the third would restrict abortions to situations where the woman's life or health are seriously endangered by the pregnancy.

B. *Compulsory Medical Treatment of Pregnant Women*

During the period under review, there was one reported attempt to resolve maternal-fetal conflict by compelling a pregnant woman to undergo medical treatment (*Re A. [in utero]*) [45]. In July 1990 the Children's Aid Society of Hamilton-Wentworth, Ontario appeared before Mr. Justice Steinberg of the district Unified Family Court to seek an order requiring the woman to submit to prenatal medical supervision. The Society had the following concerns about the welfare of the unborn child: (1) the husband had a criminal record and a violent, disturbed personality; (2) the husband and wife had lied to Society workers about the mother's prenatal care; (3) the mother was suffering from toxemia with a risk of severe medical complications at birth; (4) the parents had a history of improper care of their four previous children, who were all wards of the Crown residing in foster homes; and (5) the parents had made threats of physical violence against Society workers who had attempted to intervene. The application for Crown wardship of the unborn child was dismissed on the grounds that the definition of a "child" in the Ontario *Child and Family Services Act* (1984) does not accord an unborn child status as a "person" or a right to protection under the Act. The Judge stated that although the state's interest in the welfare of an unborn child increases as the child comes closer to being born, *parens patriae* jurisdiction is not broad enough to justify forcible confinement of a parent.

C. *Natural Childbirth*

As is evident in cases involving forced caesarian deliveries, maternal-fetal conflict can occur during as well as before birth. Another instance of this type of conflict is recourse to "natural" childbirth, in particular home delivery with the assistance of a midwife rather than a physician. The Supreme Court of Canada in 1991 rendered a verdict on a case where a delivery of this type resulted in a stillbirth (*R. v. Sullivan*) [44].

In 1985 two midwives in British Columbia, Mary C. Sullivan and Gloria J. Lemay, were charged with criminal negligence for causing the death of a fetus and bodily harm to the pregnant woman. They were unable to complete the delivery after the head of the fetus appeared. The woman was then taken to hospital, where the baby was delivered asphyxiated. The midwives were convicted of the first charge on the basis that the baby was a person at the time of its death. On appeal this verdict was reversed because, according to common law, the entire body of the child had to be brought living into the world for it to be a person in law. However, the appeal court held that the fetus was part of the mother's body and therefore the midwives were guilty under the second charge. This verdict was appealed to the Supreme Court of Canada, which heard the case on October 30, 1990 and rendered its decision on March 21, 1991. This Court upheld the decision of the Appeal Court on the first charge and overturned its conviction of the midwives on the second charge, on the grounds that their original acquittal on this charge was appropriate.

V. CONSENT TO TREATMENT AND EXPERIMENTATION

Since the landmark Supreme Court of Canada ruling in the case of *Reibl v. Hughes* (1980) [46], there has been substantial agreement on the general legal and ethical principles of informed consent to medical treatment and participation in research. In 1987 the Medical Research Council of Canada updated its *Guidelines on Research Involving Human Subjects* to ensure their conformity with current standards of consent [25]. During the period under consideration, these standards have been refined and interpreted by the courts and the Law Reform Commission of Canada.

A. *Court Decisions*

On March 30, 1990, the Ontario Court of Appeal released its judgment in the case of *Malette v. Shulman et al.* [38]. The case originated in June, 1979 when Mrs. Malette, a Jehovah's Witness, was given a blood transfusion by Dr. David Shulman after she had been seriously injured in an automobile accident. A nurse had advised Dr. Shulman that Mrs. Malette carried a signed (but not dated or witnessed) card which stated (in French), "As one of Jehovah's Witnesses with firm religious convictions, I request that no blood or blood products be administered to me under any circumstances." After her recovery, Mrs. Malette sued Dr. Shulman, the hospital, its Executive Director and four nurses for negligence, alleging that the blood transfusion had been medically unnecessary, and for battery, alleging that the transfusion was performed without her consent. When the case came to trial, Mr. Justice Donnelly of the Ontario Supreme Court dismissed the action for negligence but awarded damages of $20,000 against Dr. Shulman for battery, having decided that the card continued to represent Mrs. Malette's true wishes.

In a unanimous decision the Court of Appeal upheld the judgment of the lower court. Basing its decision on the doctrine of informed consent, the Court stated:

A doctor is not free to disregard a patient's advance instructions any more than he would be free to disregard instructions given at the time of the emergency.... The principles of self-determination and individual autonomy compel the conclusion that the patient may reject blood transfusions even if harmful consequences may result and even if the decision is generally regarded as foolhardy.... To transfuse a Jehovah's Witness in the face of her explicit instructions to the contrary would...violate her right to control her own body and show disrespect for the religious values by which she has chosen to live her life.

The Court concluded that the card carried by Mrs. Malette, even though undated, constituted a valid expression of her wishes concerning medical treatment, and that Dr. Shulman did not have reasonable grounds to doubt that this was the case. The award of $20,000 was upheld.

In February 1989, a Quebec Superior Court judge issued a ruling on one of the very few Canadian cases involving informed consent to participation in medical research (*Weiss v. Solomon*) [49]. Seven years previously Julius Weiss, a 62 year old man who had undergone cataract surgery, was asked by his surgeon, Dr. Solomon, to participate in a hospital research study on the ability of indomethacin in the form of eye drops to reduce the retinal edema that is a frequent side-effect of cataract surgery. The protocol required three fluorescein angiograms to be performed immediately after application of the eye drops. Weiss was informed that the procedure would be of no therapeutic benefit to him and would involve some minor side-effects, such as nausea and mild discomfort. The rare possibility of severe allergic reaction causing death was not disclosed. Weiss agreed to participate. Shortly after the first injection, he suffered a complete loss of blood pressure. Resuscitation attempts were delayed due to the lack of a defibrillator and electrocardiogram equipment in the angiography room. Death occurred almost immediately.

Weiss' wife and children sued for damages. The Court found (in addition to the hospital's negligence related to resuscitation) that both Dr. Solomon and the hospital's Research Ethics Committee were negligent in failing to ensure that Weiss was informed of the risks involved in the research. The hospital was held to be liable for the negligence of the Research Ethics Committee. The decision was based on the requirements of the Declaration of Helsinki, the Civil Code of Quebec, an earlier Canadian case (*Halushka v. University of Saskatchewan*), and the writings of legal scholars on the subject of disclosure of risk in medical research. The family was awarded $118,800 in damages.

The issues of informed consent and therapeutic privilege were factors in a February 1991 decision of the Ontario Court (General Division) (*Meyer Estate v. Rogers*) [39]. In 1985, a 37 year old woman, Denise Meyer, died as a result of an allergic reaction to an intravenous pyelogram, a diagnostic procedure in which dye is injected into the patient's vein and X-ray photographs are taken. Her estate sued the referring physician and the radiologist, alleging that the former was negligent in failing to communicate the plaintiff's allergy problems to the radiologist, in failing to try other methods of analysis or treatment, and in failing to suggest alternatives to the use of the dye. It was alleged that the radiologist failed to warn the plaintiff of the risks inherent in the use of the dye.

In his judgment, Mr. Justice Maloney dismissed the action. The radiologist had followed the recommendation of the Canadian Association of Radiologists that since the incidence of death from an allergic reaction to the dye that was used was approximately 1 in 40,000, the risks associated with fully informing the patient outweigh the risk of not doing so. The judge rejected the therapeutic privilege exception to the doctor's duty of disclosure. Nevertheless, he concluded that the non-disclosure of the risks associated with contrast media injections was not causative of the plaintiff's death, as the evidence established on a balance of probabilities that a reasonable person in the plaintiff's position would have consented to the procedure.

A June 1991 decision of the Ontario Court of Appeal in the case of *Fleming v. Reid* [36] struck down a provision of the Ontario *Mental Health Act* which allowed the administration of neuroleptic drugs in non-emergency situations to involuntary incompetent psychiatric patients who, while mentally competent, have expressed the wish not to be treated with such drugs. The plaintiff was confined in a mental health facility for treatment following his acquittal of a criminal offense by reason of insanity. Despite previous successful treatments, he refused further medication and became agitated, hostile, hyperactive and confused. A review board granted Dr. Fleming's request for a treatment order under the provisions of the *Mental Health Act*. This decision was affirmed on appeal to the District Court.

The Public Trustee, acting on behalf of the plaintiff and other involuntary incompetent patients, then appealed to the Ontario Court of Appeal. In its judgment this Court stated that, with very limited exceptions, every competent adult has the right to be free from unwanted medical treatment. To deprive involuntary patients of this right when they become incompetent and force them to submit to medication against their competent wishes without the consent of their legally appointed substitute decision-makers constitutes an infringement of the Charter right to security of the person. The Court found as well that the treatment orders made by the Board were arbitrary and unfair. The orders were therefore declared invalid.

B. *Legislative Reform*

In December 1989 the Law Reform Commission of Canada published a Working Paper entitled *Biomedical Experimentation Involving Human Subjects* [21]. The purpose of this document was to provide a clear survey of the current state of the law on this issue in Canada and elsewhere; to review the problematic aspects of experimentation on human subjects, such as the use of fetuses, children and prisoners; and to determine whether the current state of the law is satisfactory or whether there is need for reform.

The Commission believes that the law has a limited, but nevertheless important, role to play in the regulation of scientific research on humans, in addition to the regulatory role of bodies such as the Medical Research Council of Canada. The Working Paper recommends that laws be changed or introduced to deal with the following matters:

- nontherapeutic experimentation should be permitted only when the subject's free and informed consent has been properly obtained and there is an acceptable ratio between the risks incurred and the expected benefits;
- deception of research subjects should be permitted only under certain specified conditions;
- nontherapeutic research on children, mentally deficient persons, embryos and fetuses should be permitted only under certain specified conditions.

To implement these provisions, the Commission would have the Criminal Code amended by excluding from offenses against bodily integrity those cases of nontherapeutic biomedical experimentation in which free and informed consent is properly obtained and the risks incurred are not disproportionate to expected benefits.

The Working Paper is intended to generate discussion and feedback. The final views of the Commission will eventually be presented in a Report to the Minister of Justice and Parliament.

VI. CONFIDENTIALITY

The rapid spread of the human immunodeficiency virus (HIV) has raised questions about the long-standing principle of respect for the confidentiality of a person's health status. In order to prevent or at least inhibit transmission of the virus, many public officials and health care professionals have advocated less stringent standards of confidentiality than for other medical conditions. During the period under review, several committees have studied these issues and published their conclusions. The courts have also been required to deal with this topic.

A. *Policy Announcements*

In September 1989 the Federal/Provincial/Territorial Advisory Committee on AIDS published a summary of its report on *Confidentiality and HIV Seropositivity* [15]. The full report had been approved by the ministers of health the previous year. It addresses the issue of confidentiality in relation to HIV-antibody testing of individuals, but not in relation to HIV seroprevalence research.

The published summary consists of a preamble, definitions, basic principles and guidelines, and sections on laboratories, public health authorities, and disclosure. The principles and guidelines relating to confidentiality are as follows:

- People must undergo counselling before testing and should be notified about the limitations and implications of testing, including the nominal reporting of positive results in jurisdictions where this is required by law.

- Follow-up, including identification and tracing of contacts, is warranted if ethically justified or prescribed by law.
- A basic data set must be reported to public health authorities (this may require statutory authority). The set should include: (a) the name of the referring physician; (b) the patient identifier (code or name); (c) the patient's age and sex; (d) the reason for the test (e.g., patient is a member of a specified high-risk group, a donor of tissue, blood or organ, or for diagnostic purposes and approval for insurance); and (e) the test result (and the test methods used if these differ from one laboratory to another).
- The legal requirements for reporting procedures should be designed to protect the confidentiality of those tested and their contacts.
- Issues of confidentiality should be considered within the overall framework of testing.

The remaining sections of the document suggest how the right to confidentiality should be balanced against the need of others to know an individual's HIV-status. Reporting of HIV antibody test results by laboratories to public health authorities should be anonymous. Contact tracing should require the informed consent of the seropositive individual. Information about a person's HIV status should be disclosed only for the benefit of the person being tested, although this would include prevention of transmission of the virus to others. Specific measures are suggested to protect HIV-positive individuals, their contacts, the public, and health care workers. Employers cannot demand that employees reveal their HIV status or undergo HIV antibody testing.

In June 1990, the executive and council of the Canadian Association of Pathologists approved a document entitled *Guidelines for the Release of Laboratory Test Results, Reports, and Specimens to Patients, Interested Groups, and Other Institutions*. The scope of this document includes, but is not limited to, HIV tests. The basic principle behind the guidelines is as follows:

> The privacy and rights of the patient are primary in the health care system. All aspects of the doctor-patient relationship are confidential from the time the patient first sees the physician until all investigations are completed. The laboratory, as an intermediary in the delivery of health care, must maintain this confidentiality and ensure that the results are released to the attending physician. <u>It is the sole responsibility of the attending physician to release the test results to his or her patients</u> [2].

The document goes on to say that patients should not be misled about the degree of confidentiality that can reasonably be expected of a medical institution. Nevertheless, laboratory workers should ensure that test results should be released only in accordance with approved procedures. In particular, results should not be given directly to patients but only to their attending physician, unless the latter has specifically authorized their direct release.

In October 1990 the Federal Centre for AIDS Working Group on Anonymous Unlinked HIV Seroprevalence published its *Guidelines on Ethical and Legal Considerations in Anonymous Unlinked HIV Seroprevalence Research* [14]. The Working Group approved anonymous unlinked HIV seroprevalence research because: (a) it can provide accurate estimates of the prevalence of HIV infection since it avoids the self-selection bias inherent in all voluntary studies; (b) it is noninvasive; (c) it ensures personal privacy because individuals cannot be linked to test results; and (d) it is cost-effective. However, certain conditions must be fulfilled for such testing to be legally and ethically acceptable with regard to the preservation of confidentiality:

- Records must be permanently unlinked before testing, so that it would be impossible at any time to identify individual test results.
- No sample size small enough to identify individuals should be reported.
- No information that might lead to the identification of individuals should be used.

Other provisions in the *Guidelines* relate to communication with the public about such testing, and provision for individual testing on a voluntary basis.

B. *Court Decisions*

In 1983 and 1984 Clarence Sharpe, a hemophiliac, was given blood transfusions during surgery in an Ontario hospital. In late 1985, he was diagnosed as having AIDS. After his death, his estate brought an action against the hospital, alleging that the blood given to Mr. Sharpe was contaminated with the HIV. The hospital issued a third-party claim against the Canadian Red Cross Society, which had supplied the blood. The Society was requested to produce its records relating to donors whose blood was given to Mr. Sharpe.

The case (*Sharpe Estate v. Northwestern General Hospital*) was heard by Master Sandler of the High Court of Justice of the Ontario Supreme Court [47]. In his judgment on September 24, 1990, Master Sandler rejected the Red Cross Society's attempt to preserve the confidentiality of their blood donors. He stated that the harm to the Red Cross if this information were ordered to be disclosed would not necessarily be greater than the harm to the plaintiffs if they were deprived of the information. At the same time, he imposed several conditions on the retrieval and use of the records.

The Red Cross Society appealed this decision to the Ontario Court (General Division). On January 17, 1991 Mr. Justice Haley dismissed the appeal. He denied that any relationship of confidentiality ever existed between the Society and its blood donors. In any event, he stated, the conditions which Master Sandler imposed would protect the donors and their families from any social stigma.

A second AIDS-related case (*Jean-Pierre Valiquette v. The Gazette*) was decided in a Quebec court in 1991 [37]. Mr. Valiquette, a Montreal school teacher, was diagnosed as having AIDS in early 1986. He took sick leave on September 29, 1986 and then sought to return to work the following June. His school board would not accept a medical certificate from his doctor that he was fit to resume his work. In August 1987, as the school year was about to begin and the board had still not authorized his return to teaching, a story about the case appeared on the front page of a Montreal newspaper, *The Gazette*. Without naming Valiquette, it resulted in widespread recognition of him as the subject of the report. Although he did resume teaching, he suffered a relapse in September, was hospitalized in October, and died soon after.

Valiquette's heir, Réal Blais, subsequently sued the newspaper for invasion of privacy. In a judgment rendered March 12, 1991, the court found in his favor. Valiquette's right to privacy and anonymity had been violated. Damages were assessed at $37,500 plus interest and costs.

VII. EQUITABLE ACCESS TO HEALTH CARE

During the period under review, several provincial governments conducted major examinations of health care financing and delivery in their jurisdictions. The recommendations from these studies will likely be the subject of considerable debate before they are incorporated into legislation and regulations. The provincial and territorial health systems have been put under additional financial pressure as a result of federal legislation (Bill C-20, *Budget Implementation Act, 1991*), which reduces the federal contribution to provincial health programs each year until it eventually reaches zero.

Since universal access to health care is a basic feature of Canadian Medicare, there have been relatively few court challenges in this matter. One such case was decided in January 1990 (*Brown v. British Columbia (Minister of Health)*) [34]. Two-and-a-half years earlier, the Ministry of Health of British Columbia decided to place Azidothymidine (AZT) under the Provincial Pharmacare Plan, with the result that all AIDS patients, except those on social assistance or in long-term care facilities, were required to pay a portion of the cost of the drug. An AIDS patient, Kevin Brown, and the Vancouver Persons with AIDS Society challenged this decision on the grounds that it was discriminatory and contrary to the Canadian Charter of Rights and Freedoms. At that time British Columbia was the only province requiring patients to contribute to the cost of AZT. It did pay the entire costs of some other drugs, however, namely those administered to cancer patients through the Cancer Control Agency, and cyclosporin for transplant patients.

In his judgment, Mr. Justice Coultas of the British Columbia Supreme Court dismissed the plaintiffs' challenge to the Ministry's decision. He found that the Ministry had not shown administrative unfairness in making this decision and that the decision did not violate rights guaranteed in the Charter.

VIII. WITHDRAWING/WITHHOLDING TREATMENT FROM DYING PATIENTS

Attention to the two related issues of withdrawing/withholding treatment from the terminally ill and advance directives increased markedly during the period under review. These topics were the subject of considerable activity by legislative bodies and professional associations.

A. *Withholding/Withdrawing Treatment*

In no less that four of its publications, the Law Reform Commission of Canada has recommended that the Criminal Code be amended or revised to recognize the right of competent individuals to refuse life-sustaining medical treatment and to protect physicians who withhold or withdraw such treatment when requested by the patient or appropriate proxy or when they judge that the treatment has no therapeutic value and is not in the best interests of the patient ([16];[17];[18];[19]. The federal government has thus far not acted on these recommendations. However, a private members' bill entitled *An Act to Amend the Criminal Code (Terminally Ill Persons)* (Bill C-203) was scheduled to be debated in the House of Commons in the Fall of 1991. The intent of this bill is to incorporate into law the above-mentioned recommendations of the Law Reform Commission of Canada. Given the history of private member's bills, it is unlikely that this one will pass. However, it is expected that the federal government will introduce legislation to revise the entire Criminal Code in 1992, and that withholding/withdrawing of treatment will be debated in that context.

The current professional position on this topic is expressed in the *Joint Statement on Terminal Illness*, which was prepared and endorsed by the Canadian Nurses Association, the Canadian Medical Association, and the Canadian Hospital Association [10]. This document deals only with resuscitation of the terminally ill. It consists of a set of clinical criteria for non-resuscitation, procedural guidelines, and a recommendation that palliative care be provided to patients for whom resuscitation is inappropriate. The associations which produced the Joint Statement have acknowledged that it is outdated and needs to be expanded to include other procedures besides resuscitation. At the time of writing, the revision of the Statement is not yet complete.

B. *Advance Directives*

In July 1990 the Law Reform Commission of Manitoba released its *Discussion Paper on Advance Directives and Durable Powers of Attorney for Health Care* [23]. The substance of this document is contained in its conclusion, where readers are invited to respond to the following questions:

- Should the law be reformed to permit the creation of a mechanism which would give legally binding effect to the wishes of a person not competent to make decisions about medical treatments which were expressed at a time when he or she had such competence?
- If so, what mechanism or mechanisms should be adopted? The advance directive, the durable power of attorney for health care, or both? Is some other mechanism preferable?
- In adopting one or more of these mechanisms: (a) What should their scope be? (b) What obligations and immunities should health care professionals and medical facilities have in following them? (c) Who should be permitted to execute one? Should substituted judgment be permitted? (d) What requirements for witnesses should there be? (e) Should handwritten documents be recognized? (f) Should a statutory form be required or permitted? (g) Should substantial compliance with any formalities associated with execution (as opposed to absolute compliance) be taken as sufficient? (h) What steps, if any, should be taken to ensure that a person who is asked to give effect to such a mechanism is aware of the document? (i) Should health care providers and insurers be prohibited from requiring an individual to execute such a document as a precondition to their services? (j) How should the commencement of an attorney's authority or the effective date of an advance directive be determined? (k) What should the requirements for revocation be?
- Specifically with respect to durable powers of attorney for health care: (a) Who should be eligible for appointment as an attorney? (b) Should the decisions of attorneys be reviewable? (c) Should a person who executes a durable power of attorney be permitted to name a substitute attorney?

The Commission will take into account the comments which it receives on the Discussion Paper when preparing its final report. In accordance with *The Law Reform Commission Act*, that report will be submitted to the Minister of Justice and Attorney General of Manitoba.

One of the groups which responded to the invitation to comment on this Discussion Paper was the Manitoba Medical Association (MMA). The Association's brief does not answer all of the questions posed by the Law Reform Commission, but instead raises a number of concerns about the idea of advance directives in general and lists the safeguards which should be included in any specific legislation. The brief also contains a section on "Matters of Principle" which discusses three issues:

- Should advance directives apply only to terminal conditions? Some of the MMA sections and committees, including the Committee on Ethics, recommended that legislation should protect the right of a patient to determine future medical treatment for non-terminal conditions which would lead to what the individual patient would consider to be an unacceptable quality of life.
- Should advance directives be binding on health care providers? The MMA

wants providers to be free to question directives which appear capricious, illogical, or medically unsound.

- How should the validity of advance directives be determined? The MMA calls for an efficient means of making such determinations, to avoid prolonged disputes or indecision.

At its 1991 annual meeting, the Canadian Medical Association adopted a policy on advance directives [9]. The CMA advises physicians to discuss the advantages and limitations of such documents with their patients. The advisability of patients naming specific individuals to act as their proxy decision-makers should also be discussed, whether or not patients wish to execute advance directives. Physicians should honor these documents unless there are reasonable grounds to suppose that they no longer represent the wishes of their patients or that the patients' understanding was incomplete at the time of making the directives.

IX. ACTIVE EUTHANASIA

In its publications dealing with death and dying, the Law Reform Commission of Canada has consistently opposed the legalization of active euthanasia ([16];[17];[18];[19]). This lack of support did not prevent one member of Parliament from introducing a private member's bill in support of euthanasia. Bill C-261, *An Act to Legalize the Administration of Euthanasia Under Certain Conditions to Persons Who Request It and Who Are Suffering from an Irremediable Condition and Respecting the Withholding and Cessation of Treatment and to Amend the Criminal Code*, received first reading on June 19, 1991. As the title indicates, the bill had two purposes. The secondary aim was similar to that of Bill C-203 described above, to entrench the right of individuals to refuse life-sustaining medical treatment and to protect physicians who do not provide such treatment. The primary purpose of the bill was the legalization of active euthanasia.

Bill C-261 did not define euthanasia but rather suggested a process by which it could be administered. An individual who is suffering from an "irremediable condition" and is at least 18 years old would be able to make a written application to a "referee in euthanasia" for a "euthanasia certificate". The referee, who would be appointed by the Attorney General, would interview an applicant and his/her physicians and, if satisfied that the applicant understands the nature and purpose of the application and has made it freely, would then issue a euthanasia certificate. The certificate could be presented to a qualified medical practitioner, who would administer euthanasia. Refusal to grant a certificate could be appealed to the Attorney General. The bill included a sample application form and two medical certificate forms for transmission to the referee.

Bill C-203 was debated for its allotted time but did not come to a vote. According to the rules of the House of Commons, it will not be discussed further during the current session of Parliament.

1991 also saw the publication by the Ontario Medical Association (OMA) of a document entitled, *Euthanasia and the Role of Medicine*, which was prepared by the OMA Committee on Medical Bioethics [27]. The document is intended to serve as the first step in the development of a position on this topic. It includes an historical perspective on euthanasia, an ethical review which identifies the principal values relevant to the topic, and the conclusions of each member of the Committee.

At the annual meeting of the OMA Council in June 1991, a statement on euthanasia was adopted. The Association contends that the establishment of policy on this issue is the responsibility of society. The expertise and experience of physicians can make a valuable contribution to the societal debate, but they should not be put in the position of leading society in this matter. The statement suggests that a forum be established as soon as possible to consider the legalization of euthanasia in Ontario.

X. ORGAN DONATION, SALE, AND TRANSPLANTATION

A. *Legislation*

The shortage of organs for transplantation has prompted numerous suggestions for "required request" policies in Canadian hospitals, but to date only one province has enacted legislation in this matter. In 1989, Nova Scotia amended its *Human Tissue Gift* Act by adding the following provisions:

(1) Where a person dies in a hospital and a consent has not been given [for organ retrieval], the hospital shall, as soon as practicable after the death of the person, request permission or cause permission to be requested from the person entitled to consent...to remove tissue from the body of the deceased person to be used for transplant purposes.

(2) Permission shall not be requested...where a person designated for the purpose of this subsection by the hospital or where a physician determines that: (a) no tissue could be used for transplant purposes because of the condition of the body of the deceased person and of the tissue thereof; (b) there is no need for the use of any tissue from the body of the deceased person for transplant purposes; or (c) the emotional and physical condition of the person from whom permission is required to be requested makes the request inappropriate.

Sections (3) and (4) require that records be kept of cases where the provisions of section (2) are invoked.

B. *Policy Announcement*

In 1990 the Canadian Paediatric Society published a document entitled *Transplantation of Organs from Newborns with Anencephaly* which had been prepared by the Society's Bioethics Committee [11]. The document raises and answers the following questions:

- Should infants with anencephaly be regarded as people? Yes, and they should be treated in the same way as any other patients.

- Are they potential donors of organs? Only in a very few cases.
- If so, under what circumstances? If the organs are still usable at the time that brain death can be determined according to the medical or legal definition of death that applies to other human beings.
- Should the definition of death or of brain death be changed? No. The Committee strongly opposes any change in the definition of death for four reasons: (a) it might be extended to other groups of "near-dead" patients, such as those in a persistent vegetative state; (b) it would lead to negative effects on people's confidence and trust in physicians in general and pediatricians in intensive care units in particular; (c) it would have negative effects on staff otherwise committed to caring for these patients; and (d) it would be a further step toward the consideration of anencephalic infants simply as a means to an end.
- Should aggressive life support be used for infants with anencephaly in anticipation of brain death? No. Anecdotal experience suggests that this practice does not lead to early death but prolongs survival and therefore prolongs dying. There is also evidence of distress among the parents and the staff during such attempts.
- How should priorities be determined in neonatal units with limited resources? The Committee feels that it is clearly unethical to transfer the resources needed for neonatal intensive care to an experimental, unproven project such as the maintenance of infants with anencephaly for the purposes of organ donation.

While not condemning altogether the transplantation of organs from anencephalic newborns, this document concludes that it should be accorded very low priority and should be subject to stringent controls.

XI. OTHER ISSUES.

Two sets of issues other than those dealt with above have attracted particular attention during the period under review: genetics and professional codes of behavior. Also discussed in this section are an initiative in medical ethics education and a proposal for a national biomedical ethics council for Canada.

A. *Genetics*

In 1990 the Medical Research Council of Canada (MRCC) released the final version of its *Guidelines for Research on Somatic Cell Gene Therapy in Humans* [26]. This document, which had been distributed as a discussion paper the previous year, contains four main sections: (a) medical and scientific background, (b) ethical issues, (c) ethical and scientific review, and (d) conclusions. Its conclusions are as follows:

- For the foreseeable future, gene therapy research on humans should be considered only for diseases which meet all the following criteria: they are caused by a defect in a single gene; they cause a live-born human being to suffer severe debilitation or early death; they cannot be treated successfully by any other means.

- For the foreseeable future, there should be no attempts to undertake research in humans which involves deliberate alteration of the patient's germ line, or which involves gene transfer in human embryos.
- Research with animals and with other models in the area of somatic cell gene therapy for humans is needed and appropriate.
- Any attempt to treat an inherited disease by somatic cell gene therapy should be regarded as a research protocol, and subject to procedures and considerations as outlined in this document and in the MRCC's *Guidelines on Research Involving Human Subjects* [25].
- A National Review Committee should be formed to evaluate all proposals in Canada for research on somatic cell gene therapy in humans. A prerequisite for submitting protocols to the National Review Committee is that they should first have been accepted by their local Research Ethics Board.

In June 1991, the Science Council of Canada, the national advisory agency on science and technology policy, released a report entitled *Genetics in Canadian Health Care* [30]. It is the result of a project launched in 1986 to review the current and potential role of genetic knowledge and technologies, to examine related policy issues, and to stimulate progress in addressing them.

The report consists of seven chapters: overview, the role of genes in health and disease, genetic technologies and health care applications, genetic health care services in Canada, ethical and legal concerns, education and awareness, and research. The Council believes that ethical concerns must be fully integrated into the new genetics. In particular:

- Genetic technologies and services are, and should continue to be, evaluated, and delivered in the context of a caring society that values individuals and accepts human diversity and disability. Individual choice (autonomy), counselling, informed consent, and confidentiality are the cornerstones of the ethical delivery of genetic services.
- The primary objective of genetic applications in health care is to treat or prevent genetic disorders; the technologies should not be used with the primary goal of reducing the costs of health care or improving the human species. Genetic services should be initiated only if there are benefits to the recipient such as disease prevention or treatment, or lifestyle or reproductive choices.
- Individuals and families should have access to beneficial technologies in order to make informed decisions about their own health care and reproductive options. The decisions should be based on reliable technical information and accurate, non-directive counselling.
- Participation in genetic services should be a matter of free choice. Individual and family decisions must be respected and supported. Individuals should not

be penalized (through reduced medical or social services, for example) for their reproductive or personal health care decisions.

- Genetic technologies should be used in a manner consistent with acceptance of human diversity and disability. We must ensure that availability of genetic services does not decrease our acceptance of the disabled.

B. *Professional Codes and Guidelines*

In 1991 the Catholic Health Association of Canada published its new *Health Care Ethics Guide* [13]. This document is addressed to the owners, boards, personnel, and residents/patients of all Roman Catholic health care institutions in Canada and to Catholics who work in or are patients/residents of other health care facilities. It replaces the *Medical/Moral Guide* which had governed Catholic health care institutions in Canada since 1970.

The guide consists of six main sections, each with two parts: a statement of fundamental principles and values that underlie the treatment of the issues discussed in the section, and a series of articles that serve as formulations of a contemporary Catholic understanding of how the principles and values are applied in particular circumstances. The subjects of these six sections are the communal nature of health care, dignity of the human person, human reproduction, organ donation and transplantation, care of the dying person, and research on human subjects. Three appendices present a summary of the steps used in an ethical decision-making process, a glossary of terms, and a bibliography.

The Canadian College of Health Service Executives is a professional association of health care institution administrators. In 1991 the College adopted a new set of ethical rules for its members. The one-page document, entitled *Standards of Ethical Conduct for Health Service Executives* [4], is divided into five sections, which outline the responsibilities of members to individuals, the organization, community and society, and the profession. The fifth section deals with conflicts of interest.

The document emphasizes respecting confidentiality of information, serving the public, contributing to the improvement of the health of Canadians, and practicing with honesty and integrity. In particular, members are expected to foster informed decision making by patients, attend to the ethical dimensions of resource allocation, and report to the College when there are reasonable grounds to believe a member has violated the *Standards*. Failure to comply with the provisions of the document may be cause for termination of membership in the College.

During the period under review, two professional societies developed guidelines for relationships of their members with the pharmaceutical industry. The Canadian Society of Hospital Pharmacists (CSHP) guidelines were published in November 1990 [12]. Their stated goals are to clarify the industry/pharmacy relationship and to maximize the efficiency and effectiveness of this relationship.

The guidelines deal with the following matters: communication between industry representatives and pharmacists, including the responsibilities of each party, research, education, the role of pharmaceutical manufacturers' representatives, gifts and donations, participation in focus groups and surveys, product pricing, and hospital/clinic displays. In general, the document encourages close collaboration between pharmacists and industry. Furthermore, it is stated in an introductory note that "it remains the responsibility of users of this document to judge its suitability for their particular purpose."

At its 1991 annual meeting, the Canadian Medical Association approved its own *Guidelines for an Ethical Association between Physicians and the Pharmaceutical Industry* [8]. This document deals with physician participation in industry sponsored research, surveillance studies, continuing medical education, and marketing. While the focus is on relationships with the pharmaceutical industry, the guidelines are meant to apply to other commercial relationships as well, such as those with manufacturers of medical devices and baby food and providers of medical services.

The main goal of the CMA guidelines is to ensure that when a relationship with industry entails a conflict between the personal interests of physicians and the well-being of patients, the latter should prevail. Even the appearance of physicians receiving unearned benefits from industry should be avoided. In particular, lavish entertainment in the guise of continuing medical education is no longer acceptable.

C. *Medical Ethics Education*

In 1989 the Council of the Royal College of Physicians and Surgeons of Canada approved a proposal for incorporating biomedical ethics into all residency training programs. The proposal had been prepared by an ad-hoc committee of the Royal College and approved by the College's Biomedical Ethics Committee in May 1989. It includes sections on goals, curriculum, implementation, evaluation and recommendations to other groups within the Royal College, especially the specialty societies.

In 1990 the Accreditation Committee of the Royal College issued a report which specifies the objectives, settings and evaluation requirements for the teaching of ethics in residency programs [29]. Specific objectives are set forth under three headings: knowledge, attitudes, and skills. Suggested settings for ethics teaching include individual faculty-patient-resident encounters, team or group interactions, formal educational sessions, and ethics committees. Attention to ethics should be part of the evaluation process for individual residents and for programs.

D. *National Biomedical Ethics Council Proposal*

In response to the proliferation of national ethics and bioethics committees around the world, the Law Reform Commission of Canada decided to produce a study paper to determine the feasibility and desirability of such a committee for Canada. The document, *Towards a Canadian Advisory Council on Biomedical Ethics*, was prepared by Jean-Louis Baudouin, Monique Ouellette and Patrick A. Molinari, all members of the Faculty of Law of the University of Montreal, and was published by the LRCC in 1990 [22]. It contains the following recommendations

- A Canadian Advisory Council on Biomedical Ethics should be established.
- The primary responsibilities of this Council should be: (a) to ensure greater coordination of research and activities in biomedical ethics throughout the country; (b) to make the results of Canadian and foreign research available to all who are interested; (c) to provide advice to government and other public authorities on all matters relating to biomedical ethics and, in particular, to issue non-binding public opinions on problems of current interest; (d) to act as a national biomedical ethics think tank; (e) to provide information, both to other organizations with similar goals and to the general public, in order to make Canadians more aware of major contemporary issues; (f) to establish contacts with international bodies and organizations in other countries concerned with biomedical ethics; (g) to present Canada's position on major problems in biomedical ethics to the international community; (h) in seeking to attain these objectives, to work in direct and close cooperation with existing organizations.
- The Council should be composed of between 22 and 30 members and should represent some of the Canadian groups and organizations involved in the field of biomedical ethics, without necessarily representing the interests of all of them.
- The members of the Council should be appointed by the Governor General in Council and selected for their expertise, independence, open-mindedness and originality; they should come from various professional backgrounds and have different types of training, so that the multidisciplinary and interdisciplinary nature of the Council will be assured; in particular, physicians, surgeons, nurses, research scientists, ethicists, philosophers, theologians, psychologists, medical administrators, and members of the legal profession should be included.
- The Council should be incorporated and should for administrative purposes report to Health and Welfare Canada.
- The Council should have the necessary powers to manage its internal affairs, in particular the power to set up committees, hire experts and recruit staff.
- The Council should be funded separately from the Department under which it operates.

As of June, 1991 the federal government, to which the LRCC reports, has taken no action on this proposal.

XII. CONCLUSION

In accordance with the scope of this book, the preceding description of bioethics in Canada has dealt only with the activities of governments, the courts, and professional associations. As such, it does not present a complete picture of bioethics in this country. Omitted from consideration are the vast majority of research activities, publications, educational programs, and clinical, research, and professional ethics committee deliberations.

Despite these limitations, this report does permit some tentative conclusions about the status and role of bioethics in Canada:

- Canadians seem to have an aversion to legislative control over ethical aspects of health care. Despite the many recommendations of the Law Reform Commission of Canada to update the Criminal Code to deal with current issues in ethics, medicine and law, neither the federal government nor the general public thinks that such legislation is required or even desirable. The provinces and territories are equally uninterested in health care legislation, apart from its financial dimension.
- As a result, the regulation of ethically problematic health care issues is generally performed by para-governmental or professional bodies, such as the Medical Research Council of Canada, the National Council on Bioethics in Human Research, and the provincial/territorial Colleges of Physicians and Surgeons. These bodies prefer to issue guidelines rather than laws, leaving the application of these guidelines in large measure to subsidiary groups such as local ethics committees.
- The widely accepted principles of Canadian health insurance provide a common framework for mutual understanding and acceptance of the "rules of the game" by all participants in the health care system, including the various health care professions, administrators, and patients. Conflict is not thereby eliminated, but it seems to be reduced, at least in relation to the situation in the United States as evidenced by the much lower rate of malpractice suits in Canada than in the United States.

The rapidly increasing financial constraints on health care in Canada are threatening two of the basic principles of the system, comprehensiveness and universal accessibility. While this threat is an important ethical issue in itself, it is also likely to have major repercussions on many other issues, such as the new reproductive technologies, withdrawing/withholding treatment from dying patients, and active euthanasia. Governments may become more heavily involved in medical decision-making, if for no other reason than to control expenditures.

What role will there be for bioethics in an atmosphere of economic domination? Two scenarios are possible: (1) ethical considerations will be regarded as unaffordable luxuries and not relevant to health care policy formation; (2) the allocation of scarce resources, at the macro, meso and micro levels, will be recognized as a fundamentally ethical issue, one requiring the input of ethicists as much as economists.

At present it is impossible to predict which of these scenarios will prevail. Many factors will determine the outcome, not the least of which is the willingness and ability of the bioethics community to deal with the economic aspects of health care. The response of these individuals and groups to this challenge will be evident in the next volume of this yearbook.

Department of Ethics and Legal Affairs
Canadian Medical Association
Ottawa, ONTARIO

Center for Bioethics
Clinical Research Institute of Montreal, QUEBEC

BIBLIOGRAPHY

(Items marked with an asterisk are available in French.)

Articles and Reports

1. British Columbia Branch, The Canadian Bar Association: 1989, *Report of the Special Task Force Committee on Reproductive Technology*.
2. Canadian Association of Pathologists, Section of Clinical Pathology: 1990", "Guidelines for the Release of Laboratory Test Results, Reports and Specimens to Patients, Interested Groups and Other Institutions", *Canadian Medical Association Journal* 143, 847-848.
3. *Canadian Bar Association: 1990, *Submission of the Canadian Bar Association to the Royal Commission on New Reproductive Technologies*.
4. Canadian College of Health Service Executives: 1991, *Standards of Ethical Conduct for Health Service Executives*, Canadian College of Health Service Executives, Ottawa.
5. Canadian Fertility and Andrology Society and the Society of Obstetricians and Gynaecologists of Canada, Combined Ethics Committee: 1990, *Ethical Considerations of the New Reproductive Technologies*, Ribosome Communications, Toronto.
6. *Canadian Medical Association: 1991, *Brief to the Standing Senate Committee on Legal and Constitutional Affairs Re: Bill C-43, An Act Respecting Abortion*, Canadian Medical Association, Ottawa.
7. *Canadian Medical Association: 1991, *New Human Reproductive*

Technologies, Canadian Medical Association, Ottawa.

8. *Canadian Medical Association: 1991, *Guidelines for an Ethical Association between Physicians and the Pharmaceutical Industry*, Canadian Medical Association, Ottawa.
9. *Canadian Medical Association: 1991, *Advance Directives for Resuscitation and Other Life-Saving or Sustaining Measures*, Canadian Medical Association, Ottawa.
10. *Canadian Nurses Association, Canadian Medical Association, Canadian Hospital Association: 1984, *Joint Statement on Terminal Illness*,
11. Canadian Paediatric Society Bioethics Committee: 1990, "Transplantation of Organs from Newborns with Anencephaly", *Canadian Medical Association Journal* 142, 715-717.
12. Canadian Society of Hospital Pharmacists: 1990, "Guidelines for Pharmacists and the Pharmaceutical Industry", *Dimensions* 67 (November), 14-16.
13. *Catholic Health Association of Canada: 1991, *Health Care Ethics Guide*, Catholic Health Association of Canada, Ottawa.
14. Federal Centre for AIDS Working Group on Anonymous Unlinked HIV Seroprevalence: 1990, "Guidelines on Ethical and Legal Considerations in Anonymous Unlinked HIV Seroprevalence Research", *Canadian Medical Association Journal* 143, 625-627.
15. Federal/Provincial/Territorial Advisory Committee on AIDS: 1989, "Confidentiality and HIV seropositivity", *Canadian Medical Association Journal* 141, 523-525.
16. *Law Reform Commission of Canada: 1982, *Euthanasia, Aiding Suicide and Cessation of Treatment* (Working Paper 28), Law Reform Commission of Canada, Ottawa.
17. *Law Reform Commission of Canada: 1983, *Euthanasia, Aiding Suicide and Cessation of Treatment* (Report 20), Law Reform Commission of Canada, Ottawa.
18. *Law Reform Commission of Canada: 1986, *Some Aspects of Medical Treatment and Criminal Law* (Report 28), Law Reform Commission of Canada, Ottawa.
19. *Law Reform Commission of Canada: 1986, *Recodifying Criminal Law*, Vol. 1 (Report 30), Law Reform Commission of Canada, Ottawa.
20. *Law Reform Commission of Canada: 1989, *Crimes Against the Fetus* (Working Paper 58), Law Reform Commission of Canada, Ottawa.
21. *Law Reform Commission of Canada: 1989, *Biomedical Experimentation Involving Human Subjects* (Working Paper 61), Law Reform Commission of Canada, Ottawa.
22. *Law Reform Commission of Canada: 1990, *Toward a Canadian Advisory Council on Biomedical Ethics* (Study Paper), Law Reform Commission of Canada, Ottawa.
23. Law Reform Commission of Manitoba: 1990, *Discussion Paper on*

Advance Directives and Durable Powers of Attorney for Health Care, Law Reform Commission of Manitoba, Winnipeg.

24. Manitoba Medical Association: 1990, *Response to the Manitoba Law Reform Commission Paper: Advance Directives and Durable Powers of Attorney for Health Care*, Manitoba Medical Association, Winnipeg.
25. *Medical Research Council of Canada: 1987, *Guidelines for Research Involving Human Subjects 1987*, Ministry of Supply and Services Canada, Ottawa.
26. *Medical Research Council of Canada: 1990, *Guidelines for Research on Somatic Cell Gene Therapy in Humans*, Ministry of Supply and Services Canada, Ottawa.
27. Ontario Medical Association: 1991, *Euthanasia and the Role of Medicine*, Ontario Medical Association, Toronto.
28. Roy, D.J. and Williams, J.R.: 1987, "Canada: Conflict as well as Consensus", *The Hastings Center Report* 17 (June), Special Supplement "Biomedical Ethics: A Multicultural View", 32-34.
29. *Royal College of Physicians and Surgeons of Canada: 1989, *Postgraduate Biomedical Ethics Teaching*, Royal College of Physicians and Surgeons of Canada, Ottawa.
30. *Science Council of Canada: 1991, *Genetics in Canadian Health Care*, Science Council of Canada, Ottawa.
31. William, J.R.: 1986, *Biomedical Ethics in Canada*, Edwin Mellen Press. Queenstown, Ontario and Lewiston, New York.
32. Williams, J.R.: 1989, "Commissions and biomedical ethics: the Canadian experience," *The Journal of Medicine and Philosophy* 14, 425-444.

Cases

33. *Borowski v. Canada (Attorney General)*, S.C.R. 342, 57 D.L.R. (4th) 231 (S.C.) (1989).
34. *Brown v. British Columbia (Minister of Health)*, B.C.J. 151 (S.C.) (1990).
35. *Canadian Abortion Rights Action League Incorporated v. Attorney General of Nova Scotia*, 69 D.L.R. (4th) 241, 96 N.S.R. (2d) 284 (S.C.) (1990).
36. *Fleming v. Reid*, 4 O.R. (3d) 74 (C.A.) (1991).
37. *Jean-Pierre Valiquette v. The Gazette*, C.S. (1991).
38. *Malette v. Shulman et al.*, 72 O.R. (2d) 417 (C.A.) (1990).
39. *Meyer Estate v. Rogers*, 2 O.R. (3d) 356 (G.D.) (1991).
40. *Murphy v. Dodd*, 70 O.R. (2d) 681 (H.C.J.) (1990).
41. *Nimi v. Morris*, 64 Man.R. (2d) 319 (C.A.) (1990).
42. *R. v. Morgentaler*, 1 S.C.R. 30, 44 D.L.R. (4th) 385 (S.C.) (1988).
43. *R. v. Morgentaler*, 99 N.S.R.(2d) 293, 270 A.P.R. 293 (P.C.) (1990).
44. *R. v. Sullivan*, S.C.J. 20 (1991).
45. *Re A. (in utero)*, 75 O.R. (2d) 82 (U.F.C.) (1990).
46. *Reibl v. Hughes*, 2 S.C.R. 880, 114 D.L.R. (3d) 1 (S.C.) (1980).

47. *Sharpe Estate v. Northwestern General Hospital*, 2 O.R. (3d) 40 (G. D.) (1991).
48. *Tremblay v. Daigle*, 2 S.C.R. 530, 62 D.L.R. (4th) 634 (S.C.) (1989).
49. *Weiss v. Solomon*, 1 RJQ 1 731 (C.S.) (1989).

JOSE A. MAINETTI, GUSTAVO PIS DIEZ, AND JUAN C. TEALDI

BIOETHICS IN LATIN AMERICA: 1989-1991

I. INTRODUCTION

In our region, bioethics as an academic discipline and a public movement is still in its beginning stages. The historical changes resulting from scientific and technological advances in biomedicine and from the liberal and pluralist character of industrialized countries have barely begun to occur in the developing countries of Latin America, which remain largely "pretechnical" in their orientation. Bioethics as a secular discipline, with its principles of beneficence, autonomy, and justice, and its emphasis on the rational and free agent in the therapeutic relationship, has not yet reached Latin America [5].

Nonetheless, with the dissemination of the new medical technologies (special care, organ transplants, and assisted reproduction) and the advent of democratic governments in the region, public and academic interest in bioethical issues has been stimulated in the 1980s. On the one hand, litigation in medicine has increased, perhaps because of the distance which specialization poses between the professional and the patient; indeed, malpractice and the patient rights movement imitate the "American way" of doing bioethics. On the other hand, there has been an academic resurgence of practical moral and political philosophy which has been applied to medicine according to a pluralistic model of consensus formation, again along the lines of bioethics in the United States [7].

Within this context, it is still too early to expect major developments in bioethics at the governmental and professional levels which this Yearbook surveys. Our region will require a host of legislative and policy responses to the complex realm of today's biomedicine.

This report on major legislation, court rulings, regulatory changes, and policy announcements by governments and professional associations on bioethical topics is based on the search undertaken at our *Centro Nacional de Referencia en Bioetica* for Argentina. Data gathering from other countries of the region posed numerous difficulties, and therefore, unfortunately, the information is fragmentary. Nevertheless, the publications of the Pan-American Health Organization (to which our group has contributed significantly), as well as the regional material gathered in our Center, have been useful for that purpose.

B. Andrew Lustig (Sr. Ed.), Bioethics Yearbook: Volume 2, 83-96.

II. PROFESSIONAL-PATIENT RELATIONSHIP AND HEALTHCARE ETHICS

As with other issues in bioethics, Latin America reveals characteristic tendencies in its approaches to issues of confidentiality, consent to treatment and experimentation, and the notion of a right to health care and its costs. Although it is true that new regulations are appearing both in public policy and in professional guidelines, it is also the case that there is a wide gap between theory and practice in bioethics.

The field of bioethics, born in the United States, has tended to give priority to the principle of autonomy rather than the principle of beneficence. Thus, respect for confidentiality and the requirements of informed consent reflect the priority of autonomy. By contrast, practice in Latin American healthcare and research has tended to emphasize the principle of beneficence.

As a result, relations between medical professionals and patients, as well as health care policies, still exhibit a strongly paternalistic character, despite the ever-growing record of new legislation. If in North America there remains an opposition between theoretical autonomy and practical beneficence, in Latin America there emerges a tension between the priority of beneficence and a principle of justice that theoretically endorses the equal right to health care. Many Latin American societies, because of inadequate social and economic development, are not able to guarantee the right to basic health care.

There is no doubt, however, that recent legislative initiatives, health policy decisions, and updated professional codes are efforts to reduce the distance between beneficent practice and just policies. The paradigm of this transformation has been AIDS, both in its impact on the professional-patient relationship and on health care policy.

Having made these preliminary comments, we will now analyze recent rules concerning the ethical aspects of confidentiality, consent, equal access to health care, and the control of health expenditures.

A. *Confidentiality*

In May 1991 in Argentina, the authorities of La Plata Military District disclosed the results of a blood test among 5,407 adolescents who were to join the Army. One in every 160 persons tested positive for HIV infection. The public disclosure of these results prompted significant public debate. The National Act Number 23.798 on AIDS had been passed in 1990. Article 2 states that the law's provisions are in no case meant to:

> a) affect personal dignity; b) cause discrimination, stigmatization, degradation or humiliation; c) exceed the background of legal exceptions limiting medical secrets....; d) trespass the privacy of any inhabitant of the Argentine Nation; or e) individualize people through cards, records or databases which, for this purpose, [would] be codified [20]

When the above case occurred, this statute was not yet fully in force. Today however, Article Two is viewed as fundamentally safeguarding confidentiality, despite numerous criticisms directed at it.

The protection of confidentiality in AIDS cases has lately been the focus of numerous rules in various Latin American countries. Nevertheless, the situation is not uniform throughout the region. In some countries (Ecuador, El Salvador, Guatemala, Honduras, Nicaragua, Paraguay, the Dominican Republic, and Venezuela), a legislative vacuum exists which is likely to be filled by general legislation calling for compulsory reporting of the disease. Brazil, while emphasizing AIDS education, has also insisted on compulsory reporting without specifically attending to issues of confidentiality [24]. Legislation to protect confidentiality in AIDS cases has been passed in Argentina [20], Bolivia [22], Chile [28], Colombia [29], Costa Rica [30], Haiti [31], Mexico [34], Panama ([36];[37]0, Peru [38], and Uruguay [39].

Bolivia and Peru have paid particular attention to the effects on public opinion of breaches of confidentiality in AIDS cases. These countries have discussed adopting criminal sanctions for those who publish sensationalist material. Argentina and Colombia, in turn, rely on legislation in to safeguard patient confidentiality. We have already commented on the Argentine statute. In Colombia, Article 22 of Decree 559 asserts:

> Epidemiologic information related to HIV infection is confidential. Professional secrecy shall not be ...[an impediment] to providing such information in those cases under legal provisions and regulations [29].

Article 33 of that decree states:

> Health suppliers who know or provide health care to an HIV-infected person, either symptomatic or asymptomatic, are compelled to keep confidentiality of consultation, diagnosis and evolution of the disease. This provision shall also be observed in cases of people with risky sexual behavior whose condition is not seropositive [29].

In the broadest sense, professional codes, and those of medical ethics in particular, have traditionally respected the rule of doctor-patient confidentiality. In recent years, declarations have been adapting that rule to modern technological developments. For example, the "Declaration on Ethics in Medicine" of the Latin American Association of Academies of Medicine (Quito,1983) addressed issues of confidentiality raised by the computerization of clinical records. A second example of new rules is the Brazilian "Code of Medical Ethics" [24]. Section Nine of the Code devotes eight articles to confidentiality in general and to particular issues of confidentiality pertaining to minors and workers.

Of special regional significance is the creation of the National Genetic Data bank in Argentina, formed in the wake of the people who disappeared during the military governments of the late 1970s and early 1980s. The purpose of the

Data Bank is to store genetic information belonging to presumptive relatives in order to identify children through genetic engineering techniques. Despite the crucial importance of confidentiality in these cases, the law which created the Data Bank addresses the issue only briefly in the eighth of ten articles: "The records and files of the National Genetic Data Bank shall be kept inviolable and unalterable..." [17]. The Genetic Databank has so far proved essential in identifying fifty missing children. Yet, given the harmful consequences that could follow if such genetic information is used for other questionable purposes, there is need to guarantee confidentiality ([2];[10];[12].

B. *Consent to Treatment and Experimentation*

On his journey to Colombia, Bolivia, Chile, Argentina, and Brazil in 1990 to inform the PAHO on the state of bioethics in Latin America, the American bioethicist James Drane concluded that review committees for scientific research did not function effectively in any country, nor was informed consent generally obtained from patients or research subjects, despite some existing regulations [2]. Drane thus noted the gap between theory and practice to which we have already referred. This situation has not yet been remedied, although non-governmental organizations are making important efforts[1]. To date, only three Latin American countries have passed regulations on informed consent.

1. *Mexico*

The General Health Law and its Regulation on Health Research are the most relevant rules on the subject ([32];[35]). Less important, although more recent, are "Technical Norm Number 313", which regulates the submission of research proposals and technical reports in health care institutions, and "Technical Norm Number 314", which regulates record keeping and data-gathering in health research ([33];[34]).

Article 465 of the General Health Law imposes in prison sentences of one to eight years on those who perform clinical research on human beings without the written consent of the subject or his agent, the latter defined according to Article 100 of that statute. The punishment levied is more severe in cases involving minors, handicapped persons, the elderly, or prisoners. Article 466 sets penalties for those who perform artificial insemination without woman's consent. Article 324 articulates the need for written and expressed consent in order to procure organs and tissues [32].

Section Two of the Regulation on Health Research discusses "Ethical Aspects of Research on Human Beings". Article 14 of that regulation requires "the written informed consent of the person under research or his proxy". Articles 20-27, in turn, develop at length the concept of and the conditions for informed consent. Articles 29, 36-37, 43, 57-58, and 71, respectively, refer to informed consent for research in communities, for minors or handicapped persons, for pregnant women, for special populations, and for pharmacological studies [35].

With regard to requirements for the submission of projects and technical research reports in health care institutions, Article 15 of Technical Norm Number 313 emphasizes the need to specify "the way in which ethical precepts will be observed" [33]. Technical Norm Number 324 on record keeping and data gathering in health research sets forth specific requirements for filing information with ethics, research, and biosecurity commissions [34].

2. *Brazil*

The 1988 Code of Medical Ethics forbids physicians to engage in therapeutic experimentation without the consent of the patient or his agent [23]. It also forbids participation in any experiment on human beings for military, political, racial, or eugenic purposes. In addition, Provision Number 1 of the National Health Council discusses the ethical aspects of experimentation by stressing the need for ethics and biosecurity committees and by elaborating the requirements of consent for particular groups [27].

3. *Argentina*

Law number 11044, "The Protection of Persons Involved in Scientific Research", is the first Argentine law to deal broadly with consent [21]. As a provincial statute, it has already encouraged legislative initiatives based on its text, one at a national level and another in the province of Tucuman. According to Article 3, "All research involving study of human beings shall conform to the criteria of respect for their dignity and protection of their rights and welfare". And subsection e of Article 40 states that research involving human subjects requires "the consent of subjects under research or their respective agents through public documents which specifies the risks they face".

The requirement of consent is also emphasized in other articles of the Law. Research subjects are entitled to halt their participation in the research at any time (Article 70). The consent document shall explain "the nature of the procedures the participant will be subject to, eventual risks rising from them, the participant's free choice, and the exclusion of all forms of coercion towards him" (Article 90). In cases involving incompetent persons, consent shall be given "by the agent under authorization of a qualified judge in expeditious lawsuits" (*juicio sumarisimo*) (Article 110). Ethics and research committees are entitled to adjourn any research that may affect the psycho-physical and/or psychosocial well-being of incompetent participants (Article 120). Consent requirements are also specified for pregnant and puerperal women, newborns, fetuses and embryos (Article 140), special populations (Article 220), research where new methods of prevention, diagnosis, treatment, and rehabilitation are studied (Article 240), and pharmacological research (Article 320). Ethics committees created by this statute (Article 360) shall supervise consent (Article 360).

C. *Equitable Access to Health Care*

The right of equitable access to health care has been proclaimed a human right in Article 25 of the "Universal Declaration of Human Rights", adopted by the United Nations General Assembly on December 10, 1948. In the same year, the Organization of American States (OAS) adopted the "American Declaration of the Rights and Duties of Man", proclaiming (although vaguely) the right to health in Article XI. The "American Convention on Human Rights" (1978), signed by most Latin American countries, promises to gradually achieve full implementation of economic, social and cultural rights, among which there is a right to health, first proclaimed in the "Protocol of Buenos Aires" (OAS, 1967). In this sense, the right of equal access to health care as a "human right" has been elaborated as a matter of international law for more than four decades.

In Latin America, however, the present debate on equitable access to health care exemplifies, once again, the gap between theory and practice. Although different constitutions and statutes in Latin American countries affirm the concept, the conditions necessary to guarantee that the right can be exercised are not always present. Thus, the idea of the "progressive fulfillment" of the right to healthcare relative to the material conditions of each country has developed. In practice, of course, this idea may sometimes be used as an excuse not to fulfill a basic moral obligation. Nonetheless, the statement of norms in the delivery of health care remains fundamental, because norms help to guide the possibilities of effective social transformation. We will now describe various understandings of the right to health care in Latin America.

1. *Argentina*

The Argentine Constitution now in force, which dates from 1853, has no express reference to the right to health care. In Argentine law, the notion of police power presumes that the state has power to limit personal rights in order to protect public health. Policies of various governments, therefore, define the meaning of equal access accordingly. Two national statutes passed in 1989 have set forth a new system of social security and of national health security ([18];[19]).

Law Number 23.660 regulates social security as an important sector of the Argentine health system. The social security organizations are entities constituted by 6 percent of workers' payments charged to employers, with 3 percent charged to workers (Article 16). The purpose is to devote those resources to health services, although other social benefits should be provided for as well (Article 3). Public or private workers and retired workers (Article

8), as well as families and dependents, are compulsorily included as beneficiaries (Article 9). Social security organizations should devote at least 80 percent of their resources to health care. Those with centralized collection of funds should distribute them again according to a principle of solidarity, in order to assure equitable access to health care (Article 5). The organizations are allowed to devote up to 8 percent of their resources to administrative overhead. Social security organizations, as part of the National System of Social Security, are subject to regulative norms (Article 3).

Decree 358/90, which regulates Law Number 23.660, defines "other social benefits" as those not encompassed by the medical coverage regulated by articles 25, 26, 27, and 28 and those concordants of Law Number 23.661, which we will discuss below [14]. According to Article 3 of Law Number 23.660, social security organizations must guarantee the provision of health services according to norms set by the Health Secretary and the Social Security National Administration.

It is clear that Law Number 23.660 and its accompanying Decree are fundamental to the legal framework that assures equal access to health care in Argentina. Nonetheless, the major instrument for this purpose remains the National System of Social Security.

Law Number 23.661 created the National System of Social Security "with the purpose of assuring the full exercise of health to all inhabitants without social, economic, cultural or geographic discrimination" (Article 10) [19]. Its "fundamental aim is to provide for equitable, integral and humanized health provisions, directed to the promotion, protection, recovery and rehabilitation of health, which shall respond at the highest quality available, and which shall assure the beneficiaries [a uniform] level of services, based on a criterion of distributive justice, without any form of discrimination" (Article 20). These social organizations are agents of health security (Article 20). They must conform to Health Ministry policies aimed at coordinating social security, public health services, and private suppliers (Article 3). Security services are supplied according to national health polices based on a strategy of primary health care, decentralized operation, and freedom of choice of suppliers by beneficiaries (Article 25).

Decree Number 359/90, which regulates Law Number 23.661, broadens the concept of equal access [15]. Article 50 of the Decree states that the population shall be classified into categories according to their income levels in order to assure equity.

In Argentina, the main legal framework for egalitarian access to health care is that provided by Laws Number 23.660 and 23.661 ([18];[19]). It is important to note that the present government has proposed a law which would substantially modify this structure by creating freedom of choice in social security independently of the worker's occupation.

2. *Mexico*

The present Mexican constitution dates from 1917 and asserts the social rights posited by the Mexican Revolution. It did not initially include the concept of a right to health care, but a constitutional and amendment in 1983 added that right to Article four:

> Every person has the right to health protection. The law shall define the ways and means to provide access to health services, and shall establish the participation of the Federation and of federal agencies concerning general health, in accordance with the provisions of Paragraph XVI of Article 73 of this Constitution.

The 1984 General Health Law of Mexico, which we have already quoted, is the legislative document which regulates that constitutional right (Article 1) [32]. The National Health System, which assembles private and public sectors and social services, was created to implement the right to health protection (Article 5). This system aims at providing health services to the population and improving the quality of such services (Article 6). The Ministry of Health is in charge of coordinating the system (Article 7) and of promoting full participation of the private and public sectors as well as of social services (Article 10). Health services are classified as medical care, as public health, or as social welfare (Article 24). Expanding the quality and quantity of health services to vulnerable groups is guaranteed as a priority of the system (Article 25). Primary health care is considered to be basic health service (Article 27). The basic table of health operating costs is established (Article 28), and its permanent existence is assured by the Health Secretary (Article 29).

The Health General Law is the most important legal instrument in Mexico for the regulation of equal access to health care[3]. The essential political instrument, however, is the "National Health Program 1990-1994" of the Ministry of Health, which develops the policies required to enforce the Health General Law.

Although we have only analyzed only two countries in the region with regard to issues of access to health care, Argentina and Mexico reveal two different constitutional bases, and a different legislative development based on civil law rather than on common law (as in Canada, The United States, and the British Caribbean).

III. THE BEGINNING OF LIFE

A. *New Reproductive Technologies*

Although ethical principles have generally been observed in the practice of new reproductive technologies, (NRTs) there is no official legislation on the subject. In Argentina, however, the National Government shows increasing interest in responding to the ethical issues raised by NRTs. In 1989, the National Senate created an interdisciplinary commission to study NRTs in order to produce appropriate legislation. As a result, artificial insemination and in vitro fertilization (IVF) have recently been regulated [13]. The Penal Code adds an article which sets penalties for married women who are artificially inseminated without the consent of their husbands. In the Province of Buenos Aires, Law 11.044 guides "medical practices such as assisted fertilization of proven efficacy in human beings" [21]. Article 5 explicitly refers to informed consent and reads: "Patients will give their written consent on a pre-printed form where the methods and possible risks the proposed treatment may offer [will be listed]..." This same law requires the formation of a Provincial Ethics Committee whose basic function is to advise the Application Authority on the medical practices it oversees.

The situation in Colombia is very similar to the Argentina. According to Sanchez Torres,

> Despite the medical, legal, social, and ethical implications of these new human reproductive procedures, ... the Government of Colombia has not yet issued any standards to regulate their practice. Therefore, in dealing with related ethical issues, Colombian physicians must rely on guidance issued by the World Medical Association, as set forth in Article 54 of Law 23 (1981)... ([8] p. 511).

Recently, Mexico's General Health Law contemplates penalties for those who perform artificial insemination on adult competent women without their consent, or on those who are under age or incompetent ([35], Article 466).

B. *Abortion*

Abortion remains punishable by law in the countries of the region with two exceptions: (1) the so-called "therapeutic" abortion, when the mother's life is in danger and she consents to the abortion; and (2) cases when pregnancy results from rape. In the latter cases, legal actions – by the victim or her legal representative – should be undertaken prior to the abortion. In the last several years, Argentinian legislation has revealed a variety of attitudes with regard to the range of allowable exceptions [4].

IV. ACTIVE EUTHANASIA

Euthanasia finds no support in the law or in codes of medical ethics. The Brazilian Code of Medical Ethics stipulates that "a physician must use all the diagnostic and treatment resources at his disposal on behalf of the patient"; moreover, he should never employ any "means to shorten the patient's life, even at the request of the patient or whoever is legally responsible for him" ([23], Articles 57 and 66). Naturally, the Brazilian Penal Code penalizes homicide, and euthanasia is deemed homicide because it involves the crime of killing someone.

In Peru, controversy has arisen about whether "compassionate murder", *i.e.* euthanasia, would cause more problems than it would solve. Peruvian legislation also remains reluctant to accept "living wills", by which people can freely express their desire to be allowed to die under certain circumstances. It is feared that scarce medical resources and high costs many unduly influence persons to limit their own care. In addition, there is ongoing discussion about the rights of the family to have the physician withdraw artificial life support in such cases as persistent vegetative state. There is also debate about whether it is humane to prolong life artificially, since resuscitation does not imply brain restoration of brain function. However, the Code of Ethics of the Peruvian Medical Association clearly prohibits helping anyone to commit suicide. Passive forms of euthanasia, such as the interruption of nutrition, have also provoked discussion among Peruvian medical professionals.

Argentina still awaits a frank discussion of euthanasia. Nevertheless, the Penal Code was slightly modified to lessen penalties for those who help others, with their consent, to terminate irreversible physical suffering (Article 73).

V. ORGAN DONATION

A. *Determination of Death*

Argentina passed specific legislation on organ transplantation. Law 21.541, promulgated in 1977, not only legalized the procurement and transplantation of organs and tissues (through cadaveric or *inter vivos* donation), but also created a *National Central Unico Coordinador de Ablació e Implanters* (so called since 1990). The impact of bioethics encouraged the adoption of new standards. Article 11 of the law reads:

> The chiefs of the transplant team (...) must give thorough and clear information, ...in accordance with the sociocultural level of each patient, about the ablation and implant risks, sequels, assumed evolution and postsurgical limitations. Once the donor and the recipient have understood the real meaning of the information, [they can proceed] provided they shall make their decision, freely and at will. The fulfillment of these conditions, and the physician's evaluation of the risks of the practice, will be conveniently documented according to the established norms [16].

Article 21 of the law proposes brain death as a criterion. Actually, a person is considered dead when total and irreversible encephalic activity cessation is verified by a physician [16]. The professionals responsible for this verification are a neurologist or neurosurgeon and a clinician, and neither can be a member of the transplant team.

In Colombia, the increase in organ transplantation prompted the Government to take regulatory measures. In the National Health Code, designated as Law 09 of 1979, Title IX regulates "the donation or transfer of organs, tissues, and organic fluids from cadavers or living persons for transplantation and other therapeutic users". In 1986, the Ministry of Health issued supplemental Decree 2363, which establishes a definition of "brain death". In 1988, the National Congress passed Law 73, which further regulates the procurement and distribution of solid organs and tissues ([8], p. 511).

The criterion of brain death adopted in Argentina has also been adopted by Bolivia, Chile, Ecuador, Panama, and Peru. In some cases, the criterion of irreversible cardiorespiratory arrest is also mentioned. Costa Rica speaks of "properly tested procedures" and "death certified by methods established by the Health and Welfare Secretary", but without specifications. Similarly, Venezuela sets "procedures appropriate" to determine death legally.

B. *Commercialization of Organs and Tissues*

Not all countries in the region have adopted regulations prohibiting commerce in organs and tissues, despite the likelihood that such trafficking will disproportionately burden those who live in poverty. Moreover, the failure to distinguish commerce in organs from compensation of donors for related costs has created a significant legislative and regulatory vacuum. The absence of a legal basis for donor compensation discourages potential donors, frustrates efforts to inform and educate the public about organ donation, and leaves the matter of compensation to private agreements between donors and recipients ([5], pp. 433-434).

VI. CONCLUSION

Within contemporary medicine, areas such as research with human subjects, organ procurement and distribution, assisted reproduction, the rationing of health care, and the foregoing of life sustaining treatment involve complex ethical choices and require explicit rules and formal decision-making procedures. These rules are increasingly institutionalized in statutes, court opinions, administrative standards, and institutional protocols.

Developments that have transformed medicine in developed countries are now occurring in Latin America. As soon as bioethics is institutionalized in Latin America, the reporting and discussion of major bioethical issues will become more accurate. Toward that end, our Center of Bioethics created two important fora to improve bioethics education and information in the region. The first is the Latin American School of Bioethics (*Escuela Latinoamericana de Bioética, ELABE*). The *ELABE* of the *Fundación Mainetti* is an academic project meant to enhance governmental research on bioethics, bioethics teaching, and health policy studies. Its goal is the qualify individuals to take leadership roles in bioethics, developments in their own countries, and to establish a forum for cultural and scientific exchange throughout the region.

The second forum is the *Federation Latinoamerica de Instituaónes de Bioética* (FELAIBE), an ambitious project of transcultural and international cooperation in bioethics meant to develop a regional philosophy of bioethics which is complementary to Anglo-American approaches[1].

Faculty of Medical Humanities and Philosophy of Medicine
Universidad Nacional de La Plata
ARGENTINA

NOTES

1 In 1990, the *Escuela Latinoamericana de Bioètica* (Fundación Mainetti, Argentina) organized its First International Seminar on Ethics Committees. This was the first regional effort to train professionals from Colombia, Uruguay, and Paraguay. The antecedent of this seminar in 1989 was a Course on Health Care Ethics Committees with a national scope. In 1991, the *ELABE* organized its second International Course on Health Care Ethics Committees, taught by Professor Stuart Spicker. Hosted in Gonnet (Argentina), the *Federación Latinoamericana de Instituciones de Bioética* was created last December. One of Dr. Spicker's aims is to foster the development of ethics committees in the Region.

2 A clear example of this is the case of Hospital Moyano in Argentina. Thirty patients died of cold and malnutrition there in June 1990. Some politicians claimed that the deaths were excused by the national financial crisis.

3 For a complete and updated version of the law with all its Supplements and Technical Norms, see "Ley General de Salud", 60th edition, 1990, Editorial Porrua, Mexico.

4 See Fuenzalida-Puelma, Hernan L. *et al.* (eds), 1989, *The Right to Health in the Americas. A Comparative Constitutional Study*, Washington, Pan American Health Organization. This text analyses constitutions of all Latin American countries in connection with the right to health.

BIBLIOGRAPHY

Books and Articles

1. Cecchetto, S. *et al.*: 1990, "Premature and Disabled Newborns in Argentina", paper submitted to the "International Symposium on the Beginning of Life", University of Iowa, Iowa City, USA.
2. Drane, J.F. *et al.*: 1991, "Medical Ethics in Latin America: A New Interest and Commitment", in *Kennedy Institute of Ethics Journal* 1, 4, 325-338.
3. Fuenzalida-Puelma, Herman L. *et al.* (eds.): 1989, *The Right to Health in the Americas: A Comparative Constitutional Study*, Washington, Pan American Health Organization.
4. Garcia Maranón, E. *et al.*: 1990, *Aborto e Infanticido, Aspectos Juridicos*

y Medico-Legales, Editorial Universidad, Buenos Aires.
5. Mainetti, J.A.: 1987, "Bioethical Problems in the Developing World: A View from Latin America" in *Unitas* 60, 238-248.
6. Mainetti, J.A., *et al.*: 1990, "Bioetica e Investigación en Salud. Comentario a la Ley 11.044". Honorable Càmara de Diputados de la Provincia de Buenos Aires, *Edición de la Ley No. 11,044, La Plata.*
7. Mainetti, J.A.: 1990: "Out of America: The Scholastic and Mundane Bioethical Scene in Argentina", paper submitted to "Transcultural Dimensions of Medical Ethics", Symposium co-sponsored by Fidia Research Foundation and Georgetown University Center for the Advanced Study of Ethics. National Academy of Sciences, Washington, D.C. U.S.A.
8. Sànchez Torres, F.: 1990, "Background and Current Status of Bioethics in Colombia", in *Bulletin of the PAHO*, 24,4, 511.
9. Scholle Connor, S. *et al.* (eds): 1990, "Bioethics: Introduction to the Special Issues", in *Bulletin of the PAHO*, 24,4, vi-x.
10. Tealdi, J.C.: 1991, "Banco Nacional de Datos Geneticos: El Caso Laura/Laura", Seminar "Etica en el Principio de la Vida", II Curso Internacional de Bioetica, Escuela Latinoamericana de Bioetica, Gonnet.
11. Tealdi, J.C.: 1989, "Documentos de Deontologia Medica", en Mainetti, José A., *Etica Médica. Introdución Histórica*, Quirón, La Plata, 83-84.
12. Tealdi, J.C.: 1990, "Proyecto Genoma Humano. Quién Mide a Quien Nos Mide?", paper submitted to *II Jornadas Marplatenses de Bioética*, Mar del Plata.

Statutes
13. Argentina: *Código Civil*, Arts. 243 and 234 bis.
14. Argentina: 1990, *Decreto 358/90*. Supplement to Ley #23.660.
15. Argentina: 1990, *Decreto 359/90*. Supplement to Ley #23.661.
16. Argentina: 1977, Ley #21.541.
17. Argentina: 1987, Ley 23.511, "Banco Nacional de Datos Geneticos. Creació a Fin de Obtener y Almacenar Información Genetica que Facilite La Determinación y Esclarecimiento de Conflictos Relativos a La Filiación". Boletin Oficial.
18. Argentina: 1989, Ley #23.660.
19. Argentina: 1989, Ley #23.661.
20. Argentina: 1990, Ley Nacional #23.798, "Ley SIDA".
21. Argentina, Provincia de Buenos Aires: 1990, Ley #11.044/90.
22. Bolivia: 1989, *Resolución Bi-Ministerial 0415/89.*
23. Brazil: *Código Penal*, Article 122.
24. Brazil: 1988, *Código de Etica Médica*, Diàrio Oficial da Uniào.
25. Brazil: 1988, *Decreto #95721*, Ley #7649.
26. Brazil: 1985, *Resolución de Portaria.*
27. Brazil,Ministerio da Saúde, Conselho Nacional de Saúde: 1988, Resolución

#01.

28. Chile: 1987, *Resolución 759/87.*
29. Colombia: 1991, *Decreto #559/91.*
30. Costa Rica: 1988, *Decreto #18454-S-J/88.*
31. Haiti, Ministerio de Salud Pública y Población: 1987, *Memoràndum.*
32. México: 1984, *Ley General de Salud*, Diario Oficial de la Federación.
33. México: 1988, *Norma Técnica #313/88.*
34. México: 1988, *Norma Técnica #324/88.*
35. México: 1987, *Reglamento de la Ley General de Salud*, Diario Oficial de la Federación.
36. Panamà: 1985, *Circular 1713-DGS-VE-85.*
37. Panamà: 1987, *Decreto 346/87.*
38. Perú: 1987, *Decreto Supremo 013-87-SA.*
39. Uruguay: 1988, *Decreto 233/988/88.*

JOS V.M. WELIE AND HENK A.M.J. TEN HAVE

BIOETHICS IN A SUPRANATIONAL EUROPEAN CONTEXT: 1989-1991

I. INTRODUCTION

Although recent political changes and civil wars have increased the number of European countries from just over thirty to more than forty, at the same time Europe is moving towards increased unity. Twelve countries in western Europe, united in the European Community, are about to open their mutual borders for traffic of people and goods, there are plans for a single European currency, and political and social union are being debated. The blue flag with the twelve golden stars is flying everywhere and every businessman adds the prefix "Euro" to his brand names. One would expect that contemporary bioethical issues, which receive ever increasing attention from the mass media and which, consequently, are being debated not only by health care workers but by all sorts of other professionals and laymen as well, would have been long tackled by the European Community and similar supranational European organizations. Indeed, in the past ten years, bioethical dilemmas have not gone fully unnoticed. Several supranational official bioethics committees have been established, a host of documents have been published, and various international associations for health care ethics have been founded. But in comparison with other prominent political issues, the bioethics debate has not received much systematic, organized, or well-funded support. Consequently, our review, too, will lack a clear structure and many of our remarks will reflect the "ad hoc" nature of contemporary supranational European bioethics.

II. THE POLITICAL CONTEXT

A. *The Council of Europe*

The ideology of a united Europe has been reemerging ever since the fall of the western Empire in 476 B.C. Until recently, however, force was the only means to obtain such unity, and force always turned out to be partial and temporary. In this century, between the two World Wars, the idea of a United Europe grew stronger again, and it was the second World War that motivated the various European countries to set aside feelings of national sovereignty. Winston Churchill, in a 1946 lecture to the University of Zürich (Switzerland),

B. Andrew Lustig (Sr. Ed.), Bioethics Yearbook: Volume 2, 97-126.

called for a united Germany and France to form the heart of this new Europe. His enthusiasm inspired many other politicians, and plans were launched for a European declaration of human rights, to be protected by a European court, and for European parliament which would actively draft policies for political and economic union. On May 5, 1949 the Statute of the Council of Europe was signed. Originally, 10 European countries signed the treaty; currently 21 countries are members of the Council of Europe. But anxieties about supranational intervention in national politics resulted in a European parliament that was granted little power. Its members were not voted for directly but delegated by the national parliaments. No legislative powers were granted. Its task as a parliamentary or consulting assembly is to offer recommendations to the much more powerful Committee of Ministers that decides about implementation. Only by means of vaguely expressed "resolutions" can the Assembly speak directly to the outside world. In addition to the Committee of Ministers, made up of ministers of foreign affairs, there are various conferences of specialized ministers. The Committee of Ministers, in turn, can install specialized committees to deal with particular problems, such as judicial affairs, culture and bioethics. Finally, the Council has a Secretary General, the executive office that administrates most of the daily activities of both the Assembly and the Council. It is located in Strasbourg, a city currently on French territory, but which has been German and French in the course of Europe's history.

The powers of the Council of Europe are limited. The Council may not deal with military affairs, and most economic issues are dealt with by other supranational European institutions. Many of the treaties and protocols drafted in the areas of law, culture, and health care are not signed by all members of the Council. The two most important treaties undoubtedly are two early agreements. On November 4, 1950, the European Convention for the Protection of Human Rights and Fundamental Freedoms was signed and, in accordance with the Convention, a European Commission of Human Rights was established to oversee the implementation of these freedoms and rights by the various European national authorities. The European Court of Human Rights was founded to deal with matters concerning the interpretation of the Convention and the Court's judgments are final. In 1961, the European Social Charter was signed, dealing with social human rights. However, unlike the Convention, this Charter contains no instructions about the establishment of a special commission to oversee its implementation.

In the area of bioethics, the two most common types of documents produced by the Council of Europe are resolutions and recommendations. Originally, according to common international legal jargon, a resolution was considered to have more binding power than a recommendation. However, as a matter of fact, unlike a convention ratified by a national parliaments (such as the two treaties mentioned above) a resolution or recommendation lacks genuine binding power because neither one can force member states to adjust their national laws to comply with its contents. Consequently, the term "resolution" has become more or less synonymous with "recommendation" and the latter term is used now most often.

Although member states are not legally bound to adhere to the recommendations of the Council of Europe, the political impact of recommendations is obvious, especially if they are adopted by the Committee of Ministers, the highest political organ in the Council of Europe. Similarly, when a member state has voted in favor of a particular recommendation, though such vote has no legal consequences, it has a "moral" effect in that it is generally considered unfair to break a promise. Finally, a recommendation may have a legitimizing effect in undertaking across-the-border interventions, and even more so, in justifying national legislation. All these effects can be further increased when the recommending organization or other powerful international organs refer to the recommendation or repeat its message.

B. *The European Community*

The failure of the Council of Europe to become a powerful supranational institution which could lead the way to a truly united Europe, motivated various politicians to look for other strategies. One of them was French foreign minister Robert Schuman, who in 1951 was instrumental in passing the treaty which established the European Community for Coal and Steel. That treaty proved successful, and on March 25, 1957 in Rome, two more treaties were signed, one which established Euratom, for atomic energy issues, and the other which established the important European Economic Community. In 1965, the political institutions of these three communities were merged to form "the" European Community, which currently has 12 members (Belgium, Denmark, France, Germany, Great Britain, Greece, Ireland, Italy, Luxembourg, Netherlands, Portugal, and Spain).

Similar to the Council of Europe, the European Community has four institutions: a parliament, a cabinet of ministers, an executive commission, and a Court. The European Parliament, with over 500 members who are voted for directly by the citizens of the participating nations, is more powerful than the Assembly of the Council of Europe, but it still is not a genuine democratic parliament with legislative powers. It has only limited means to control the Council of Ministers. The Council of Ministers (or Deputy-Ministers) is the most powerful institution in the Community. Depending on the topic on the agenda, the Council may consist of different ministers. For example, if all ministers of research study issues, the Council is called the Research Council; when the Council consists of all the heads of governments, it is called the European Council (not to be confused with the Council of Europe). The Council of Ministers makes all substantive decisions. The European Commission has 17 members, who are not to represent their nations, but are appointed based on their individual political and administrative expertise. The administrative staff of the Commission is divided into 20 Directorates-General. Finally, the Court of Justice, has thirteen judges and may be consulted not only by nations, but also by individuals.

The main difference between the Council of Europe and the European Community is that the latter, unlike the former (or any other international organization), can bind member states without the need for national ratification of its conventions. Furthermore, new legislation at a national level is not binding when it is inconsistent with existing European Community law. The member states have surrendered part of their sovereignty; and in that sense, the European Community is not so much an international but rather a supranational organization, similar to that of a budding (con)federation.

III. BIOETHICS ON A SUPRANATIONAL EUROPEAN LEVEL

A. *Bioethics and the Council of Europe*

At their Brussels meeting in 1974, the Ministers of Justice of the member states of the Council of Europe, concerned about the legal problems surrounding transplantation surgery, proposed to examine the ethical and legal aspects through an inter-governmental program. Under the joint direction of the Steering Committee Public Health and the Steering Committee for Legal Cooperation, a multidisciplinary expert committee was set up which drafted a resolution on removal, grafting and transplantation of human substances. In May 1978, the Committee of Ministers adopted this resolution (R (78) 29), and a recommendation on the international exchange and transportation of human substances followed in 1979. In the early 1980s, two more recommendations were passed, one on the protection of persons suffering from mental disorders (1983) and another on recombinant DNA research (1984).

During its March 1985 meeting in Vienna (Austria), the European Ministerial Conference on Human Rights adopted Resolution Number 3, in which the Council of Europe was asked to become the focal point and clearing house for information, opinions, and where appropriate, joint international action with regard to biomedicine. A report by the French Minister of Justice, "The Protection of Human Beings and their Physical and Intellectual Integrity in the Context of the Progress Being Made in the Fields of Biology, Medicine, and Biochemistry", had drawn attention to the fact that innovative medical developments are challenging the traditional legal order. In his report, the French minister called on European legislators to undertake a supranational effort to safeguard the integrity of human subjects. In June 1985, the European Ministers of Justice at an informal meeting in Edinburgh (United Kingdom), discussed a note of the Spanish Minister on human fertilization and embryology. They concluded that any exclusive national regulation in this field would run the risk that certain techniques, prohibited by some nations of the Council, might be applied in others.

In response to these two meetings, the Committee of Ministers on June 28, 1985 decided that bioethical issues, which until then had been discussed by no less than four different committees, should be dealt with by a single specialized committee. For this purpose, under the auspices of the Directorate of Legal Affairs, the "Ad Hoc Committee of Experts on Ethical and Legal Problems Relating to Human Genetics", which had been established by the Committee of Ministers in 1983 in response to Recommendation 934 (1982) of the Parliamentary Assembly on genetic engineering, was renamed Ad Hoc Committee of Experts on the Progress of Biomedical Sciences. In 1989, this committee was renamed again and became the current Ad Hoc Committee of Experts on Bioethics (CAHBI). The CAHBI is a multidisciplinary body of lawyers, biologists, physicians, ethicists, and human rights experts representing the member states and several other countries and organizations. The CAHBI meets twice a year to discuss topics submitted by its Secretariat. Under the auspices of the CAHBI, in recent years, various subcommittees have been established to prepare CAHBI documents, such as the special Working Party on Genetic Testing and Screening (CAHBI-GT-GS), the Select Committee of Experts on the Use of Human Embryos and Fetuses (CAHBI-R-EF), and the Bioethics Convention Study Group (CAHBI-CO).

The texts of the CAHBI intend to "reaffirm major principles and values which must guide any regulation on bioethics and also indicate which limits must at all cost be respected" (CAHBI/INF (90) 1). The Committee is intended to work toward harmonizing the policies of member states as far as possible, and, if desirable, framing appropriate legal instruments. The main problem facing the CAHBI is how to achieve international consensus on such delicate matters. Two draft recommendations, one concerning artificial insemination (which was voted down by little Liechtenstein), and one concerning abortion, could not be adopted by the Council of Ministers because the vote was not unanimous. Although two recommendations were passed in 1990, the CAHBI has explored alternative means to promote constructive dialogue in bioethics, for example, by publishing reports and studies.

To promote the further exchange of ideas and information, the CAHBI has also begun to organize international symposia. The first was held in Strasbourg, France, from December 5-7, 1989. Participants met in four workshops, which dealt with teaching and research, research in bioethics, information and documentation, and ethics committees. The second symposium is scheduled to take place in 1992.

Despite the 1985 concentration of bioethics into the CAHBI, in the Council of Europe other bodies which deal with bioethical issues remain. Four Conferences of Specialized Ministers, dealing with human rights, justice, health, and family affairs have taken stands on bioethical issues. Each of these conferences have installed expert committees to deal with bioethical issues. For example, in addition to the CAHBI's Working party on genetic testing and screening, the European Committee on Crime Problems (CDPC) has installed a Working Party on Genetic Testing for Police and Criminal Justice Purposes, in which the CAHBI participates. The Working party is developing a draft recommendation on this subject (CAHBI/CDPC-GT). The Committee of Experts on Family Law (CJ-FA) has been investigating bioethical questions pertaining to the rights of children and the responsibilities of parents. There is also a Public Health Committee which has been studying the issue of organ donation.

The Council of Europe's Parliamentary Assembly is another forum that has dealt with bioethical issues. Indeed, the Assembly has preceded the Committee of Ministers by more than a decade by adopting various resolutions and recommendations on bioethical issues in the early 1970s.

Finally, the European Commission for Human Rights, established to ensure, together with the European Court, observance of the Convention, has dealt with bioethical issues. For example, Mr. C.A. Nørgaard, as President of the European Commission of Human Rights, in 1989 asked the Steering Committee for Human Rights (also called the Committee of Experts for the Development of Human Rights [DH-DEV]) to consider the scope of Article 8 of the Convention in light of new medical techniques particularly artificial insemination. Christian Byk (who is also a member of the CAHBI, did not yet write the document as representative of the CAHBI), drafted this report, which was published in 1990.

B. *Important Documents on Bioethics of the Council of Europe*

The following provides a partial list of important documents on bioethics from the Council of Europe since 1976, organized by year.

1. *1976*
   a. *Resolution 613 (1976)*: On the rights of the sick and dying. (Adopted by the Parliamentary Assembly on January 29, 1978, 24th Sitting).
   b. *Recommendation 779 (1976)*: On the rights of the sick and dying. (Adopted by the Parliamentary Assembly on January 29, 1976, 24th Sitting).
2. *1977*
   a. *Recommendation 818 (1977)*: On the situation of the mentally ill. (Adopted by the Parliamentary Assembly on October 8, 1977, 12th Sitting).
3. *1978*
   a. *Resolution R (78) 29*: On harmonizing the legislation of member states relating to removal, grafting and transplantation of human substances. (Adopted by the Committee of Ministers on May 11, 1978 at the 287th meeting of the Ministers' Deputies).
4. *1979*
   a. *Recommendation R (79) 5*: From the Committee of Ministers to member states concerning international exchange and transportation of human

substances. (Adopted by the Committee of Ministers on March 14, 1979 at the 301st meeting of the Ministers' Deputies).

5. *1982*
   a. *Recommendation 934 (1982)*: On genetic engineering. (Adopted by the Parliamentary Assembly on January 26, 1982, 22nd Sitting).
6. *1983*
   a. *Recommendation R (83) 2*: From the Committee of Ministers to member states concerning the legal protection of persons suffering from mental disorders committed as involuntary patients. (Adopted by the Committee of Ministers on February 22, 1983 at the 356th meeting of the Ministers' Deputies).
7. *1984*
   a. *Recommendation R (84) 16*: From the Committee of Ministers to member states concerning notification of work involving recombinant deoxyribonucleic acid (DNA). (Adopted by the Committee of Ministers on September 25, 1984 at the 375th meeting of the Ministers' Deputies).
8. *1986*
   a. *Recommendation 1046 (1986)*: On the use of human embryos and fetuses for diagnostic, therapeutic, scientific, industrial and commercial purposes. (Adopted by the Parliamentary Assembly on September 24, 1986, 18th Sitting).
   b. *CAHBI/INF (86) 1*: Ad Hoc Committee of Experts on Progress in the Biomedical Sciences: Provisional Principles on the techniques of human artificial procreation and certain procedures carried out on embryos in connection with those techniques. Strasbourg, March 5, 1986.
9. *1987*
   a. *CAHBI-Opinion*: Opinion given by CAHBI at the request of the Government of The Netherlands on the question of "terminating the life of a patient at his express request" (this document remains confidential).
   b. *CM (87) 112; Addendum*: Draft recommendation on artificial procreation; prepared by CAHBI, but not adopted by the Committee of Ministers.
10. *1988*
   a. *CAHBI-Opinion, No. 7244*: Opinion of the CAHBI on Parliamentary Assembly Recommendation 1046 (1986) concerning the use of human embryos and fetuses for diagnostic, therapeutic, scientific, industrial and commercial purposes. May 1988.
11. *1989*
   a. *Recommendation R (89) 14*: From the Committee of Ministers to member states on the ethical issues posed by HIV infection in the health care and social settings. (Adopted by the Committee of Ministers on October 24, 1989 at the 429th meeting of the Ministers' Deputies).
   b. *Recommendation 1100 (1989)*: On the use of human embryos and foetuses in scientific research. (Adopted by the Parliamentary Assembly

on February 2, 1989, 24th Sitting).

c. *CAHBI-R-EF (89) 1 Revised 2*: Report on the use of human fetal, embryonic and pre-embryonic material for diagnostic, therapeutic, scientific, industrial and commercial purposes; prepared by Dr. Anne McLaren, CAHBI-Select Committee of Experts on the Use of Human Embryos and Fetuses (CAHBI-R-EF), May 1989; revised and supplemented by the author March 1990.

12. 1990

a. *Recommendation R (90) 3*: From the Committee of Ministers to member states concerning medical research on human beings. (Adopted by the Committee of Ministers on February 6, 1990 at the 433rd meeting of the Ministers' Deputies).

b. *Recommendation R (90) 13*: From the Committee of Ministers to member states on prenatal genetic screening, prenatal genetic diagnosis and associated genetic counseling. (Adopted by the Committee of Ministers on June 21, 1990 at the 442nd meeting of the Ministers' Deputies). For the explanatory Memorandum, see CAHBI/INF (90) 3.

c. *CAHBI (90) 1 revised*: Genetic diagnosis and screening of children and adults: an information document by Dr. H.J. Müller.

d. *CAHBI/INF (90) 1, 2nd revision*: The work of the Council of Europe in the field of bioethics; Information Note prepared by the Directorate of Legal Affairs.

e. *CAHBI/INF (90) 3*: Information document containing the text of Recommendation R (90) 13 and an Explanatory Memorandum to this Recommendation.

f. *CAHBI-GT-GS (90) 1*: The applicability of principles of prenatal genetic screening to the problems of genetic testing and screening in general.

g. *CAHBI-GT-GS (90) 2*: Working party on genetic testing and screening; Report of the First meeting, Strasbourg, France, September 26-28, 1990. Appendix contains preliminary draft recommendation on genetic testing and screening for health care purposes.

h. *CAHBI-Gt-GS-Document Müller*: Genetic testing and screening for medical purposes.

i. *CAHBI/CDPC-GT*: Summary report of the first meeting of the joint Working party on genetic testing for police and criminal justice purposes, September 1990. Contains draft Recommendation.

j. *DH-DEV (90) 12*: Medical and biological progress and the European Convention on Human Rights. Report drawn up for the Steering Committee for Human Rights as requested by Mr. C. A. Nørgaard, President of the European Commission of Human Rights. By: Christian Byk, Member of the CAHBI. Strasbourg September 27, 1990

C. *Bioethics and the European Community*

Compared with the Council of Europe, until very recently the European Community (EC) had neglected the area of bioethics. Indeed, the Community has been developed first of all as an economic union; free trade and a single currency are typical topics on the agenda of the Community meetings. But the Community is not limited to such issues. Political union and a social supranational policy are being debated, the EC also deals with health care. There is a united fight against various kinds of diseases such as AIDS, cancer, and cardiovascular disease. And during the past decade, subsidies have been made available for biomedical research projects between different universities and research groups in European nations. Nonetheless, it was not until the late 1980s that the EC identified bioethics as an area in need of special attention.

By contrast, the European Parliament had already engaged itself in bioethics many years earlier. During the 1980s, a considerable number of debates took place in the Parliament about a variety of bioethical issues such as the Human Genome Project, care of the dying and euthanasia, post-mortem experiments in hospitals, the free choice of physicians, mental health care, AIDS, and the possibility of European harmonization of bioethical issues. Various individual members of parliament have commissioned studies, such as the comparative study of existing legislation in EC member states on experiments with embryos, which was mandated by the Socialist fraction of the EC and published in 1991 [7]. In addition, many member states have drafted resolutions on bioethical issues. However, the European Parliament still lacks the necessary power significantly and directly to influence European Community policy. Consequently, the political impact of these activities is very limited. The real power still resides in the Council of Ministers, which did not get involved until the end of the eighties.

In response to the development of German research projects in genetics, at the instigation of the German Minister Riesenhuber at the Kronberg conference of the European Community Ministers for Research on June 29, 1990, the Research Council decided to install two special bioethics committees. The existing preliminary (ESLA) Committee for Ethical, Social, and Legal Aspects was formalized into an official Working Group on the Ethical, Social and Legal Aspects of Human Genome Analysis. The second committee investigates research with embryos.

Unlike the Council of Europe's CAHBI committee, these two EC committees are primarily research committees. Actually, within the EC there is no one with functions similar to those of the CAHBI or of any other special bioethics working groups in the Council of Europe. Most of the EC's scrutiny of bioethical issues is limited to assessing the framework of biomedical research. For example, the Commission of the EC mandated a short-term analytical inventory of possible ethical problems posed by modern biomedical research in the leading and research centers in the European Community. The final report is to be submitted in December 1991. Furthermore, as part of the BIOMED program for international biomedical research, special funds were reserved for in 1991 research in the area of bioethics.

D. *Biomedical and Health Research*

Since 1983, the European Community has coordinated its research and technical development (RTD) activities through multiannual Framework Programs. These Framework Programs are implemented through specific RTD programs on selected areas of research. In April 1990, the Council of Ministers adopted the Third Framework Program, which tenure of 5 years, a budget of 5.7 billion ECU (approx. 7.3 billion US dollars), and which contains 15 specific RTD programs. One of these, the research program in "Biomedicine and Health" (BIOMED 1) was adopted by the Council in September 1991, with a budget of 131.67 million ECUS [3]. BIOMED has two objectives: a) to contribute to improving the efficacy of medical and health research and development in the member states, in particular by coordinating of their research and development activities and applying the results through EC cooperation and resources pooling; and b) to encourage basic research in the field of biomedicine and health throughout the community.

BIOMED is unique, because it is the first RTD program in the field of biomedicine to include bioethics as a research concern. The program focuses on four areas:

- *Area 1: Development of coordinated research on prevention, care and health systems*. This area attempts to harmonize methodologies and protocols in epidemiological, biological, clinical and technological research. Its key targets are: a) drugs and the administration of medicines; b) risk factors and occupational medicine; c) biomedical technology; and d) health services research.
- *Area 2: Major health problems and diseases of great socioeconomic impact*. This area covers major health problems and economically and socially significant disease groupings, in particular AIDS, cancer, cardiovascular disease, mental illness and neurological disease, the aging process, and age-related health problems and handicaps.
- *Area 3: Human genome analysis*. This area covers research aimed at the completion and the integration of the genetic and physical maps, research on the genetic basis for biological functions, and research on the genetic component of multifactorial conditions such as Alzheimer's disease. In addition, Area 3 includes efforts to establish a mechanism to coordinate gene sequencing among various centers.
- *Area 4: Research on biomedical ethics*. This area will involve problems relating to the research carried out in the first three areas of the program. It may also be linked to the possible applications of the research results. An extensive list of research objectives and topics is suggested. For example,

that list includes research "[to] define the ethical and legal issues arising from the removal, grafting and transplantation of human cells, tissues and organs" (Area 1); "[to safeguard] confidentiality of medical records" (Area 2); and "[to explore] implications of the human genome initiative [for] the concepts of personal identity, responsibility, determinism, reductionism, health and disease" (Area 3). A separate topic is the impact of research on medicine and society. For example, one of the research tasks suggested is the: "[c]omparative studies of local ethical committee practice in participating countries, and also of the variation over time of the acceptability of certain practices as well as strategies for coping with this variation".

Most of the research in BIOMED 1 will be implemented either through "shared-cost" research projects (Community participation is 50 percent of the total project costs) or "concerted" research actions (the Community does not participate in the costs of the research itself, but reimburses only coordination costs, such as meetings and travel). The budget is 27.5 million ECUS for Area 1, 72 million ECUS for Area 2 (including 25 million ECUS for AIDS research), 27.5 million ECUS for Area 3, and 4.67 million ECUS for Area 4.

Bioethics research will be accomplished primarily through concerted actions or research networking, *i.e.*, the coordination of individual research activities. Such networks must contain a minimum of two teams, each from a different country, but much larger groups can also be established. Choosing this method will contribute to the general objective of the BIOMED program, *viz.*, transnational collaboration and mobility of research workers. Without itself initiating new research programs or developing new research areas, the Community will make efficient use of existing research activities in various countries, by transferring skills and expertise from advanced to less advanced research centers, harmonizing research methods and standardizing procedures to facilitate the translation of results into practical applications within the Community, and optimizing the capacity and economic efficiency of health care efforts beyond the specific research topic. In general terms, a concerted action must contribute to strengthening the Community's economic and social cohesion.

A first call for proposals was published by the European Commission in October 1991 (with a closing date of January 31, 1992) [2]. A second call for proposals for all areas has a deadline on January 10, 1993. The selection of the first proposals is scheduled for June 1992, and funding of research activities will start at the end of that year.

Area 3 (Human Genome Analysis) is not included in the first call for proposals, but it will be included in the BIOMED program during the second call. A separate call was issued late 1991, specifically targeted on studies on the Ethical, Social and Legal Aspects of Human Genome Analysis. From 42 proposals received, 18 have been selected for EC funding, after consultation with the Working Group on the Ethical, Social and Legal Aspects of Human

Genome Analysis. Support, however, will be received only for 12 months. Consequently, most projects are likely to be pilot-studies. The range of topics is very wide, from "Valuating Cystic Fibrosis Carrier Screening Development", "The Permissibility of Genetic Tests in Life Insurance", and "Experimental Genetic Therapy: Ethical Aspects" to "Women's Perspectives on the Ethical, Social and Legal Applications and Implications of Human Genome Analysis" and "The Standard Model for Production of Educational Materials of Genetics for Secondary Schools" [6].

IV. NON-POLITICAL SUPRANATIONAL ORGANIZATIONS

In the late 1980s, increasing international contacts among ethicists, philosophers, and interested physicians resulted in the founding of several European societies and associations.

A. *European Association of Centers of Medical Ethics*

This Association (EACME) was founded in 1985 through the initiative of the Bioethics Centers of Barcelona (Spain), Brussels (Belgium) and Maastricht (The Netherlands). Its statutes were officially agreed upon in September 1986. The aims of the Association are: (1) to provide its members with information on ethical and health problems; (2) to provide research tools; (3) to coordinate collaborative research and to exchange results; (4) to contribute to the European debate on bioethics; and (5) to promote the teaching of ethics among those involved in health care.

Full members of the EACME are institutions, departments, or units which have provided evidence of actual experience in the field of medical ethics, collaborative work, contribution to research, and publications. Membership of EACME includes presently some thirty Centers in Europe (Belgium, Germany, France, Greece, Italy, the Netherlands, Spain, Sweden, Switzerland and the United Kingdom).

The Association holds one or two plenary meetings every year. Past conferences have included such topics as: "Documentation in the field of Bioethics: How to Achieve Computerized Exchange of Data Between the European Centers" (Lyon, January 1986); "Collaboration in the Field of Bioethics: Toward a European Research Network" (Maastricht, September 1986); "The European Guidelines in Medical Ethics" (Rome, September 1987); "Bioethics and Human Rights: Convergences and Dissent" (Lyon, March 1989); and "Bioethics and Research on Human Subjects" (Geneva, September 1989). As this list of topics indicates, activities of EACME have the very practical focus of promoting the interest of the participating centers. A great deal of effort has been spent developing a European documentation and databank system using several languages. Moreover, in cooperation with the French "Association Descartes", the EACME is establishing a network of centers, organizations, laboratories, committees and "resource persons" to deal with bioethics in the twelve countries of the European Community.

B. *The European Society for Philosophy of Medicine and Health Care*

This Society (ESPMH) was founded in 1987 by individual physicians, philosophers and ethicists interested in establishing mechanisms for international cooperation, particularly within Europe. The goals of ESPMH are: (1) to stimulate and promote the broad development of knowledge and methodology in the field of philosophy of medicine and health care; and (2) to become a center of contact for European scholars in this field and to promote international contacts between members in the various countries of Europe. The Society has approximately 300 members (including 3 institutional members), and the majority of its members are practicing physicians. Most members are from northwestern Europe, but since 1991 new members are increasing from the eastern Europe.

Since its establishment, ESPMH has organized annual conferences on various themes: "Medicine and the Growth of Knowledge" (Maastricht, The Netherlands 1987); "Values in Medical Decision-Making and Resource Allocation in Health Care" (Aarhus, Denmark 1988); "From Brentano to Bieganski: European Traditions in Philosophy of Medicine" (Czestochowa, Poland 1989); and "Consensus Formation and Moral Judgment in Health Care Ethics" (Maastricht, The Netherlands 1990). The latter conference was initially to be held in Bochum (Germany), but because of anti-bioethics activists who sought to prevent the conference, it was relocated to Maastricht. The 1991 conference in Oxford (United Kingdom) on "Philosophy and Mental Health" will be held without such disturbances.

By selecting the most interesting and scholarly conference presentations for elaboration into chapters, books have been developed which focus on the annual conference themes. Rather then being proceedings of the conference, these volumes seek to address themes in new, coherent and systematic ways. Thus far, two volumes have appeared [5].

Another activity of the Society is the publication of a newsletter three times a year. This newsletter includes contributions, book reviews, and news about bioethical and philosophical activities throughout Europe. Its main purpose is to be a medium for sharing information between members.

C. *International Association of Law, Ethics, and Science*

This association, established in 1989 under the auspices of International Institute for Ethical and Legal Studies on the New Biology (The ISENB, directed by Professor Marco Milani-Comparetti), is known as "Milazzo Group", after the town in Italy where the ISENB is located and its meetings are held. The headquarters of the Association is in Strasbourg (France).

UNESCO and the Council of Europe are members of the governing board.

The Association has the following aims: (1) to create an international network of persons interested in ethical, legal and social questions raised by scientific developments; (2) to encourage the exchange of information and experience on these questions; (3) to pursue activities designed to improve understanding of the relationships among man, science and society; (4) to contribute to the activities of national and international organizations and institutions; and (5) to improve the multidisciplinary study of bioethical issues in order to enhance public discussion and debate.

Thus far, the Association has held two meetings. The first, in 1989, focused on the theme "artificial procreation"; the second in 1991, emphasized issues in genetics. The Association intend to meet every other year in Milazzo. A third conference, on "Bioethics and Power", is planned for 1993.

The membership of the Association is not open. Members are persons and organizations invited to participate in the first meeting, and those accepted by the Governing Board after nomination by three members from at least two countries. The Association is now represented in 40 countries.

Starting in 1990, the ISENB and the Association are publishing the *Journal International de Bioéthique/ International Journal of Bioethics* (editor-in-chief: C.Byk). The journal aims to encourage a multidisciplinary debate in bioethics at an international level. It publishes articles in English as well as French, with numerous, contributions from the Milazzo Group. The journal has four parts: contributions on the theme of the biennial conference, other bioethical issues, presentations of research in bioethics, and a newsletter section with brief information on relevant events in science and ethics.

V. RECOMMENDATIONS AND REPORTS ON VARIOUS BIOETHICAL ISSUES

A. *Preliminary Remarks*

As discussed earlier, only during the last few years has bioethics become the topic of systematic, organized research and deliberation at a supranational European level. Over the years, members of the European Community's Parliament have often raised questions on various subjects of bioethical relevance, but, to our knowledge, no consensus documents with significant authority over member states have yet been produced. The two committees installed in 1990 at the request of the Community's Research Council have met only a few times, and neither has published public reports. However, the various bodies of the Council of Europe that have shown an increased interest in bioethics have published a number of documents. The topics covered touch on many important contemporary bioethical issues, including reproductive technologies, AIDS, and organ donation. But many equally important issues have been left unaddressed, such as abortion and equitable access to health care. Because of this silence, we are left with no choice but to skip those topics in the following review.

Since the status of these documents differs considerably – ranging from research reports of single officials to recommendations passed by the Council of Ministers – we will focus our discussion on documents with an indisputable authoritative status, *i.e.*, primarily resolutions and recommendations passed by either the Parliamentary Assembly or the Committee of Ministers. Furthermore, interest in the areas of reproductive technology and genetic medicine has grown rapidly during the past few years, both in the Council of Europe and the European Community. Much of our attention, therefore, will be focused on these topics. Finally, in our discussion of these documents, we will not refer to those articles and statements that have become "common moral standards" in the areas of medicine and health care, *e.g.*, the principle of "informed consent" for any and all research that is undertaken with human subjects. Similarly, the repeated calls for high quality, care and caution in the area of medicine will not be repeated, in spite of their undeniable continued importance.

B. *New Reproductive Technologies and Genetic Medicine*

1. *Genetic Diagnosis and Treatment*

The Committee of Ministers of the Council of Europe adopted, Recommendation R (90) 13 "On Prenatal Genetic Screening, Prenatal Genetic Diagnosis, and Associated Genetic Counseling" on June 21, 1990. The CAHBI began to prepare this recommendation in 1988. They began with the belief that ethical principles common to all States of Europe and based on their moral and spiritual values should govern prenatal genetic screening, diagnosis and counseling. The Recommendation therefore formulates 14 principles, and urges governments of the member states to adopt legislation conforming to these principles. CAHBI started its preparatory work by conducting a survey of existing and foreseen applications of prenatal diagnosis and genetic screening as well as associated genetic counseling. The survey included several findings: (1) prenatal genetic screening as a public health service appeared to be available throughout Europe (although not all countries were equipped at the same level); (2) no member state had compulsory prenatal screening or diagnosis; and (3) the general level of understanding of and information about these services is inadequate.

According to the Recommendation, recent progress in genetics requires a consideration of the optimum balance "between, on the one hand, the dignity of the human person and the corresponding inviolability of his body, and on the other hand, the continuous and irreversible advance of science". Such consideration is urgently needed, because growing number of genetic diseases can be identified and diagnostic procedures can help people at risk to make procreative decisions. However, the increasing availability and accuracy of these procedures also raise serious problems, for example, "concern in some countries that the increasing use of these procedures will encourage an attitude according to which all available means should be used to avoid the birth of handicapped children".

In drafting its principles, the CAHBI appears to proceed from four general assumptions. A first, somewhat implicit, assumption is that genetic diagnosis and screening should almost exclusively focus on the interest of the persons (couple and/or women) involved. If persons do not understand diagnostic information, they should be helped to make an autonomous decision. This assumption may explain why genetic counseling is emphasized in Principle 1 of the Recommendation: "No prenatal genetic screening and/or prenatal genetic diagnosis tests should be carried out if counseling prior to and after the tests is not available". In such counseling sessions, both members of the couple should participate whenever possible: "the participation of both partners in the counseling sessions should be encouraged" (Principle 5). However, the problems that may arise in this context, for example, in cases of disagreement are not adequately addressed. The majority of CAHBI experts did not recommend requiring the consent of the future father prior to testing the woman. The woman, however, cannot be examined without her consent. In this controversy, CAHBI recommended leaving the adjudication of disputed cases to national law.

Second, the general moral right to information is applicable to genetic diagnosis; thus a woman has the right to receive information concerning her pregnancy. Since information about genetic diagnosis and screening is inadequate, the CAHBI emphasized that fuller and more accurate information about these procedures should be available. Principle 14 in fact requires that "where there is an increased risk of passing on a serious genetic disorder, access to preconception counseling and, if necessary, premarriage and preconception screening and diagnostic services should be readily available and widely known". Every effort should be made to inform couples about such services.

Furthermore, Principle 8 explicitly confirms that the information given during the counseling "must be adapted to the person's circumstances and be sufficient to [enable the person to] reach a fully informed decision". On the one hand, this requirement ensures that the woman can make a decision with full knowledge of the pertinent facts. It also explains why "this information should in particular cover the purpose of the tests and their nature as well as any risks which these tests presents". On the other hand, access to the information revealed through the testing should be restricted and carefully controlled. Principles 12 and 13 make clear that because genetic data are particularly sensitive, so that confidentiality must be assured. Principle 12 is most explicit: "Any information of a personal nature obtained during prenatal genetic screening and prenatal genetic diagnosis must be kept confidential".

Principle 13 states that the "right of access to personal data collected pursuant to prenatal genetic screening and prenatal genetic diagnosis should be given only to the data subject", but leaves it to national law and practice to determine the manner in which the data subject may have access to [her] information. In some countries, access to medical information may only be given through the intermediary of the data subject's physician. In other countries, direct access is the norm. The Explanatory Memorandum discusses certain problematic situations are where application of the right of access to genetic data may lead to difficulties, for example, concerning the question of who is entitled to exercise this right. For instance, when information has been collected on genetic relatives without their knowledge, the question arises as to whether those relatives should be informed and whether they have a right of access to data. The experts recommend the resolution of such difficulties to national law, although they propose that advising relatives of the existence of relevant data should not be made mandatory. The text of the Principle, however, refers to another situation: "Genetic data which relate to one member of the couple should not be communicated to the other member of the couple without the free and informed consent of the other".

Third, one of the basic values common to the member states is moral respect for the diversity of the personal opinions and convictions of individuals. This implies that every person should decide freely whether or not to undergo medical screening and diagnostic tests, but also that pregnant women or couples should be free to make decisions in the light of the results of diagnosis and the associated counseling. Principle 4 states: "The counseling must be non-directive; the counsellor should under no condition try to impose his or her convictions on the persons being counseled but inform and advise them on pertinent facts and choices". This principle reflects well-established rules and practices in European and other democratic countries. Concordant with this statement is Principle 6: "Prenatal genetic screening and prenatal genetic diagnosis may only take place with the free and informed consent of the person concerned". Principle 7 reiterates that informed consent is necessary even where tests are routinely offered. In fact, Principle 6 makes explicit that the woman is the only person who can give or refuse consent to tests. In a subparagraph, the issue of legally incapacitated persons is also addressed. Such persons should not be denied access to genetic services; rather, their legal representatives should be consulted on their behalf. But the same principle also states that "prenatal genetic screening or prenatal genetic diagnosis should not be carried out when the person to undergo tests objects". The Explanatory Memorandum clarifies that this statement refers to the legally incapacitated person who is capable of understanding. When a person is capable of expressing his views and refuses to undergo these tests, they should not be carried out, regardless of the prior consent given of his legal representative. What procedure to follow if the legally incapacitated person is unable to express his views remains unanswered; the wording of the principle leaves open whether of not the legal representative must be consulted or give consent.

Fourth, genetic diagnosis and screening are medical procedures. This has two implications: only physicians should be responsible for these tests (Principle 3), and tests should never be used for non-medical purposes (Principle 2). The latter principle stipulates that the tests "should be aimed only at detecting a serious risk to health of the child". The Explanatory Memorandum explains that the principle refers to "those conditions in the child to be born for which treatment can be given during pregnancy or immediately after the birth". Moreover, tests should not be allowed for abnormalities which are of minor significance (although in the Recommendation no measure of significance is provided) or for detection of characteristics which have nothing to do with congenital abnormalities (*e.g.*, sex determination).

Principle 9 states that the woman should not be compelled by the requirements of national law or administrative practice: "In particular, any entitlement to medical insurance or social allowance should not be dependent on undergoing these tests". Principle 10 states that no discriminatory conditions should be applied to women who do or do not seek these tests. Consistent with principle 6, which requires the woman's freedom of choice, both principle 9 and 10 specify that her freedom should not be reduced by either direct or indirect influences. From its survey, CAHBI learned that the Greek Orthodox Church in Cyprus demands a certificate proving that a genetic screening test for thalassemia has been carried out before celebrating a marriage between Greek Cypriots. This, however, was apparently not regarded as an infringement on the freedom of choice, since the results of this test remain confident and are not revealed to the church authorities.

Finally, Principle 11 requires that in genetic tests "personal data may only be collected, processed and stored for the purposes of medical care, diagnosis and prevention of disease and research closely related to medical care".

2. *Genetic Engineering*

In the early 1980s, concern about the new recombinant DNA techniques led to a number of activities. The Parliamentary Assembly of the Council of Europe adopted Recommendation 934 "On Genetic Engineering" in January 1982. The Assembly made the following relevant recommendations to the Committee of Ministers. The Committee should:

[a] "provide for explicit recognition in the European Convention on Human Rights of the right to a genetic inheritance which has not been artificially interfered with, except in accordance with certain principles which are recognized as being fully compatible with respect for human rights (as, for example, in the field of therapeutic applications)";

[b] "provide for the drawing up of a list of serious diseases which may properly, with the consent of the person concerned, be treated by gene therapy (though certain uses without consent, in line with existing practice

for other forms of medical treatment, may be recognized as compatible with respect for human rights in the probability of a serious disease being transmitted to a person's offspring)";

[c] "lay down principles governing the preparation, storage, safeguarding and use of genetic information on individuals, with particular reference to protecting the rights to privacy of the persons concerned in accordance with the Council of Europe conventions and resolutions on data protection".

Responding to these recommendations, the Committee of Ministers of the Council of Europe established a new Ad Hoc Committee of Experts on Genetic Engineering (CAHGE) in 1983. This committee was charged to study the problems posed by genetic engineering, in order to harmonizing the policies of member states, and, if desirable, to frame an appropriate legal instrument. The work of this committee resulted in Recommendation R (84) 16, adopted by the Committee of Ministers in September 1984. The Recommendation requires all researchers doing recombinant DNA work to inform the competent national authorities, so that any work which may pose dangers to human beings or the environment can be stopped or revised. The Recommendation also requires safeguards for the confidentiality and intellectual property of any information given to those authorities.

D. *Use of Human Embryos*

Further elaboration of Recommendation 934 (1982) on genetic engineering resulted in Recommendation 1046 on the use of human embryos and fetuses for diagnostic, therapeutic, scientific, industrial and commercial purposes. This recommendation was adopted by the Parliamentary Assembly of the Council of Europe on September 24, 1986. In the text of the Recommendation, the Assembly notes that recent progress in reproductive medicine, particularly the technique of in vitro fertilization, makes it possible, in principle, to control human life in its earliest stages. However, the legal status of the embryo and the fetus at present is not defined by law nor do adequate provisions governing the use of living or dead embryos and fetuses exist in the member states. The Assembly also recognizes that any exclusively national regulation runs the risk of being ineffective, because research activities could be transferred to another country which does not enforce the same regulations. These considerations indicate the need for European cooperation, the need to define the extent of legal protection of developing human life, and the need to develop clear ethical guidelines which take into account the balance between the principles of freedom of research and of respect for human life.

The Assembly also recognizes the need for further recommendations to the governments of the member states. In its discussion, the Assembly develops several ideas:

- As in the earlier Recommendation 934, the Assembly recognizes the right to a genetic inheritance; this right should not be artificially interfered with except for therapeutic purposes;

- The use of human fetal and embryonic materials and tissues must be limited to purposes which are clearly therapeutic and for which no other means exist;
- Human embryos and fetuses must be treated in all circumstances with a respect which comports with human dignity.

Although a definition of the status of the embryo is not given, the Recommendation states that "human life develops in a continuous pattern, and.... it is not possible to make a clear-cut distinction during the first phases of its development..." (p. 2). This text apparently reflects a compromise, for it remains unclear what concept of the moral status of the human embryo is being adopted.

The recommendations made are heterogeneous, and in fact, are an enumeration of restrictions on the research and use of embryos and fetuses. The more important recommendations are paragraphs 14.A.iii and A.iv. Paragraph 14.A.iii calls on governments "to forbid any creation of human embryos by fertilization in vitro for the purposes of research during their life or after death". Paragraph 14 A iv calls specifically for a ban on the undesirable use of reproductive technologies, referring to:

a) the creation of identical human beings (e.g., by cloning);
b) the implantation of a human embryo in the uterus of another animal or the reverse;
c) the fusion of human gametes with those of another animal (making an exception for the hamster test);
d) the creation of embryos from the sperm of different individuals;
e) the fusion of embryos or any other operation which might produce chimeras;
f) ectogenesis;
g) the creation of children from people of the same sex;
h) choice of sex by genetic manipulation for non-therapeutic purposes;
i) the creation of identical twins;
j) research on viable human embryos;
k) experimentation on living human embryos, whether viable or not;
l) the maintenance of embryos in vitro beyond the fourteenth day after fertilization.

Further recommendations concern procedural and organizational issues, such as the creation of national registers of accredited medical centers authorized to carry out reproductive techniques, the creation of national multidisciplinary committees or commissions on artificial human reproduction, and the preparation of a European convention regulating the use of human embryonic and fetal tissue for scientific purposes.

Recommendation 1046 has been submitted to the Committee of Ministers but has not been agreed upon. The CAHBI was invited by the Committee to express an opinion on Paragraph 14 of the Assembly recommendation. This "Opinion of the CAHBI (CAHBI, No.7244)" was forwarded in May 1988 to the Assembly via the Committee of Ministers. In its review of Paragraph 14, the CAHBI agrees with the Assembly that there should be limits to the use of embryos and fetuses, but it proposes a more coherent and limited set of principles:

a) any creation of human embryos by fertilization in vitro for research purposes should be forbidden;
b) only the use of dead embryos or fetuses may be permitted;
c) such use may only concern the therapeutic and diagnostic purposes;
d) the use must be with the consent of persons concerned;
e) supply of embryonic or fetal material may never be the subject of commerce or trade.

The CAHBI agrees with most of the more specific prohibitions mentioned in paragraph 14.A.iv, except the prohibition of research on viable human embryos. The CAHBI would not totally exclude all such research, but would permit research that is intended to benefit the embryo in question. The CAHBI also questions whether an embryo can be "viable" in the generally accepted meaning of the term. With regard to the prohibition against maintaining embryos in vitro beyond 14 days, the CAHBI observes that this prohibition in fact concerns two matters: (1) the duration of maintenance of embryos in vitro; in that case the CAHBI agrees that the maximum duration should not exceed 14 days; and (2) the duration of storage of embryos by freezing; in that case, national law should set a time limit for such storage.

On February 2, 1989 the Parliamentary Assembly adopted Recommendation 1100 on the use of human embryos and fetuses in scientific research. This recommendation is a new version of the earlier Recommendation 934, which takes account of intervening criticisms. The most significant change is that only procedural and organizational issues are addressed in the text itself, including recommendations to establish national or regional multidisciplinary bodies and draw up national or regional registers of accredited and authorized centers. More substantial ethical issues concerning the use of embryos and fetuses have been referred to the appendix, entitled "Scientific Research and/or Experimentation on Human Gametes, Embryos and Fetuses and Donation of Such Human Material". The status of this appendix is not clarified, but it contains normative rules ordered according to potential research materials: gametes, live preimplantation embryos, dead preimplantation embryos, post-implantation embryos or live fetuses in utero, post-implantation embryos or live fetuses outside the uterus, and dead embryos or fetuses. Although some of the prohibitions issued in Recommendation 934 are reemphasized, the rules formulated in Recommendation 1100 appear to be less stringent. The earlier absolute prohibition of any creation of human embryos by in vitro fertilization for the purposes of research during their life or after death (Paragraph 14.A.iii, R 934) is not reissued in Recommendation 1100. The absolute prohibition of research on viable human embryos (Paragraph 14.A.iv, R 934) is mitigated in Recommendation 1100, where it is stated that "investigations of viable embryos in vitro shall only be permitted – for applied purposes of a diagnostic nature or for preventive or therapeutic purposes – if their non-pathological genetic heritage is not interfered with". The same is true for the prohibition of research on living human embryos. That prohibition has been qualified by four conditions. As it now reads,

> research on living embryos must be prohibited, particularly if the embryo is viable; if it is possible to use an animal model; if not foreseen within the framework of projects duly presented to and authorized by the appropriate public health or scientific authority or, by delegation, to and by the relevant national multidisciplinary committee; if not within the time-limits laid down by the authorities mentioned above".

The last two conditions can be read in a positive sense, and seem to allow different opportunities in different member states. Moreover, the relative importance of these conditions is unclear. It is obvious that "viable" has a different meaning than in other contexts. A "viable" embryo is defined here as an "embryo that is free of biological characteristics likely to prevent its development"; this definition, however, is open to various interpretations.

In accordance with the Opinion of CAHBI, it is explicitly stated that "Investigation of and experimentation on dead embryos for scientific, diagnostic or other purposes shall be permitted subject to prior authorization".

The Parliamentary Assembly concludes that the successive proposals in Recommendations 934 (1982), 1046 (1986) and 1100 (1989) provide a framework of principles from which national laws or regulations can be developed in a uniform manner. Recommendation 1100 has been submitted to the Committee of Ministers, but thus far without definitive results.

The most recent document follows from the Committee of Ministers' invitation to the CAHBI to offer its opinion on Recommendation 1046. To review the scientific state of the art, a Select Committee of Experts on the Use of Human Embryos and Fetuses (CAHBI-R-EF) was established. Under the auspices of this Committee, Dr. Anne McLaren wrote a "Report on the Use of Human Fetal, Embryonic and Pre-Embryonic Material for Diagnostic, Therapeutic, Scientific, Industrial and Commercial Purposes". Initially finished in May 1989, a second, revised version of the Report was completed in March 1990 (CAHBI-R-EF 89-1 rev-2). The aim of the Report is to review the biomedical science and the use of human fetal, embryonic and pre-embryonic material practicable now or in the immediate future. The report is a scholarly explication of the various uses of such material in different contexts and for various purposes.

E. *Consent to Treatment and Experimentation*

In Recommendation R (90) 3, the Committee of Ministers first discusses the need to harmonize legislation throughout various member states, because existing legal provisions are either divergent or insufficient. The recommendation then sets forth a number of principles. It is argued, for example, that "in medical research the interests and well-being of the person undergoing medical research must always prevail over the interests of science and society" (Principle 2.1). Thus, the argument that large numbers of (future) patients will greatly benefit from experiments does not justify inflicting even the slightest harm on the persons undergoing the research. It follows from this principle that, as a general rule, "a legally incapacitated person may not undergo medical research unless it is expected to produce a direct and significant benefit to his health" (Principle 5.1) Obviously, if in exceptional circumstances allowed for by national law, a legally incapacitated person is participating in experiments, his legal representative may not receive any form of renumeration (Principle 13.2). One should note, however, that mentally competent research subjects may not be offered any financial benefits either (other than those covering expenses) (Principle 13.1). The Council apparently interprets the "interests and well-being of the person undergoing medical research" which are to prevail over other interests not to include financial gain. The individual is granted the right to refuse (further) participation at any time, but not to "sell" his time and or to undergo certain health risks as a way of "earning a living".

The protection of the mentally handicapped against the risks of medical experiments had been a concern to the Committee of Ministers some seven years earlier. According to Recommendation R (83) 2 "Concerning the Legal Protection of Persons Suffering from Mental Disorders Placed as Involuntary Patients", non-therapeutic clinical trials of products and therapies on committed mentally handicapped persons should be forbidden (Article 5.3). A treatment that is still considered experimental may be given only if the attending physician considers it indispensable and if the patient, after being informed, has given express consent (Article 5.2). The same applies to treatments which present a serious risk of causing brain damage or adversely altering the patient's personality (Article 5.2). The question arises as to whether committed patients have a right to refuse (by withholding consent) any medical treatment, or only risky treatments. Article 5.1 states that "a patient put under placement has a right to be treated under the same ethical and scientific conditions as any other sick person". Since competent patients have a right to refuse any treatment, it follows logically from Article 5.1, that committed patients have the same right. Furthermore, it seems paradoxical to assign a "right" to be treated and, at the same time, to assume a "duty" to be treated. However, Article 3.b suggests a different conclusion: a patient may be placed when, "because of the serious nature of his mental disorder, the absence of placement would lead to a deterioration of his disorder or prevent the appropriate treatment being given to him". It is common (though not undisputed) practice to justify committing mental patients who pose a serious risk to themselves. However, to commit someone because treatment deemed appropriate by his physicians otherwise cannot be given implies that treatment may be forced upon the mentally handicapped patient. Moreover, the Council not only sanctions involuntary treatment of committed patients, but also allows mentally handicapped patients to be committed in order to be treated. We stress that this conclusion is not explicitly reached in the recommendation; on the contrary, other articles in this and other recommendations tend to emphasize the unconditional status of the principle of informed consent. But such delicate matters as involuntary medical treatment do require a very careful phrasing that leaves no room for dubious interpretations.

The Parliamentary Assembly of the Council of Europe has issued a Recommendation 818 (1977) "On the Situation of the Mentally Ill". Unlike the recommendations of the Committee of Ministers, the Assembly's recommendations reflect discussions among parliamentary members on such philosophical issues as the nature of mental disease. Recommendation 818, for example, states that the Assembly has reached certain conclusions "considering that the definition of mental illness is extremely difficult, since criteria change with time and from place to place, and since a whole new range of psychological disturbances have emerged, linked with the working rhythm, stresses, and the sociological patterns of modern life" (§ 2). Furthermore, the Assembly warns that tragic errors and abuses have occurred with the internment of mentally handicapped patients (§ 5), and that improved medical and psychotherapeutic technology can sometimes constitute threats to the rights of patients (§ 8). In accord with these considerations, the Assembly recommends that the governments of member states review their legislation by redefining basic concepts such as "dangerous" and by reducing to a minimum the practice of compulsory detention for an "indeterminate period" (§ 13.I.i). The Council recommends establishing a Council of Europe Working Party to redefine "insanity" and "mental abnormality" and to reassess the implications of such redefinitions for civil and criminal law (§ 13.I.vi). Overall, these recommendations are primarily concerned with the legal or societal mistreatment of the mentally handicapped. The Assembly is convinced "that the concept of the criminally insane implies a contradiction in terms, in that an insane person cannot be considered responsible for criminal actions" (§ 7). Consequently, most of the Assembly's suggestions apply to the protection of the mentally handicapped and less to the relationship between health care providers and their mentally handicapped patients.

A final note should be made about the recent guidelines issued by the Commission of the European Community's Working Party on the efficacy of medicinal products. This so-called "note for guidance" pertains to good clinical practice for trials on medicinal products in the European Community. The guidelines in this 46-page document, implemented on July 1, 1991, are directed primarily towards the pharmaceutical industry, but also to all involved in the generation of clinical data for inclusion in regulatory submissions for medicinal products ([1], Foreword). The document discusses the protection of trial subjects and consultation of ethics committees; the responsibilities of sponsors, monitors and investigators; the handling of data; and the requirements of careful research design. As expected, informed consent is considered an absolute, requirement. It is interesting, in light of the Council of Europe Committee of Ministers stance on research with mentally handicapped, that the EC apparently takes a slightly more liberal position:

If the subject is incapable of giving personal consent (e.g., unconsciousness or severe mental illness or disability), the inclusion of such patients may be acceptable if the Ethics Committee is, in principle, in agreement and if the investigator is of the opinion that participation will promote the welfare and interest of the subject. The agreement of a legally valid representative that participation will promote the welfare and interest of the subject should also be recorded by a dated signature. If neither signed informed consent nor witnessed signed verbal consent are possible, this fact must be documented with reasons by the Investigator (§ 1.13).

D. *Care of the Dying*

The oldest bioethical document of the Council of Europe, Resolution 613 (1976) issued by the Parliamentary Assembly, was entitled "The Rights of the Sick and Dying". Contrary to its name, this short recommendation pertained mostly to the definition of death. It did not, however, offer a particular definition, rather, it expressed concern for the consequences of death-definition-debates for the dying patients. The Assembly maintained that "unnecessary anguish may be caused by uncertainty over the most appropriate criteria for the determination of death" (§ 3) and that "no other interests may be considered in establishing the moment of death than those of the dying person" (§ 4). Existing criteria for the determination of death have to be critically examined by the responsible bodies in the medical professions of member states (§ 5), and an attempt should be made to harmonize and practical standards (§ 6). In an ensuing Recommendation 779 (1976) with the same name, adopted on the very same day, the Assembly focused directly on the care, the sick, and dying. The Assembly noted that there is a "tendency for improved medical technology to lead to an increasingly technical – sometimes less humane, treatment of patients" (§ 2). In large modern hospitals, sick persons may find it particularly difficult to defend their own interests (§ 3). Consequently, the Assembly recommends increasing the training of health personnel (§ 10.1.a) and improving opportunities for patients to adequately prepare for death (§10.1.c). Special reference is made to the right of the sick to full information about their illness and proposed treatment (§ 10.1.b).

With regard to the issue of euthanasia, the Assembly maintained that "the doctor must make every effort to alleviate suffering", but he has "no right, even in cases which appear to him to be desperate, intentionally to hasten the natural course of death" (§ 7). National multidisciplinary ethics committees must establish rules for the treatment of dying patients, and discuss any national legal consequences resulting from unintentional hastening of death in the care of dying patients (§ II).

F. *Organ Donation and Transplantation*

Organ donation and transplantation were of concern to the Council of Europe very early on. The earliest Committee of Ministers' Recommendation R (78) 29 is "On Harmonization of Legislation of Member States Relating to Removal, Grafting and Transplantation of Human Substances". The recommendation stresses the importance of anonymity, which is to be respected except where there are close personal or family relations between donor and recipient (Article 2.2). The unique status of family relationships is also reflected in the recommendation's tenet that organ or tissues removal of which presents a foreseeable risk to the life or health of the donor may only be permitted exceptionally, when it is justified by the motivations of the donor, the medical needs of the recipient, and the family relationship between donor and recipient (Article 5). This article also reveals the Council's strong reluctance to grant individuals a right to decide about their own health and course of life. Self-sacrificing altruism is allowed only within the limits of familial relationships; donating organ or tissue to earn money is prohibited unconditionally by recommendations in two separate articles (Articles 9 & 14). On the other hand, there should be no financial barriers to organ and tissue donation: the (potential) donor should be fully compensated for any medical and social costs (Article 9). Moreover, in a separate Recommendation R (79) 5 concerning international exchange and transportation of human substances, the Council recommends "to exempt[ing] substances and their containers from all duties and taxes at importation and exportation" (§ I.4).

The Council's concern for the well-being and protection of the mentally handicapped (see also the earlier discussion of consent to treatment and experimentation), is reflected in Recommendation R (78) 29. Article 6.2 maintains that "the removal of substances which cannot regenerate from legally incapacitated persons is forbidden". Only in exceptional cases, when (a) the donor has demonstrated his ability to understand the issue and given his consent, (b) his legal representative and an appropriate authority have authorized removal, and (c) the donation presents no foreseeable substantial risk to the life or even the health of the donor, may a mentally handicapped person donate non-regenerative organs to a recipient who genetically is closely related (Articles 6.2 & 6.3).

G. *AIDS*

The Committee of Ministers, managed as early as 1989, to agree on Recommendation R (89) 14 "On Ethical Issues of HIV Infection in the Health Care and Social Setting". AIDS had been of concern to the Council many years earlier. In June 1983, Recommendation R (83) 8 on preventing the transmission of AIDS from blood donors to patients receiving blood products was passed, and the subsequent Recommendation R (85) 12 pertained to the screening of blood donors for the presence of AIDS. However, both recommendations, and the 1987 Recommendation R (87) 25, concern more technical issues made agreement relatively easy. It is therefore amazing that the Council passed Recommendation R (89) 14 in 1989, since this recommendation is rather explicit in setting definite ethical standards for the care of AIDS patients. The recommendation begins by arguing that "in the light of present knowledge, voluntary testing, integrated into the process of counseling, is the approach which is most effective from the public health view, and most acceptable ethically and legally" (Appendix to R (89) 14, Principle 15). Test results and counseling should be confidential (Principles 17 & 18) and any reporting of AIDS cases to the health authorities should be anonymous (Principle 37). The Council also accepts as a "general rule", to be broken only in "extreme cases", that partners of HIV-positive patients may not be informed without consent of the patient. On the other hand, systematic screening is deemed contrary to the rights of individuals when carried out without informed consent of the screened individuals (Principle 29). There is no justification for screening in the work place or educational setting (Principle 87). Employers may not compel a prospective employee to submit to an HIV test (Principle 89). Moreover, if an employee chooses not to reveal positive HIV status, sanctions cannot be imposed (Principles 90). Finally, compulsory screening, is considered "unethical, ineffective, unnecessarily intrusive, discriminatory and counterproductive" (Principle 33).

These principles constitute a clear reaction against any public hysteria but also against any discriminatory protectionism in the area of health care. The recommendation explicitly argues that compulsory screening may not be introduced for any particular "captive" population group, such as prisoners, military recruits, or immigrants (Principle 34). The freedom of HIV-positive travellers or immigrants to move between member states may not be restricted (Principle 47), nor may governments place such patients in any kind of quarantine (Principle 48). On the other hand, health care workers are thought to have an obligation to care for AIDS patients (Principle 69). Yet, when a health care worker is HIV-positive, he should refrain from medical practices that entail the slightest risk to the patient (Principle 66). Counseling and financial assistance should be available for such health care workers (Principle 67), but the Council clearly places a very heavy and unequal burden on health care providers.

It should be noted that the emphasis on voluntariness is not merely reflected in the Council's negative opinion of any compulsion. At various places, the importance of education, training, research, and safe-handling techniques is stressed. Furthermore, the recommendation urges that voluntary testing facilities should be easily accessible and free of charge (Principle 17). Prevention should be stimulated by non-coercive approaches, such as making sterile syringes and needles available to drug users and condoms available to prisoners (Principle 23). The latter precepts reveal the Council's eye for the pragmatic issues this deep concern for the needs of those in danger of acquiring AIDS as well as HIV-positive patients. Characteristic of this concern is the Recommendation's conclusion, that "national authorities should consider studying insurance possibilities for HIV-infected individuals" (Principle 112).

VI. CONCLUSION

The preceding review of the bioethics activities undertaken by the various European organizations has made clear the increasing interest in international aspects of health care ethics among both scholars and politicians. Many projects have been undertaken and many recommendations passed during the preceding decade. However, many plans have not been developed beyond proposal. For example, the Secretary General of the Council of Europe has proposed setting up a "European Ethics Committee", under the auspices of the Council of Europe. However, the Directorate of Legal Affairs has argued that, at the present time, there are too few national ethics committees to establish an encompassing international committee (CAHBI/INF (90) 1). Indeed, this proposal of the Council of Europe has not yet been implemented, nor is there information about tangible plans for the committee. What remains, rather, is merely an agreement among the CAHBI members that a thorough study of national and local ethics committees in member states (CAHBI (90) 3 should be undertaken. Similarly, during the conference of the Ministers of Justice in Istanbul in June 1990, it was suggested that a "Convention on Bioethics" be developed which would articulate common general standards for the protection of the human person in the context of biomedical services. The CAHBI is studying the proposal, but once again, there are no indications of significant further development.

These failures may have many different causes of which we are unaware. But various impediments which hinder a coordinated, systematic, and in-depth approach can be identified. First, there is the problem of fragmentation. Nowadays, the Parliamentary Assembly of the Council of Europe is being invited to participate in the CAHBI, and the European Community is represented. Similarly, in international conferences, such as the 1989 conference in Strasbourg (France), many international organizations as well as bioethics research institutes and university departments have participated. But the large number of organizations, each with its own political history and goals, its own structure and body of officials, its own funds and obligations, obstructs coordinated cooperation. Furthermore, many bioethical endeavors at the political level, both in the Council of Europe and in the European Community, seem to have a rather "ad hoc" nature. Recently, two European Community bioethics committees were established, each with a different political structure and status, and neither embedded in a larger framework of bioethics activities. Even the oldest, most authoritative, and best organized bioethical body formation, the Council of Europe's CAHBI, reflects in its name this "ad hoc" status: Ad Hoc Committee of Bioethics Experts.

In addition to the problem of fragmentation, cooperation between the academic and the political world is lacking. Existing committees, such as the CAHBI, have a mixed membership of administrators and scholars. But this mixture seems accidental and may not yield optimal results. On the one hand, each member is supposed to represent his or her own member state, but when the professional backgrounds of the representatives differs, a rather amalgamous political outcome is to be expected. Furthermore, the mandates given to the CAHBI members by their national governments appear to differ considerably. On the other hand, those members who are professional bioethicists or health lawyers cannot be expected to bear the burden of continuously supplying the committee with systematic, in-depth research studies about topics on the agenda. Instead, there is an obvious need for close cooperation between well-equipped scientific bioethics research groups which can provide thorough studies to be implemented by supranational political and administrative bodies. It should be noticed, however, that currently bioethics researchers are hindered severely by the lack of a European databank and information system for relevant bioethics documents. This very chapter is bound to be incomplete because it lacks access to a comprehensive literature review.

Of course, all these problems are of minor importance compared with the single most significant obstacle: the lack of power. As mentioned earlier, the recommendations of the Council of Europe lack any and all formal binding power. The Council asks member states to comply with rules and guidelines and to inform the Council periodically of actions taken on the recommendations. But member states are free fully to disregard those recommendations. There are indications that some influence is exerted by the Council's recommendations on national legislation, but to our knowledge, no research has been done to assess exactly how much.

The European Community is able to exert more power on its member states. Although no attempts have been undertaken to establish an explicit European bioethical policy, it is expected that once the two committees on genetics and embryo research publish their results, these reports will have a political impact. Obviously, much current EC policy in the area of health care implies a particular bioethical policy. On the other hand, there are indications that some countries have undertaken attempts to have bioethics identified as a national rather than supranational concern. Since the EC, unlike the Council of Europe, exerts increasing binding influence over the various members states, particularly in the area of economics, agriculture, trade and finances, "immaterial" issues such as bioethics, appear to have become the only areas left to express national diversity.

It is always easy to criticize, particularly as outsiders who are not established in the bureaucratic network of a vast supranational organization such as the Council of Europe or the European Community. It cannot be denied that there is a rapidly increasing interest in supranational bioethical issues, both among academic scholars and among politicians and officials. However, the practical realization of new ideas and theories always takes time. It is to be expected that many fragmentary endeavors will slowly merge to yield a more coordinated approach. After all, the CAHBI itself was established to coordinate the undertakings of four different committees, and existing international bioethics associations are still increasing their membership. Closer cooperation between the political and the academic scenes is being stimulated as well. Both the Council of Europe and the European Community have begun efforts to compile list of European bioethics scholars. In addition, the European Community has provided a considerable amount of money for bioethics research which should help to generate academic international bioethics networks. Although there is still a long way to go, a promising start has been made.

Catholic University of Nijmegen
Department of Ethics
Philosophy and History of Medicine
Nijmegen
THE NETHERLANDS

BIBLIOGRAPHY

1. CMPM Working Party: 1990, *Good Clinical Practice for Trials on Medicinal Products in the European Community*. Commission of the European Communities, Directorate-General for internal market and industrial affairs. III/3976/88-EN, 1990.
2. EC Commission: 1991, "Research and Technology Development in the Field of Biomedicine and Health: Call for proposals". *Official Journal of the European Communities* October 25, 1991, Number C 278/9-11.
3. EC Council: 1991, Council Decision of September 9, 1991, Adopting a Specific Research and Technological Development Program in the Field of Biomedicine and Health (1990 to 1994)". *Official Journal of the European Communities* September 24, 1991, Number L 267/25-32.
4. Have, H.A.M.J.ten, Kimsma, G.K. and S.F.Spicker (eds.): 1990, *The Growth of Medical Knowledge*. Kluwer Academic Publishers, Dordrecht.
5. Jensen U.J. and Mooney G.(eds.): 1990, *Changing Values in Medical and Health Care Decision-Making*. Wiley & Sons, Chichester.
6. *Newsletter Biomedical & Health Research*. European Community, Number 3, 1991, p. 9-10.
7. Trappenburg, M. and Cortel, M: 1991, *De Frankenstein-cirkel. Rechtsvergelijkende notities over regelgeving rond embryo-experimenten*. Rapport in Opdracht van de Socialistische Fractie in het Europees Parlement, Leiden.

DAVID GREAVES, MARTYN EVANS, DEREK MORGAN,
NEIL PICKERING AND HUGH UPTON

BIOETHICS IN THE UNITED KINGDOM AND IRELAND: 1989-1991

I. INTRODUCTION

This report provides both a summary and discussion of recent developments in bioethics in the United Kingdom (UK) and Ireland on a wide range of topics. Those developments have occurred both formally, in legislation and case law, and less formally, in scholarly discussion and popular debate. This report, therefore, draws upon a variety of sources in its analysis of issues. With regard to legislation, for example, it discusses the recently passed Human Fertilization and Embryology Act in Sections II and III, the National Health Service and Community Care Act in Sections VII and IX, and the Human Organ Transplant Act in Section XIII. The report also comments extensively on recent court decisions in a number of areas. Cases of maternal-fetal conflict are discussed in Section IV, those involving severely ill newborns in Section IV, cases involving issues of informed consent in Section VI, and a case of alleged active euthanasia in Section XI. In their discussion of cases, the authors also speculate upon the possible implications of these cases for future developments in bioethics. Finally, the report draws upon scholarly exchanges, as well as accounts in the popular media, to present a fuller picture of the volatile debates on certain issues. Although it is impossible to review completely all recent developments, the report provides a comprehensive summary and analysis of the most important recent developments in bioethics in the United Kingdom and Ireland.

II. NEW REPRODUCTIVE TECHNIQUES/PRACTICES

The introduction of legislation to regulate aspects of the "reproduction revolution" – controlling embryo research and licensing the provision of infertility treatments such as IVF and services using donated gametes – was achieved in the Human Fertilization and Embryology Act of 1990 of the United Kingdom [37].

The medical profession exercised an enormous influence over the legislation as it relates to treatment services, and it may be that the operation of the supervisory body, the Human Fertilization and Embryology Authority (HFEA), will be to a large extent dependent not only on the good will but on the

B. Andrew Lustig (Sr. Ed.), Bioethics Yearbook: Volume 2, 127-152.

cooperation of the profession.

The Act has several purposes. The first is to regulate certain infertility treatments which involve keeping or using human gametes and to regulate the keeping of human embryos outside the human body. The Act deals with only four of the treatments currently in use: artificial insemination or gamete intra-fallopian transfer (GIFT) using donated gametes, egg donation, embryo donation, and in vitro fertilization (IVF). The second purpose is the statutory regulation of embryo research. Thirdly, there is a prohibition on the creation of hybrids using human gametes, the cloning of embryos by nucleus substitution to produce genetically identical individuals, and genetic engineering to change the structure of an embryo. The fourth purpose is to effect changes to the Abortion Act of 1967. These latter three purposes are probably the most controversial aspects of the legislation; they were the most hotly debated and aroused the most public comment.

The main vehicle for conveying these changes is the Human Fertilization & Embryology Authority (HFEA), which was established on November 7, 1990. The Authority is a statutory agency charged with a wider range of responsibilities than its predecessor, the Voluntary (later Interim) Licensing Authority, established following the publication of the Warnock Committee Report by the Royal College of Obstetricians & Gynecologists and the Medical Research Council. HFEA has to operate within the terms of the 1990 Act, and is primarily a licensing body.

If we ask, however, why the legislation was introduced, we can gain a clearer understanding of the place of the Act in the regulation of clinical practice, and research. One of the strongest arguments heard in the Parliamentary debates was the need for regulation of research on hyman embryos, but also and importantly, to protect the integrity of reproductive medicine and to safeguard scientists and clinicians from legal action and sanction. The statutory scheme has been introduced to ensure that these sensitive issues of moral and legal complexity are dealt with in a clear framework. It seeks to balance what are the sometimes conflicting interests of scientific research, clinical practice and public concern. Similarly, it seeks to mediate between the families who may benefit from research into the causes of genetically inherited disease or chromosomal abnormalities. In all cases, the broader social, moral, and philosophical interests which disclose fundamentally different ways of conceiving of the world are seen potentially to be in conflict. In short, the Act is one important manifestation of who we are and who we say we want to become; the question which it raises is "who is to be or not to be?" In providing an answer, the medical profession is not able to claim, nor with few exceptions does it, that scientists or clinicians have a special expertise.

III. ABORTION

On this issue there was one event of major importance in the time period being considered, namely, the changes to the law on abortion which were made in the 1990 Human Fertilization and Embryology Act. The legislation was preceded by extensive campaigning by those who regarded the current law as too liberal and by those who feared the loss of hard-won rights for women. Attention was particularly focused on the question of the time limit for abortions, largely because of the belief that a fetus could be viable earlier than the 28-week limit then in force under the 1929 Infant Life (Preservation) Act. Under that Act, it was illegal to abort any fetus capable of being born alive, and 28 weeks was held to mark the point when this would certainly be the case.

Since this part of the 1929 Act was unaffected by the legalization of some abortions under the 1967 Abortion Act, it could continue to be invoked in court. Hence, in 1990, the mother of a child with spina bifida who had a scan at 25 1/2 weeks sued the health authority and a radiologist for not giving her the diagnosis that would have enabled her to choose an abortion. The judge, however, upheld the defense case that abortion would not have been a legal option at that time because the baby would have been capable of being born alive, where this meant breathing by its own lungs independently of the mother [4].

Nevertheless, despite the presence of the viability condition, there was pressure to recognize the effect of advances in medicine since the 28-week limit was set, with many claiming that 24 weeks would better represent the point of viability. Others wanted even greater restriction, the most prominent campaign being for the 18-week proposal in a private members bill put forward in 1988 by David Alton, only to be lost through being "talked out".

When the time came for a vote on the government's legislation, the limit was reduced but only by 4 weeks to 24 weeks, in line with medical opinion regarding viability. This was to be the limit for abortions carried out because of risk to the physical or mental health of the woman or of her existing children. There was, however, to be no limit under the new Act for abortions carried out because of a risk of grave permanent injury to the woman, a risk of her death, or of serious handicap for the child.

What then, was thought to be the general effect of the new act? J. Warden held that overall the law had been relaxed, given the absence of any time limit for the exceptions to the 24-week rule [53]. This apparent permission for abortion in some circumstances up to the time of birth (however infrequently one might expect doctors to take that option) caused alarm to some. The fear on the opposing side was that the 24-week limit would be too constraining, because there were serious fetal abnormalities (presumably other than those that would constitute exceptional cases) that could not reliably be detected within that time. Whether or not this is the case, there is clearly a problem of principle here. If the legal time limit is tied to viability, and medical science is able to advance the point of viability, will diagnostic skills keep pace? If not, at least part of the rationale for legalized abortion would have to be reassessed.

Finally, while the law relating to abortion was being reviewed and changed, attention was also given to the conscience clause in the 1967 Act. This clause gave those doctors and nurses who objected to abortions the right not to be involved in them. Some commentators expressed concern that there was discrimination against those who chose to exercise this right, (*e.g.*, [53]), and a recommendation was made by the Social Services Committee that medical schools be discouraged from asking candidates for their views on abortion. It was also proposed that the right to invoke the conscience clause be extended to staff.

IV. MATERNAL-FETAL CONFLICTS

In the wardship case of *Re F (in utero)*, a 36-year old woman (F), became pregnant early in 1987 [61]. She was under the care of a psychiatrist, having a history of severe mental disturbance and periodic drug usage. Local social workers had some years before taken her 10 year old son into care. Her access to him had been terminated as part of long-term fostering arrangements. Just before the expected date of the birth, F disappeared. The local authority applied for the fetus to be made a ward of court; this would terminate all parental rights, duties and powers and vest them in the court. Directions could be made about F's continuing antenatal attendance, the welfare of the fetus, and the obstetric regime to which F would ultimately have had to submit.

In the United States, even more draconian orders have been sought, and in a limited number of cases, obtained and enforced. These have required some women to undergo a caesarean operation to which they would not consent despite the advice of their obstetricians, concerning the supposed benefit or welfare of the fetus [28]. Cases of this type have been inevitable since the urging of the recognition of the fetus as a patient with its own rights and interests from the early 1970s onwards. Both the High Court and the Court of Appeal refused to issue the wardship order or any ancillary order.

The Court of Appeal concluded that the court cannot extend the wardship jurisdiction to protect the interests and welfare of the fetus, although Lord Justice May indicated that this was a case in which such a maneuver would have been justified. All the judges who heard the case agreed that, until birth, the fetus has no civil legal personality, an approach confirmed judicially in 1990 in *Rance v. Mid Downs Health Authority* [58]. Whatever its maturing moral status, the fetus enjoys no legal protection beyond that indirectly afforded through restrictions on lawful abortion. English law draws a fundamental distinction between a newborn infant and a fetus. Since an unborn child has, *ex hypothesi*, no existence independent of its mother, the only purpose of extending the jurisdiction to include a fetus is to enable the mother's actions to be controlled. Giving respect to the autonomy of the pregnant woman outweighs any claims which might be advanced on behalf of the fetus.

The Court dismissed arguments that the fetus could be the subject of protection under the European Convention on Human Rights. Article 2 (1) of the Convention provides that "Everyone's right to life shall be protected by law." According to Lord Justice Balcombe, extending this to apply to the fetus might mean curtailing the right to life of the pregnant woman; in an extreme case a woman might have to sacrifice her life for the benefit of the developing fetus. (There are some philosophical traditions which would see this as the true meaning of sacrifice.) All the judges were concerned that to allow wardship in a case such as this, however reprehensible each judge felt the woman's behavior to be, would open the door to much greater demands for the prenatal regulation of the behavior of pregnant women. Lord Justice Balcombe was alarmed that a court might later be asked to direct a woman's health regimen (relating to the consumption of tobacco or alcohol), social life ("any activity which might be hazardous to the child"), or any obstetric preference which in the opinion of the relevant physician might cause injury to the fetus. The judges agreed that the issue of the wardship jurisdiction for the benefit of the fetus was a matter for Parliament and not the courts. Several jurisdictions in the United States have passed "fetal abuse" or protection statutes, under which a pregnant woman can be arraigned on criminal penalty where her life style or preferences might damage the fetus [12].

These concerns and cases, of which *Re F* is an example, reflect and are representative of a more general mood which sees the fetus emerging, both figuratively and literally, from the shadows of moral concern. They are part of a broader movement to accord enforceable legal rights to the unborn (against accidental injury, being aborted, and so on) which is paralleled in concerns about human embryo research. British courts are becoming more closely drawn into reproductive and obstetric conflicts; but there are at least two reasons why this temptation has been correctly resisted in *Re F*; the "slippery slope" argument and the "therapeutic conflict".

These difficult and troubling cases can be recognized to be at the top of a tortuous and slippery slope. The "horrible result" which lies at the bottom is the reduction of pregnant women to no more than "fetal containers", seeing them as a resource to be used to the ends of another [1]. Alternatively, the "arbitrary result" once on the slippery slope includes such issues as the following: (a) determining the age at which legal protection should be accorded the fetus; and (b) to what sorts of threat must the fetus be subject and what likelihood of harm is necessary before the court would entertain wardship? [54]

Lawyers should be hesitant to rush into the delivery room. The relationship between a pregnant woman and her developing fetus, or between the woman and her obstetric advisers, is not a straightforward one. Creating an adversarial relationship between the woman and her fetus will not clarify either. Physicians often disagree about the appropriateness of, say, an obstetric intervention, but most women, when faced with the risk of death or serious injury to their fetus, submit to their obstetrician's judgment, even if grudgingly. While English law is quick to accept medical uncertainty as a justification for physician error, we seem in danger of being less ready to recognize the implications of this notion for patient self-determination. If there is a course to be set between the Scylla of treating women as fetal containers and the Charybdis of regarding their fetuses as uterine cargo [34], the price which we must be prepared to pay for protecting the integrity of all competent adults is the occasional risk of death or serious injury to an unborn fetus. Professional obstetricians alone cannot be allowed the license to make the occasional mistake.

V. CARE OF SEVERELY ILL NEWBORNS

As often happens, a case at law has provided the chance of considering the moral as well as the legal aspects of a medical problem. In April 1989, the Court of Appeal heard the case of *Re C (a minor)* concerning a handicapped, premature baby suffering from unusually severe hydrocephalus which had resulted in the brain being poorly formed [60]. Because, for independent reasons, she had been made a ward of court the doctors asked for the court's assistance regarding her treatment.

Expert medical opinion was that the baby could not be saved and that the aim of medical interventions should be to ease suffering rather than to produce a short prolongation of life (although antibiotics, intravenous infusions or nasal gastric feeding were not ruled out). The response from the lower court had been an order from Mr. Justice Ward that was based explicitly on the idea of putting C's interests first. While this idea was supported throughout as being the correct policy, certain items in the court order raised concerns that were usefully clarified later. First, Justice Ward's initial order permitted the hospital to "treat the minor to die". This caused anxiety that some sort of active euthanasia was being allowed, whereby staff could act to bring about the death of C. The amended order from Justice Ward made it clear that active euthanasia was not intended, but rather that the hospital was at liberty to allow the baby's life to come to an end peacefully and with dignity. In other words, what was permitted was (to use a term many reject as misleading) passive euthanasia.

Another concern regarding the court order was its direction that the hospital should relieve C's pain, suffering and distress but that "it should not be necessary to prescribe and administer antibiotics to treat any serious infection which she might contract or to set up intravenous infusions or nasal gastric feeding regimes" [60]. Naturally, a court ruling on a substantive matter of treatment would be of great significance, but this passage was deleted by the Court of Appeal as being inconsistent with another part of the order, *viz.*, that treatment should continue in accordance with the report of the expert pediatrician mentioned above. The correct understanding of this report was held to be that such measures could be adopted if necessary to relieve suffering, but not if the aim was only to prolong the baby's life.

The case was clearly an important one. Morgan remarks of Justice Ward's order that it was "the first time that an English court has acknowledged *and condoned* the pediatric practice of managing some neonates towards their death, rather than striving to "save" or "treat" at all "cost" ([35], p. 14). Equally, of course, it is clear (and not surprising) that the law still upholds as crucial the distinction between knowingly acting with a view to a certain result occurring, and knowingly allowing that result to occur earlier than it would have had some intervention had been made. Philosophical skepticism regarding this distinction has yet to have much effect on the practice of law. Nevertheless, it is significant enough that the Court of Appeal decided that a baby in C's condition was a special case requiring special treatment. While a social worker had proposed that treatment should be the same as for a child without handicap, the court upheld the view of the local authority's legal department that treatment should be appropriate to C's condition. How different this would be in practice presumably remains an open question, since no doubt conditions that would be life-threatening will often also cause suffering and thus be treated under the latter description.

In all this, the idea of a baby's best interests remains vital but problematic, raising questions both of what counts as meeting the requirement and of who is best placed to make the assessments. The courts again had recourse to the idea in the decision of the Court of Appeal in *Re Baby J* [59]. J, born very prematurely, had severe brain damage, being apparently blind, probably deaf, and unlikely to develop speech or even limited intellectual abilities. On the other hand, he will probably feel pain as would a normal child and could live to his late teens. He has required a ventilator following severe breathing failures, as well as a drip and antibiotics. Since J was a ward of court, it was the court's responsibility to give or refuse consent for further treatment in the event of another collapse.

The lower court had ruled that it would not be in J's best interests to be put on the ventilator again in the event of breathing failure, unless the doctors deemed it appropriate in particular circumstances. Antibiotics and maintenance of hydration would be used in the event of chest infection, but not prolonged manual ventilation.

The appeal against this ruling was made by the official solicitor, who argued that consent to life-saving treatment should never be withheld by a court. This was rejected by the Court of Appeal as absolutist. Instead, the court held that while there was a strong presumption in favor of prolonging life, account had to be taken of the quality of life prolonged; and there could be cases where causing increased suffering without commensurate benefit would not be in the child's best interests.

Once again, then, the idea that not everything need be done to prolong the life of very severely handicapped neonates was upheld, though, of course, the problem of deciding just when the quality of life will be too poor remains. In addition, there is the problem of providing the best account of what it is that we are seeking. *Re Baby J* refers to the customary notion of "best interests", which is hard enough to determine, but there is also a reference to the idea that the court must decide what the patient would choose if he were able to make a sound judgment. This test, obscure at the best of times, seems hardly to make little sense in the case of a severely brain-damaged infant.

VI. CONSENT TO TREATMENT AND EXPERIMENTATION

Perhaps the main focus of attention in this area has been provided by a case where treatment appeared to be required, yet valid consent was regarded as unobtainable. The case of *F v. West Berkshire Health Authority* took up the question of whether an adult woman with a mental handicap which rendered her legally incompetent could have an operation for sterilization [57]. The case was significant because F, not being a minor, was not a ward of court. In addition, though she was for many years a resident in hospital, this was not compulsory. In this sense (though doubtless not in others) she was a voluntary patient.

F was 36 years old but with a mental ability regarded as that of a four or five year old child. Her verbal capacity was that of a child of two, a fact that was said to restrict her to indicating likes and dislikes and to expressing rather idiosyncratic emotions. The case arose when it was discovered that she had begun a sexual relationship with a male resident of the hospital, entailing a risk of pregnancy. It was believed that her lack of understanding of this condition, and of labor and delivery, would render her unable to cope with these events; thus, sterilization was regarded as the best way of avoiding the situation.

There was, it seems, no dispute in the courts that sterilization was indeed the best solution, but the matter was taken to the highest court, the House of Lords, as a test case. The judgments in several courts were that the operation was permitted (and desirable, as being in the best interests of the patient) despite the fact that F's consent could not be obtained. Yet the fact of consensus does not solve the problem of how we should determine the substantive best interests of an incompetent patient.

The test favored by the Lords was the established idea of avoiding negligence, where this is understood in terms of the Bolam Test [36]. Under this test, a doctor will not be negligent if "he or she establishes that they acted in accordance with a practice accepted at the time by a responsible body of medical opinion skilled in the particular form of treatment in question". However, one might think that this test allows practitioners too great a license and that a requirement of support for the treatment from the *majority* of relevant opinion would be a more cautious and wiser rule. This kind of test was favored earlier in the case's progress by the Court of Appeal.

Another important question is that of the role of the courts in these decisions. The House of Lords found that the courts had no powers to make orders regarding such operations as sterilization, but decided it was desirable that in cases involving incompetent adults, that a court opinion be obtained to the effect that the operation would be in the patient's best interests.

One practical result of this judgment was the issuing by the Official Solicitor of Guidance of the appropriate procedure to follow in sterilization cases, whether of minors or mentally incompetent adults [45]. This document is suitably cautious in setting out in detail what must be established before the operation is sanctioned. In broad terms, it discusses the determination that the patient is, and will remain, incompetent; that there is a real danger of pregnancy occurring; that pregnancy would indeed be traumatic (and worse, in this report, than sterilization); and that other less intrusive treatments are unavailable and unlikely to become available in the near future. It would seem to remain true, though, that these requirements are to be assessed according to the relatively weak Bolam test mentioned above.

Groups that have taken an interest in this problem include the Royal Society for Mentally Handicapped Children and Adults, which proposed that ethics committees could be used to render decisions, particularly when there was a conflict of medical opinion and invariably where any serious intervention such as sterilization was being considered. This proposal was important, for it suggested that such decisions are more than purely medical ones, going beyond the question of whether or not we have secured agreement among medical professionals (or some portion of them) in favor of the treatment. The Society produced its own working party report, *Competency and Consent to Medical Treatment*, in March 1989 [48].

The Law Society issued a discussion document, *Decision-Making and Mental Incapacity*, on the general problem of deciding for the incompetent. The report stressed that the aim should be, as much as possible, for the dependent person to make decisions, and that any necessary alternative to this, with regard to medical treatment or admission to residential care, should involve a multidisciplinary group.

In 1990 the Code of Practice that was required under the 1983 Mental Health Act was finally published. Although the code is too long for summary, it is noteworthy in its emphasis on the idea that we should resist generalizing about mental illness and the incapacity to consent, but instead should concentrate on the capacity of an individual at a particular time and according to an appropriate form of explanation. Where treatment is deemed necessary in the absence of consent, the Code naturally follows the ruling of the House of Lords in the case of *F v. West Berkshire Health Authority* described above.

VII. CONFIDENTIALITY

There have been two recent acts of Parliament in the United Kingdom which have substantial implications for confidentiality. The National Health Service (NHS) and Community Care Act was implemented in April 1991. Although it is not directly concerned with confidentiality, its administrative provisions pose a potential threat to the traditionally confidential nature of the doctor-patient relationship. The problem arises because the new system of purchaser and provider relies on providers to bill health authorities for treatment, and for each individual invoice details are required about the patient's age, sex, date of birth and ethnic origin as well as about diagnosis, treatment and medical history. As an editorial in the *British Medical Journal* pointed out, "This clearly breaks the ethical principle that identifiable personal health information should be kept confidential and released only when the patient has given explicit consent or when strictly necessary for treatment". This potential problem, probably an oversight by the legislators, could be overcome by the introduction of a code which could be broken only in particular agreed circumstances.

The Access to Health Records Act was passed in 1990 and implemented in November 1991. It gives patients right of access to their manually recorded health records for the first time, although it still allows the doctor discretion to prevent access when he believes that it could cause serious harm to the patient. Similar access to personal computer-held health records is already permitted under the 1984 Data Protection Act. These two acts are important, because they overturn the long-held principle that doctors should normally keep medical records from patients, and introduces instead the notion that patients have a right of access to their records except when the doctor can show good grounds on which to deny it. Many individual doctors say they have no objection to patients seeing their own records, and some allowed their patients to do so previously. However, the main bodies representing the medical profession, *e.g.*, the British Medical Association, continue to oppose this change and especially its enactment in law, claiming that in general it is not in patients' best interests.

In November 1989 the Government of the United Kingdom announced that it was to introduce anonymous HIV testing to determine the number and monitor the distribution of people who are HIV-positive for the purpose of health-care planning [39]. Testing was to be carried out by taking random samples of the residue of specimens of blood already obtained for diagnostic purposes in other health care situations. Although there is an opting-out procedure, whereby individuals may request not to be included in the testing program, serious ethical issues still remain and these have led to considerable controversy. The most worrying is that an individual identified as HIV-positive has no possibility of learning this result since the testing, whose purpose is solely epidemiological, has been organized so that the results are not attributable to a given individual. Therefore, the most serious personal medical information will be denied to particular people, not because it is deliberately withheld from them, but because the program has been established so as to make it unattainable. This may appear to defuse the potentially explosive issue of confidentiality through anonymity procedure of anonymity, but, alternatively, it might be interpreted as a way of masking the issue improperly. An Institute of Medical Ethics Working Party recognized this as a harm that such a program would entail, but concluded that "not only are there no serious ethical objections to anonymous testing for HIV, but its introduction should be welcomed [24]". The issue of confidentiality alone would seem sufficiently serious to make such a judgment contentious.

Recently, there have been two important court cases in the United Kingdom concerning confidentiality, one in England and the other in Scotland. The English case involved W, a mental patient who was compulsorily detained in a special hospital and who sued a psychiatrist, Dr. Henry Egdell, for an alleged breaching confidentiality [56]. W wished to prevent certain information about himself, which had been obtained by Dr. Egdell in the preparation of a special report, from being forwarded to the hospital where he was held. Dr. Egdell requested permission from W's solicitor to send this report, and despite the fact that permission was refused, he proceeded to send the report anyway. The allegation was that to have sent the report without such permission constituted a breach of confidence, but the case was dismissed by the Court of Appeal in November 1989. Lord Justice Bingham concluded that, because W had killed several people some years previously,

"A psychiatrist who became aware, even in the course of a confidential relationship, of information that led him to fear that decisions might be based on inadequate information and with a real risk of danger to the public was entitled to take reasonable steps to communicate his concern to the responsible authorities. Dr. Egdell's conduct was necessary in the interests of public justice and the prevention of crime [56]".

The second case came before the Court of Session in Edinburgh in 1989. Mr. A-B, who claimed to have become HIV-positive following blood transfusion, went to court in an attempt to gain access to his blood transfusion records. However, the Secretary of State for Scotland intervened in the hearing on the grounds that disclosure of the blood donor's name would put the national blood supply at risk. The hearing was then terminated, since there was no precedent for the court overriding a ministerial objection to disclosure of information [55].

VIII. EQUITABLE ACCESS TO HEALTH CARE

The new arrangements introduced in the United Kingdom in April 1991 under the NHS and Community Care Act promised to provide increased patient choice. However, in practice, health authorities are organizing block contracts with particular hospitals and other health care facilities, and making extra-contractual referrals (ECRs), *i.e.*, referrals outside these block contracts, exceptional, thus preventing patients from exercising choice. For example, one district health authority (DHA) has given guidance about the management of ECRs as follows:

One of the reasons for separating purchasers from providers is to help purchasers to exercise greater choice about how the limited health care budget for the resident population is spent.... By stressing clinical priority within the ECR authorization policy, patient choice, a corner-stone of the stated benefits of the reorganization, may be reduced.

Arrangements such as these, however, may tend to reduce equity in access to health care because they disadvantage those who, because of poverty, disability, or lack of personal transport, have difficulty travelling to the selected health care facilities.

Within the NHS the impression has lingered that those who are financially better off obtain a greater share of NHS funding per capita than those who are poorer. However, recent research has led to a different interpretation of the evidence on which this was founded. With better methods of standardization, it is now been claimed that the previously supposed relationship is actually reversed and that the poor receive more NHS funding per capita. But because the poor have higher levels of morbidity, when this is taken into account, the extra resources they receive are not sufficient to compensate for their worse level of health. Thus, the poor are at a disadvantage, but much less so than had been thought before [6].

If the work of this single study is confirmed, it represents an important finding. It has long been assumed that richer middle class patients are able to command access to more health care resources, because they are more articulate in presenting their case and hence more successful in utilizing the health services. The class structure has therefore been seen as a barrier to equitable access to health care which can only be overcome by positive measures to favor poorer working class people. The findings of the above study suggest that such systematic inequities have already been largely overcome by such positive measures, or are far less pronounced than had been supposed.

IX. ETHICAL ISSUES POSED BY COST-CONTAINMENT MEASURES

By far the most significant change in health care provision in the United Kingdom in recent years has been the government's introduction of an internal market within the NHS aimed at effecting competition among all elements of the service, to improving efficiency, and to reducing costs per unit of health care provided. This major change in philosophy, set forth in the NHS and Community Care Act, has given rise to the most heated debate about the direction of British health care since the introduction of the NHS in 1948.

The heart of the controversy concerns the question whether the reforms are about efficiency and cost-containment within the NHS, or whether there is an underlying political interest in fostering the development of a two-tier system of health care on the American model, where private care is encouraged and containment applies only to state expenditures on health care. The United Kingdom spends about 6 percent of its GNP on health care, a figure which has changed very little in the past decade. In most other western countries, this figure has been rising to higher levels (most notably in the United States, where it now stands at approximately twice the U.K. percentage). Thus, as one American commentator has observed:

> ....the restructuring of a system (the NHS) that delivers more services per million pounds than any other on the grounds that it does not do so seems insane. Were the cost of the NHS £45bn and rising at 5 percent over inflation it would be understandable why the government thought it had a problem, but with NHS costs at £30bn, rising at 2 percent over inflation, the question is why do British leaders think they can treat the nation's ills for a third less than anyone else? [30]

The question remains as to whether the changes that have been introduced will lead to an expansion of the private sector, thus restricting benefits to those who are able to pay, or whether there will be a major injection of capital into the NHS which will discourage the further expansion of private health care while spreading the benefits more equitably throughout the population. What strains credibility is that the NHS will become more efficient overall. In fact, the reverse appears more likely; perhaps the most serious argument against the present changes is that they will lead to increased administrative costs, without any parallel increase in benefits to patients, so that more money will provide the same amount of total health care as before.

Behind the political furor that these reforms have raised is the question of what constitutes a just distribution of total health care resources for a developed nation such as Britain. Because the focus has been almost entirely on the internal distribution of health care resources and efficiency, this wider question has been largely obscured. There seems to be a reluctance in the United Kingdom, as elsewhere, to confront this most thorny of ethical issues, which faces all governments and policy-makers of whatever political persuasion. In Britain it seems as if political wrangling serves to distract from confronting the need to deal with the expansionary pressures on health care for the benefit of all citizens. There is a common anxiety about voicing the problems which we all share but fear to acknowledge. In the meantime, we continue to limit health care in a piecemeal fashion according to the political expediency of the moment. The present contortions of the NHS demonstrate this process within one nation and on one political stage, but it is being played out in different ways throughout the countries of the developed world.

In the Republic of Ireland, there has also recently been controversy about health care provision, but against a rather different background. A decade ago, health care expenditures reached a peak of over 8 percent of GNP, but has declined dramatically to around 7 percent today. The reduction in the Irish health care budget reflects the serious economic difficulties that the country has experienced in the 1980s. Not surprisingly, these cuts have not gone uncontested. One response was the establishment by the Department of Health of a Commission of Health Funding, which reported in 1989. The main targets of its criticism were the management and financial systems of health care delivery, rather than the amount of overall funding. These conclusions mirror a strong political theme in the United Kingdom, where current difficulties in health care are identified as problems of management and economics.

In the United Kingdom, there has been, since 1983, a directive to all health authorities in the NHS requiring them to effect cost-improvement programs every year. By combining these and costing them nationally, the government has claimed that more resources are available for the service on a year-by-year basis, because of improvements in efficiency. However, an independent study in 1989 showed that many of the apparent savings are in reality spurious, because accountants have responded to government requests for cost-improvements in ways which exaggerate the benefits to be gained. For example, some cost-improvement schemes can be effected only once, but are budgeted as if they applied recurrently [27]. However, figures are then used by the government to claim improvements in efficiency, which in part, at least, are imaginary. Such deceptive statistics exemplify the sort of distortion likely to occur when cost is made the paramount measure of health care activity, and cost-containment the overriding goal, as is the current trend.

X. WITHDRAWING/WITHHOLDING TREATMENT FROM DYING PATIENTS

One of the sharpest debates in the public arena recently has been over do-not-resuscitate orders. A.K. Marsden has provided the following guideline: "Resuscitation should be started on all patients with sudden loss of consciousness and absent breathing or pulses. It is inappropriate, however, to attempt to resuscitate those patients whose lives are drawing naturally to a close because of irreversible disease" [33]. Here what is appropriate is linked to the specific condition of the patient, though what exactly should count is left vague.

A different link is established in a report on "Resuscitation Decisions in a General Hospital" by Stewart and his colleagues. The report assumes that "cardiopulmonary resuscitation should be restricted to those *most* likely to benefit" (our emphasis); thus, "a relatively high proportion of patients designated not for resuscitation is appropriate in general wards" [52]. This might be taken to mean that at *any* time a certain percentage of patients in such wards should be given do-not-resuscitate (DNR) orders. What is appropriate, in here, however, surely should apply to individuals, and not to kinds of wards.

The paper also notes that "there should also be discussions and consultation with patients... when making this decision" [52]. What remains unclear is whether or not the patient's views (as compared with the medical prognosis) will be the deciding factor. A further issue is posed by patients with senile dementia. The study showed that six such patients were given DNR orders, but the authors fail to comment on this. The author also concluded that "eight [patients] with moderate or severe functional impairment after stroke" should have been given DNR orders, but were not. It is worth noting that for patients more than 75 years old, geriatricians were more likely to write DNR orders (21 patients out of 33) than were general physicians (13 patients out of 33).

The attitudes of Stewart and his colleagues are especially interesting in the light of a survey taken by doctors of the Imperial Cancer Research Fund [50].

The survey pointed out differences in the attitudes to chemotherapy among doctors, nurses, unaffected members of the general public, and patients actually faced with the conditions which required the treatment. Patients confronting decisions were far more likely to choose chemotherapy, despite a poor prognosis and the side effects of the treatment, than those not suffering from cancer. The writers ask whether only those actually affected "can evaluate such life and death decisions", and whether the views of those unaffected are deemed correct only insofar as they correspond. Both assumptions appear to clash with that of Stewart *et al.* quoted above.

XI. ACTIVE EUTHANASIA

Public pronouncements on euthanasia have become very important because the subject is firmly in the public mind. This public awareness results from the growing number of AIDS victims, the growth of the hospice movement, the activities of the Voluntary Euthanasia Society, the release of a British Medical Association Report in 1988, and the publication of a book by a well known broadcaster [26]. Familiar ethical and conceptual issues have arisen in the course of the debate. Distinctions are drawn between active and passive euthanasia, and between attempts to alleviate pain and deliberate killing.

These two distinctions were illustrated in two legal cases. The first concerned the parents who failed to call for medical assistance for their daughter, who had deliberately taken an overdose, because they had agreed with her beforehand that they would not intervene. The daughter had a rapidly deteriorating condition, with only a few weeks to live. The parents were found guilty, but were sentenced to probation. The decision not to punish them more severely was based upon the fact that they had not *actively* participated in their daughter's suicide [41].

In the second case, a murder charge against a Senior House Officer was dropped [10]. Dr. Stephen Lodwig had given a terminally ill cancer patient a mixture of lignocaine and potassium chloride. The prosecution offered no evidence, because (i) other trials were proceeding elsewhere and the results were "encouraging"; (ii) the time the patient had remaining to live was, perhaps, just minutes rather than days or months; and (iii) death could have been the result of morphine or other complications. For the defense lawyer, the crucial question was whether or not the doctor intend to kill the patient? Other issues considered were the agonized condition of the patient and pressure on the doctor from relatives to "do something". A doctor is allowed to give doses of pain killer that he knows may kill, if his aim is to relieve the patient's distress. Of course, this does not appear to rule out the possibility that killing the patient may be the only way to relieve his distress. Nonetheless, any *deliberate* attempt to kill technically remains murder.

Concern about euthanasia was heightened by reports coming from the United States of America of a doctor who had invented a "suicide machine" and had allowed a patient to use it [9]. The woman was reported to be "no longer able to spell or play the piano,...[but] well enough to win at tennis and understand the consent forms". In addition, opposition to the option of euthanasia for AIDS patients was voiced at a London Medical Group conference in February 1989 [31]. There it was argued that something could always be done to palliate pain, and that active euthanasia was an inappropriate expression of medical power.

In other places, efforts to increase attention to care of the dying were encouraged. In 1990, the UK government guaranteed to match voluntary funding for the running expenses of the 145 English hospices [21].

On the other side of the debate, the Voluntary Euthanasia Society carried out a Public Opinion Poll in April 1989 [39]. Questions were asked about a situation in which someone was suffering from "an incurable physical illness that is intolerable for them" and had "previously requested [euthanasia] in writing". The word "euthanasia" did not appear, and was replaced by "immediate peaceful death". The role of the health care system was characterized as "medical help". Respondents were asked whether the law should be changed to allow adults to receive such "medical help"? 32 percent strongly agreed and 43 percent agreed that the law should be changed. Among members of the major religious denominations, majorities agreed also. Citing the poll results, 15 Members of Parliament signed an Early Day Motion supporting the introduction of legislation to legalize euthanasia [40].

An Institute of Medical Ethics working party published a report on euthanasia in 1990 [23]. The report views "assisting death" to be morally justifiable (though not, it appears, obligatory). The patient must have expressed "sustained wishes" for assisted death, and the benefits of ending life must "greatly outweigh" those of prolonging it. The doctors and the clinical team must be satisfied that no other means of relieving pain and distress could work. These means include "pharmacological, surgical, psychological, [and] social" resources. Thus, the doctors must consult other relevant experts and persons – including the family – before acceding to the patient's wishes. These stipulations seem to make the doctor and the clinical team the decision makers because they judge the relative strengths of all arguments in any particular case. An interesting further point is the use of the words "assisting death" rather than "euthanasia", "assisted suicide", "homicide upon request" or "killing". "Euthanasia" is rejected because different people attach different meanings to it. (One might think that the problem could be overcome by their saying what they meant by it, rather than abandoning its use). "Killing" is rejected because it fails to imply "a gentle act of merciful clinical care". If the aim is to justify "assisted death", then "killing" certainly does seem the wrong word. Another conclusion of the paper is that, in some cases, there is no morally significant difference between killing and letting die.

The Voluntary Euthanasia Society continued to argue their point, and produced a proposed advance directive, essentially a Living Will, though one not yet having any legal standing in the United Kingdom [42]. The directive suggests that a person has the right to revoke the advanced directive at any time. What is unclear is whether a person can revoke the directive at a time when his or her mental condition is not such as to allow him or her to have made such a directive. At such a time, should they be reminded of their previous decision? Moreover, the directive makes it clear that no civil liability is to be accorded to those carrying out the advanced directive. However, the question of criminal liability remains. As yet, it is not known whether such liability would arise. The preamble to the directive notes that little has changed since the 1988 report produced by Age Concern and The Center of Medical Law and Ethics of King's College, London [2]. However, whether this moral *status quo* should be regarded as deplorable or praiseworthy, remains unclear.

XII. THE DEFINITION OF DEATH

There have been no recent significant governmental pronouncements on the definition of death in the United Kingdom. In England and Wales, the law continues to view one as dead when a qualified and competent doctor certifies one to be; so no particular interpretation of the definition of or criteria for human death are implied in the law as it stands.

By implication, however, the law generally upholds the application of the criteria for "brainstem death" in appropriate cases; thus it is not unlawful for a qualified doctor to declare as dead someone who fulfills the criteria for "brainstem death". Moreover, by implication, the Government's pronouncements on the procurement and supply of organs uphold the concept and application of "brainstem death", because the practice of organ transplantation relies on such a definition of death in the great majority of cases. The Government in effect advocates and encourages the dissemination of such a definition when it identifies "misunderstandings" among the public concerning the circumstances under which organs are obtained, as a barrier to organ donation.

Perhaps the most striking affirmation of the definition of "brainstem death" is to be found in the policy of "elective ventilation". This policy has been largely pioneered in Exeter [11], though its adoption is being actively considered elsewhere, *e.g.*, at the University Hospital of Wales, Cardiff. The policy involves identifying patients who are about to die on general medical wards, but who could in principle be used as organ donors if maintained in a suitable condition. To accomplish this, they are transferred to Intensive Care and ventilated prior to the declaration of death. They are then declared "brainstem dead" before being eviscerated. Since this practice has not been deemed to be unlawful, and since the Government has been silent on it, the practice may be assumed to enjoy at least the tacit approval of both the law and the Government. Until these tacit approvals are challenged they seem unlikely to be withdrawn. It remains to be seen whether any such challenge will be forthcoming from the ethical debate of the present definition of death and its associated practices.

XIII. ORGAN DONATION, SALE AND TRANSPLANTATION

The past two years have seen a significant heightening of interest in transplantation matters among the British public, arising primarily from the much-reported activities of three British doctors allegedly involved in the use of commercially-traded human kidneys. The case was so notorious, and the public concern and hostility so intense, that the Government very rapidly passed legislation against the transplantation of commercially-obtained human organs. Such legislation had been considered four years earlier at the time of a previous scandal, but had then been thought inappropriate and had thus lain dormant. The new law, with its surrounding circumstances, has dominated organ transplantation law and ethics in recent years; accordingly most of this section will be devoted to considering it.

Although the circumstances of the most recent case doubtless gave grounds for great disquiet, much of the outcry was unreflective, stirred up as it largely was by sensationalized newspaper headlines. The condemnation of the buying and selling of human organs as "abhorrent" or "repugnant" was virtually universal. The wrongness of such trading was taken as axiomatic, but little attention was paid to whether such trading was wrong in itself, or wrong because in practice it preyed upon the most vulnerable, *viz.*, those in poverty sufficiently to prompt choices most of us would not even contemplate. The superficial character of the resulting debate extended even to Parliament, and, following some precipitous remarks by Ministers, the Government very swiftly brought to Parliament the Human Organ Transplantation Bill, which received the Royal Assent on 27th July 1989, a mere six months after the scandal which inspired it was first reported.

The haste with which the legislation was assembled and brought is reflected in features of the Bill (now the Human Organ Transplant Act [HOTA]) which can be criticized. Essentially, the new law sought to accomplish two things: (1) to make it unlawful to use, obtain or deal in commercially-procured human organs; and (2) to restrict transplantation of organs from living donors to recipients genetically related to the donor. At first sight, these two aims appear unconnected; indeed the law's difficulties stem from the weakness of the connection on which it relies. It is assumed that if commercial transactions in human organs are more likely to occur between donors and recipients who have no genetic or family connection, then restricting transplantation of live-donor organs to exchange within families will reduce or remove the risk that commercial exchanges will be attempted. Since genetic relationships can be established scientifically, such a restriction may be enforced straightforwardly. On this basis, the new legislation would both outlaw the commercial use of organs, and put in place a practical mechanical inhibition to such use.

However, the connection between the law's two aims should be regarded with great skepticism. It takes an unrealistic view of the nature of familial and genetic relationships by making several assumptions: (1) that those who are unrelated genetically to the organ recipient are especially vulnerable to commercial exploitation by those acting to procure the organ for such a recipient; (2) that those who are genetically related are not vulnerable to such exploitation; and (3) that financial exploitation is the only, or the most pernicious, kind of exploitation. Each assumption seems highly questionable. The emotional pressure that can be brought to bear upon a reluctant juvenile to donate a kidney to a sibling is only the most obvious of many kinds of exploitation that have nothing to do with commerce, but everything to do with vulnerability.

The key to preventing exploitation – commercial or otherwise – lies in establishing the genuine voluntariness of the donor's willingness to donate. This is, of course, principally a matter of the information available to and understood by him, and of his free and intentional choice to act as he does. By contrast, his voluntariness has no logical connection with his genetic relationship to the recipient, although it is true that family ties may in any individual case be decisively important. But such ties simply provide the donor with important reasons for acting in the way he does. Thus, establishing the donor's voluntariness is a matter of asking the right kind of question about his reasons for donating, rather than a matter of identifying his genes.

Paradoxically, the HOTA appears to recognize this by admitting the possibility of exceptions to the strict rule about genetic relatives. Such exceptional candidates must satisfy a statutory body, the Unrelated Live Transplant Regulatory Authority (ULTRA), that their wish to donate an organ is informed, considered and voluntary. It is conspicuous that the new law envisages statutory ethical scrutiny of live organ donation, but only in a small minority of cases. ULTRA is charged with establishing that the donor's decision is free of coercion or inducement, including financial inducement. Since there are plainly other kinds of pressure, such as emotional pressure, and since emotional pressure seems *prima facie* to be at least as possible within a family setting as outside it, one can view the new law as asking all the right questions, but in only very few of the right cases. It is, of course, true that the normal clinical responsibilities for obtaining informed consent remain in place for those doctors dealing with related living donors. But the HOTA provides for no statutory ethical scrutiny of clinical practice regarding related living donors.

In December 1988, the Department of Health issued health circular HC (88) 63, which instructed Health Authorities to ensure that relevant hospitals have explicit procedures for identifying potential organ donors and notifying the appropriate transplant coordinator. These procedures were designed to identify potential organ donors among suitable patients. Suitable patients would be those in whom a diagnosis of brainstem death had been made, and who exhibited no medical contraindications to organ donorship. The procedures required that the appropriate transplant unit or coordinator be contacted, and that the consent of relatives be obtained.

In June 1990, the Secretary of State for Health told Parliament that a special Health Authority should be established to manage the United Kingdom Transplant Service [20]. On April 1, 1991, the new Authority assumed its role.

On March 28, 1991, Stephen Dorrell, junior Minister of Health, outlined to Parliament the Government's "educationalist" approach to increasing the supply of donor organs: the public must be encouraged to understand the benefits from transplantation. He explicitly ruled out implementation of the so-called "opting out" schemes of donor identification, on the grounds that no one has a right to anyone else's organs. He further referred to a misunderstanding among NHS patients, and society as a whole, of the circumstances in which organs can be taken from a dead donor, identifying this misunderstanding as an important barrier to donorship [22].

A survey in south and west Wales showed that only 26 out of a potential 188 suitable patients (in a survey of 9840 hospital deaths) were used as organ donors. Potential donors were identified as those with severe cerebrovascular accidents and in the age range of 50-69 years. The study noted that elective ventilation had doubled the organ donation rate in Exeter, the district [49].

A confidential audit of deaths in Intensive Care Units (ICUs) in England has been carried out since 1989. The first six months' report showed that of 5,803 patients dying in ICUs in that period, 497 were confirmed as brainstem dead with no medical contraindication to organ donation. The main reason for failure to procure organs from this group of patients was identified as the refusal of relatives in 30 per cent of cases.

XIV. OTHER ISSUES

A. *AIDS*

Because the increasing incidence and fear of AIDS has raised many important ethical issues, this section will be devoted to them. The known epidemiological facts about AIDS in the United Kingdom underscore its significance both to society and to the health care profession. The number of HIV-positive individual in England and Wales up to June 1989 was 9,065, with 2,258 cases of AIDS itself. In Scotland, the equivalent figures were 1668 and 102. The monthly rate of increase was 1.75 percent from June 1988 [17]. The cumulative number of reported AIDS cases at the end of 1990 was 4,098. 1276 new cases were reported in 1990, a 51 percent increase over 1988 [25].

The questions of anonymous screening for AIDS and confidentiality has already been considered in an earlier section, but there are many other issues raised by AIDS to be considered.

Education and training are significant preventive measures. The Council of Europe in its recent recommendations, urges that children should be taught about AIDS in their usual lessons, and that teachers should be given special training [44]. The London Medical Group, meeting in February 1989, recognized the need to change the behavior of potential victims by emphasizing education [31].

The Health Education Authority placed advertisements in 201 selected magazines and newspapers in 1989 [16]. A survey in the same year revealed that more than a third of school children had engaged in sexual intercourse before they entered the fifth form (at about 15 years of age); of these, only 35 percent had used condoms on all occasions. In response to this survey, the British Medical Association produced a game for children called "AIDS and You" [43].

Testing positive for HIV infection can have serious repercussions on individuals. A debate already under way in the pages of the *British Medical Journal* made clear that a positive HIV test would be regarded as a reason for not offering insurance. During that debate, a correspondent from the Friends Provident Life Office said that the most common reason for requiring an HIV test is that the extent of coverage would be substantial [47]. In these cases a request would be made in order to protect the funds of other policy holders. The customer would be informed, on request, if the test was negative. If positive, the customer's nominated doctor would be informed and left to deal with the matter. With regard to the effects of simply having been tested, the following conclusion was reached: "The Association of British Insurers has indicated that having an HIV test as part of a recognized screening program, such as screening before blood donation...will not adversely effect insurance policies provided that the test is negative" [25].

Two attempts were made during 1989 to add provisions to the Employment Bill which was then in Parliament: first, a clause protecting prospective employees from demands by employers for evidence of negative HIV status; and second, a clause preventing discrimination against HIV-positive employees. Both failed to be made statutes. The legal position in such cases is given in *AIDS: A Guide to the Law* [15]. While doctors should not normally disclose information to third parties, a general practitioner may be obliged legally to break confidentiality by telling a sexual partner that his or her mate is HIV-positive, where the partner is also his patient.

With regard to Special Housing Policy and AIDS, a number of guidelines have been proposed. One commentator has noted that they "promise to shift the balance of the special housing debate away from its implicit preoccupation with containment and control towards the more ambitious ideals of disease prevention and health promotion" – a direction she approved of [51].

With regard to prisons, it was announced that the Home Office would not supply condoms to prisoners [18]. The reasons given were that condoms do not offer significant protection against HIV infection from anal sex, and that issuing them may give that false impression, thus encouraging anal sex and increasing infection. However, a report in the following year by the Prison Reform Trust offered six criticisms of the Home Office policy concerning HIV and AIDS in prisons [46]. The criticisms focused on the following issues: (1) the failure to identify prisoners who were HIV-positive (a failure because confidentiality was not guaranteed, *i.e.*, those discovered to be infected were often segregated); (2) segregating high risk prisoners to try and force them to accept testing; (3) the failure to establish special care units (despite promises); (4) preventing research into high-risk behavior; (5) insufficient education of prisoners and prison staff; and (6) refusal to issue condoms. The report applauds the more enlightened attitudes of the Scottish prison system, which set up a multidisciplinary AIDS management group. However, the Scottish system does not issue condoms either.

With respect to AIDS and health care professionals, two issues are identified by the Council of Europe. One is the risk to patients from HIV-positive health workers. The other is the risk to health care professionals from HIV-positive patients, and what follows from the presence of that risk. The Department of Health issued guidelines on the latter in 1990 which gave advice on how health service workers could protect themselves. The aim was to minimize the risk, but the guidelines left unanswered questions of whether or not the risk justifies the withholding of treatment from known HIV-positive patients, or is a reason which justifies testing those about to undergo operations.

The Royal College of Surgeons of Edinburgh argued that those patients suspected of being HIV-positive should be tested for HIV before operations, with their consent whenever possible, but without consent required in emergencies. The Royal College recognized that ethical problems might arise, since such testing would be done for the benefit of staff rather than patients. Moreover, the judgment of which individuals are members of high risk groups could not be disinterested. It was therefore necessary to inform patients who did not consent to HIV tests that they would receive precisely the same care as if they had consented [14]. Two other issues were also raised. First, doubt was expressed about the possibility of judging whether or not someone was in a high risk group, and about the point of doing so, since the notion of high risk groups was being undermined by the spread of the infection into the general population. Second, the risk to surgeons was deemed more serious than it first appeared, since over a thirty year period it could be calculated to be as high as 26 percent.

Within the United Kingdom the introduction of the idea of an internal market, spelled out in *Working for Patients*, has significant consequences for HIV and AIDS [8]. In a commentary on these ramifications, a number of threats to the existing level of services were identified, including those to the funding of genitourinary clinics (GUC) and to their successful working [3]. Preventing the spread of AIDS depends on providing a free and open access to specialist services that guarantee confidentiality. But funding problems arise from such arrangements, because AIDS patients have long-term needs and relations with a particular health care professional, very different patterns from situations of accident or emergency. In the plans, however, both situations are funded on the same basis. On the one hand, genitourinary medicine is accepted by the government as a core service which will be supported, but in a Department of Health circular, "District Allocations", it is unclear how the extra costs of providing an open service will be met [8]. The underlying problem is that patients frequently refer themselves to well-known clinics which are outside their areas of residence, even when their district health authority has no contractual relation with such clinics. Moreover, those districts that do have contracts may refuse to pay for the current level of service. Further problems are likely to arise because as many as eight contracts may be needed to cover the present range of services. To overcome such problems, funding needs to be made uniform at the regional or national level.

In addition, a patient's anonymity maybe threatened if the patient's district of residence has to be revealed, which may dissuade patients from early presentation. Moreover, the fact that many patients move to be nearer a particular medical center, and that some are homeless, adds to the complexities.

Still another problem is posed by "outreach" services, *e.g.*, those provided by the Terence Higgins Trust, a charitable organization. Because such bodies encourage the use of services and high expenditures, tightly budgeted districts are likely to review their relationships with outreach groups.

A more general criticism of funding proposals for AIDS and HIV infection is that they lack long-term planning. Without such planning, large inequalities in spending will remain, especially, in the amount spent per patient by different regional health authorities (varying between £16,000 and £80,000 per annum). For example, the Lothian Health Board, which deals with 60 percent of cases in Scotland, received only 15 percent of the total funding (1988-1989 figures). In addition, short-term planning has thus far taken account of the likely growth in numbers of AIDS victims.

The government tried to reassure those worried on these various counts, when Baroness Blatch spoke on two specific points. In reference to funding, she stated that "[this]...will have to take into account...self referral services for all comers, and not just for residents". With respect to confidentiality, she said, "treatment will begin and end in one local authority. There will not be a cross boundary flow of information back to the home authority of the patient who is receiving services".

Centre for the Study of Philosophy and Health Care
University College Swansea
Singleton Park, Swansea
WEST GLAM

BIBLIOGRAPHY

Articles and Books

1. Annas, G.: 1986, "Pregnant Women as Fetal Containers", *Hastings Center Report* 16, 13.
2. Age Concern/King's College London, Center of Medical Law and Ethics.: 1988, *The Living Will*, Edward Arnold, London.
3. Bentley, C. *et al.*: 1990, "Choice Cuts for Patients with AIDS", British Medical Journal 301, 501-502.
4. "Capable of being born alive", 1990, *Bulletin of Medical Ethics* 56, 5.
5. Center for Health Economics, University of York.: 1990., *Discussion Paper 70 - Government Funding of HIV-AIDS Medical and Social Care.*
6. Center for Health Economics, University of York.: 1991, *Discussion Paper 85 - An Empirical Study of Equity in the Finance and Delivery of Health Care in Britain.*
7. "Comment on Extra-Contractual Referrals", 1991, *Bulletin of Medical Ethics* 68, 6-7.
8. Department of Health: 1990, *Working for Patients: Contracts for Health Services: Operating Contracts*, HMSO, London.
9. Dunea, G.: 1990, "Letter from Chicago: Slippery slopes", *British Medical Journal* 301, 1094-1095.
10. Dyer, C.: 1990, "SHO Has Murder Charge Dropped", *British Medical Journal* 300, 768-769.
11. Feest, T.G. *et al.*: 1990, "Protocol for Increasing Organ Donation After Cerebrovascular Deaths in a District General Hospital". *Lancet*, 335, 1133-1135.
12. Field, M.A.: 1989, "Controlling the Woman to Protect the Fetus", *Law, Medicine and Health Care* 17, 114-129.
13. "Figures from PHLS Communicable Disease Surveillance Center, and Communicable Disease (Scotland) Unit": 1991, *British Medical Journal* 302, 197.
14. Gazzard, B.G. and Wastell, C.: 1990, "HIV and surgeons", *British Medical Journal* 301, 1003-1004.
15. Haigh, R and Harris, D.: 1989, *AIDS: A Guide to the Law*, Routledge, London.
16. *Hansard*, 10.1.1989, cols. 605-608.
17. *Hansard*, 12.7.1989, cols. 530-531 and 13.7.1989, cols. 604-605.
18. *Hansard*, 12.2.1990, col.7.
19. *Hansard*, 15.6.1990, cols. 1356-1357.
20. *Hansard*, 11.6.1990, col. 35.
21. *Hansard*, 9.11.1990, cols. 304-310.
22. *Hansard*, 28.3.1991, cols. 1139-1144.
23. Institute of Medical Ethics Working Party Report: 1990, "Assisting

Death", *Lancet* 336, 610-613.

24. Institute of Medical Ethics Working Party Report: 1990, "HIV Infection: The Ethics of Anonymised Testing and of Testing Pregnant Women", *Journal of Medical Ethics* 16, 173-178.
25. Kell, P.D. *et al.*: 1991, Correspondence: "Testing for HIV antibodies", *British Medical Journal* 302, 660-661.
26. Kennedy, L.: 1990, *Euthanasia: Counterblast 13*, Chatto and Windus, London.
27. King's Fund Institute.: 1989, *Efficiency in the NHS*.
28. Kolder, V. *et al.*: 1987, "Court-Ordered Obstetrical Interventions", *New England Journal of Medicine* 316, 1192.
29. Leiberman, J.R. *et al.*: 1979, "The fetal right to live", *Obstet. Gynecol.* 53, 515-517.
30. Light, D.W.: 1991, "Observations on the NHS Reforms: An American Perspective", *British Medical Journal* 303, 568-570.
31. London Medical Group Conference: 1989, "Review: AIDS, Sex and Death", *Bulletin of Medical Ethics* 48, 18-21.
32. Mansfield, S.: 1990, "Patient Power", *British Medical Journal* 301, 347.
33. Marsden, A.K.: 1989, "Basic Life Support Revised Recommendations of the Resuscitation Council (UK)", *British Medical Journal* 299, 442-445.
34. Meeker, W.: 1987, Correspondence "Protecting the Liberty of Pregnant Patients", *New England Journal of Medicine* 317, 1224.
35. Morgan, D.: 1989, "Letting Babies Die Legally", *Bulletin of Medical Ethics* 50, 14.
36. Morgan, D.: 1989, "Sterilization and Mental Incompetence", *Bulletin of Medical Ethics* 53, 20-21.
37. Morgan, D. and Lee, R.: 1991, *Blackstone's Guide to the Human Fertilization and Embryology Act 1990*, Blackstone Press, London.
38. Nathanson, V. and Macdonald, N.: 1991, "Contracts and Confidentiality", *British Medical Journal* 302, 1291.
39. News Report, 1989, *Bulletin of Medical Ethics* 50, 7.
40. News Report, 1989, *Bulletin of Medical Ethics* 53, 8.
41. News Report, 1990, *Bulletin of Medical Ethics* 55, 5.
42. News Report,, 1991, *Bulletin of Medical Ethics* 65, 4-5.
43. Notes, 1989, *British Medical Journal* 299, 1407.
44. Official Statements, 1990, *Bulletin of Medical Ethics* 58, 8-11.
45. Practice Note (Official Solicitor: Sterilization), 1989, *New Law Journal* 139, 1980.
46. Prison Reform Trust.: 1991, *HIV AIDS and Prison: Update*.
47. Robb, G.H.: 1990, Correspondence "AIDS Stigma in Insurance Market", *British Medical Journal* 300, 190.
48. Royal Society for Mentally Handicapped Children and Adults.: 1989, *Competency and Consent to Medical Treatment*. (A Working Party Report).

49. Salih, M.A.M.. *et al.*: 1991, "Potential Availability of Cadaver Organs for Transplantation", *British Medical Journal* 302, 1053-1055.
50. Slevin, M.L. *et al.*: 1990, "Attitudes to Chemotherapy: Comparing Views of Patients with Cancer with Those of Doctors, Nurses, and General Public", *British Medical Journal* 300, 1458-1460.
51. Smith, S.: 1990, "AIDS Housing and Health", *British Medical Journal* 300, 243-244.
52. Stewart, K. *et al.*: 1990, "Resuscitation Decisions in a General Hospital", *British Medical Journal* 300, 785.
53. Warden, J.: 1990, "Letter from Westminster: Abortion and Conscience", *British Medical Journal* 301, 1013.
54. Williams, B.: 1985, "Which Slopes are Slippery?", in Lockwood, M. (ed.): *Moral Dilemmas in Modern Medicine*, Oxford University Press, Oxford, pp.126-137.

Cases and Statutes

55. The Case of A-B, reported in the Bulletin of Medical Ethics 1989, 54, 3-4.
56. The Case of W, reported in the Bulletin of Medical Ethics 1989, 54, 3-4
57. *F v. West Berkshire Health Authority*, [1989], 2 All ER 545.
58. *Rance v. Mid Downs Health Authority*, [1991], 1 All ER 801.
59. *Re Baby J.*, Law Report, The Independent, 23.9.1990.
60. *Re C (A Minor)* (No.2), N L J Law Reports, 613 (1989).
61. *Re F (in Utero)*, [1988], 2 All ER 193

ANNE FAGOT-LARGEAULT

BIOETHICS IN FRANCE: 1989-1991

I. INTRODUCTION

In France, the awareness of problems in the field of bioethics intensified in the early 1980s. The National Consultative Ethics Committee for Life and Health Sciences (CCNE) was created in 1983 and became operational in 1984. The issues most discussed at that time were those raised by new modes of procreation: hired wombs, substituted or multiple parenthood, and the status of frozen embryos. During the period 1989-1991, discussion of those issues has been less volatile, largely as a result of an ongoing public debate. At the same time, other less spectacular issues have been raised, such as preservation of genetic resources within the biosphere and the legitimacy of cognitive research on human beings. In the meantime, the French Parliament has engaged in legislative action.

This paper is meant to be informative rather than argumentative. Each of the four sections is devoted to a major aspect of French bioethics discussion: (1) implementation of recent legislation on human experimentation; (2) the report to the Prime Minister, resulting in a set of recommendations for further legislative or regulatory action; (3) the reflections and recommendations of the CCNE during the period; and (4) the convening of the third medical ethics congress of the French Medical Association (*Ordre des Medecins*). A fifth section covers a variety of miscellaneous other items, and gives suggestions for further reading. For more information, one should contact the documentation center of the (CCNE).[1]

II. EXPERIMENTING ON HUMAN SUBJECTS: IMPLEMENTATION OF NEW LEGISLATION

Until 1988, biomedical research on human beings had no legal standing in France. Thus, casualties resulting from research procedures were often concealed or relabeled as "accidents of therapy", for fear of legal sanctions. Indeed, some lawyers had deemed any kind of human experimentation unconstitutional. In December 1988, a bill was passed by the French Parliament which legalized scientific research on human beings in the fields of biology and medicine, and stated the conditions under which it would be permitted. The bill entitled "The Protection of Persons Undergoing Biomedical Research"[39]

B. Andrew Lustig (Sr. Ed.), Bioethics Yearbook: Volume 2, 153-190.

was amended in January 1990 [39]. The Decrees of application appeared between September 1990 and May 1991 [18], as did Regulations by the Ministry of Health [4]. The regional committees charged to oversee research projects that were installed between July and December 1991, and it is expected that the new legislation will be fully enforced from 1992 onward.

The text of the law is divided into a preamble and five headings or parts (*titres*). The preamble (Art. L. 209-1) states principles and definitions. Part I (Art. L. 209-10) concerns "consent". Part II (Art. L. 209-11 through 13) sets forth "administrative provisions". Part III gives "specific provisions concerning research without direct benefit to the individual". Part IV (Art. L. 209-19 through L. 209-21) specifies "penalties". Part V (Art. L. 209-22 and 209-23) lists "sundries". Part VI (L. 564 through L. 605) concerns "pharmacy". Since English translation mentioned in the bibliography [39] is not authoritative, I shall quote it with possible minor corrections when needed.

The preamble is important because its first sentence officially recognizes the legitimacy of biomedical research on human beings: "Clinical trials or experiments organized and conducted in man with the objective of developing medical or biological knowledge are hereby authorized ..." (L.209-1). There has been discussion about what types of research are covered by the law. The original version of the text read "clinical trials, studies or experiments". Epidemiologists protested that observational studies should not be as stringently regulated as more aggressively invasive research. In response, the law was amended. Nonetheless, ambiguities persist. It is clear that behavioral research conducted by psychologists, and social science research, more generally are not included. It is fairly clear also that experimental research conducted outside the health care system (*e.g.*, in academic, military, leisure or athletic settings) is also included to the extent that the objectives of research have a relation with human health. In addition, Phase IV clinical trials are included, although they are genuine research, such as research on cosmetics. However, therapeutic innovation on an individual patient is excluded, as well as research conducted on the cadavers. The status of research on human embryos and fetuses remains unclear [96].

The preamble then distinguishes between two kinds of research: "Biomedical research, of which a direct benefit is expected for the person undergoing such research; and research involving either sick or healthy persons shall be described as without direct benefit to the individual" (L. 209-1). That echoes the older distinction between "therapeutic" and "non-therapeutic" research (Declaration of Helsinki, 1964). It should be noted that the law allows non-therapeutic research conducted on medical patients, which has provoked significant controversy and led to efforts to pass further legislation to prohibit such research.

Part I of the law states general conditions under which biomedical research may be conducted on human subjects. Research should be based "based on the latest scientific knowledge and on sufficient preclinical testing"; should involve no risk "out of proportion" with the expected benefit for subjects or with the interest of expected results; and should contribute to further scientific knowledge (L. 209-2). Concretely, the law stipulates the conditions under which the research should be conducted: (1) "under the direction and supervision" of an experienced physician; and (2) within an adequate technical setting and with proper scientific rigor and safety (L.209-3). Restrictions are placed on the recruitment of three categories of vulnerable subjects: (1) there should be no research "without direct benefit" on pregnant or nursing women unless it "carries no foreseeable risk" and is "of value to scientific knowledge on pregnancy or lactation"; (2) there should be no research on "persons deprived of their freedom by a judicial or administrative decision" unless "a direct and major benefit to their personal health is expected"; (3) there should be protection from unjustified non-therapeutic research for persons who lack or may lack autonomy such as children or hospital patients. The law elaborates this third condition as follows:

> [M]inors, adults under guardianship, persons in medical or social establishments and patients in emergency situations may be solicited for biomedical research only where a direct benefit to their health is expected. However, research which does not have direct benefit to the individual is permitted if the following three conditions are satisfied: (a) where such research presents no foreseeable risk to their health; (b) if it is of value to persons possessing the same age, illness or handicap characteristics; and (c) if it cannot be conducted otherwise (L. 209-6).

Article L. 209-7 concerns insurance coverage and responsibility. It states that the sponsor of the research is responsible "for the compensation of any harmful effects of the research" on subjects, "even when not at fault". Article L. 209-8 sets forth the general compensation principle: "biomedical research [should] not generate any direct or indirect financial gain for the persons undergoing it over and above the reimbursement of expenses incurred". Such a principle is to be understood in the light of the French doctrine of "non-profit" (see below, sections II and III).

Part II of the law states that there shall be no biomedical research on any human subject without the explicit consent of the subject (or of his parents or guardian if the subject is under guardianship). Consent should be given "in writing" (or, if that is impossible, should be witnessed by an independent party), after the subject has been fully informed. Provision is made, however, for the physician to withhold "information related to the diagnosis" in "exceptional cases where in the sick person's own interest, the diagnosis of his illness has not been revealed to him". However, the possibility that such information may be withheld must be specified in the protocol, in order to allow review by an ethics committee.

Here again, the contested case is that of research conducted on emergency patients, because the law allows initiation of research prior to informed consent, provided that the ethics committee has approved that possibility.

With regard to general provisions on subject's consent; the law reads as follows:

Prior to the carrying out of biomedical research on any person, free, informed and expressed consent must be obtained from such a person after the investigator or physician designated to represent him has informed this person of the following: (1) the research objective, methodology and duration; (2) the anticipated benefits, the limitations and risk associated, including in case of premature termination of the research; and (3) the opinion of the Committee mentioned in Article 209-12 ... (L. 209-9).

The exceptional provision for the case of emergency patients is expressed in this fashion:

In case of biomedical research to be carried out in emergency situations, preventing prior consent of the person to undergo such research, the protocol to be submitted to the Committee established by Article 209-11 of this code may provide that such consent need not be obtained and only the consent of the person's next of kin, if present, is to be solicited under the aforementioned conditions. The person concerned shall be informed as soon as possible and his consent requested for the possible continuation of the research (L. 209-9).

With regard to provisions for the consent of minors or adults under guardianship; the law states:

Where biomedical research is conducted in minors or adults under guardianship: (1) consent shall be given, according to the rules specified in Article L. 209-9 of this code, by persons exercising parental authority over minors. For minors or adults under guardianship, consent shall be given by the competent authority in case of research with direct benefit to the individual which does not pose a serious foreseeable risk, and, in other cases, by the guardian designated by the family council or the judge responsible for guardianship decisions; (2) consent of a minor or adult under guardianship shall also be requested where such a person is able to express his own will. Overruling his refusal or the withdrawal of his consent is prohibited (L. 209-10).

Part III of the law includes provisions for the review of research protocols and for the enforcement of reviewing procedures, Article L. 209-11 institutes review boards. These are not called ethics committees, but labeled "consultative committees for protection of persons in biomedical research" (*Comites Consultifs de Protection des Personnes dans la Recherche Biomedicale*; in short: CCPPRBs). Article L. 209-12 explains how protocols shall be reviewed by CCPPRBs, while Article L. 209-13 stipulates that medical inspectors and pharmacist inspectors from the Ministry of Health are in charge to enforce both legal dispositions and subordinate regulations.

There must be at least one CCPPRB per region (France has 26 regions, including overseas territories). The Decree Number 90-872 (Title 1, Articles R. 2001 through R. 2020) meticulously sets forth how committee members are to be chosen, how committees are to be organized, and how they are to function. Accompanying regulations allow for the creation of a total of 87 CCPPRBs, with 40 to be located in the Paris area. At the end of 1991, 56 committees have been established, of which 13 are located in the Paris area. Each committee has jurisdiction within its region (L. 209-11). Investigators are required to submit research projects to one of the CCPPRBs of their region. In case of multicentric trials, the protocol is submitted only once to a CCPPRB located in the region in which the coordinating investigator of the project has his/her activity (L. 209-12). Each committee is composed of twelve members, with twelve substitute members. Of the twelve, four (including at least three medical doctors) must be "qualified and experienced in matters of biomedical research"; one must be a general practitioner; two must be qualified in pharmacy (with at least one working within a hospital or clinic); one must be a nurse; and four should be lay persons, respectively qualified in ethics, social work, psychology, and law. Committee members are chosen in the following manner (Art. R. 2003). The government representative (*prefet*) in each region consults with professional authorities and associations, and establishes lists of qualified persons in each category. Committee members are replaced every three years. No one may be a member of more than one CCPPRB. Committee members elect their president and vice-president. If the president is chosen within the first category of four (above), the vice-president must belong to another category, and conversely (R. 2008). CCPPRBs must be empowered by statute, and be authorized by the Ministry of Health, before they are allowed to function. Their operating costs will be "financed by the revenues of a fixed duty paid by the sponsors for each biomedical research project submitted" (L. 209-11). The fixed duty is currently 9500 francs, and non-profit organizations get a reduced rate of 900 francs [4]. Duties are determined by the Ministry of Health.

Decree Number 9-872 (Title III, Art. R. 2029 through R. 2031) specifies what kind of information on the research project must be conveyed by investigators to CCPPRBs. Committees must examine proposals and answer the investigator within five weeks after a protocol has been submitted. In case additional information or substantial modifications are required from the investigator, the process is extended another four weeks (R. 2028). Deliberations are not public. Committee members must "maintain strict secrecy regarding the information to which they may have access as a result of their function" (L. 209-11). Moreover, members are not paid for that function (R. 2013). They are reimbursed for justified commuting expenses, and modestly compensated for reporting on a protocol. While they may seek advice from experts, CCPPRBs are merely consultative. They express an "opinion", which should be "unambiguously" favorable or unfavorable. When an opinion is unfavorable, investigators must wait two months before the experimental part of the research may be undertaken.

With regard to the criteria by which committees should assess projects, the Decree states:

The Committee delivers its opinion of the conditions necessary to ensure the validity of the research, notably the protection of the persons undergoing it, their information, how their consent is to be received, any sum to be paid to them, the general relevance of the project, and the appropriateness of the means used, in view of the objective sought, as well as the qualification of the investigator(s) (L. 209-12).

The Ministry of Health is to be informed of any negative opinion on a proposal. A period of two months is allowed to "suspend or prohibit" the implementation of the proposal (L. 209-12). A positive opinion expressed by the CCPPRB does not relieve the sponsor of the research of his responsibility (L. 209-12).

After a proposal has been reviewed, the sponsor informs the Ministry of Health that the experimental part of the research is about to start ("Letter of Intent", Title IV), and notifies the directors of the hospitals or other establishments in which the research is to be conducted. In a drug trial, the pharmacist of the establishment(s) is entrusted with the drug supply. Investigators must be medical doctors. They are responsible for the recruitment, information (written and oral), and the properly informed consent of subjects. If subjects are medical patients recruited for a trial deemed "directly beneficial", they must be made aware of therapeutic alternatives, be reassured that refusal to participate won't imply any therapeutic neglect, and be informed what will happen in the event that they drop out of the trial (L. 209-9 [44]. When subjects are healthy volunteers, or patients volunteering for a non-therapeutic trial, investigators must ensure that research candidates are covered by the National Health Insurance (*Securite Sociale*), and have them registered with the National File of Volunteers kept at the Ministry of Health. Volunteers must, of course, be informed of the existence of the data bank. Investigators are also responsible for writing out a report on the trial, for "any trial may yield a report" (R. 5127).

Part IV of the law is entirely devoted to specifying the conditions under which research "without a direct benefit" for subjects may be undertaken. Such research "should not carry any serious foreseeable risk for the health of persons undergoing it" (L. 209-14). Potential candidates are submitted to a preliminary medical examination. The requirement that subjects be covered by the national health insurance amounts to excluding the long-term unemployed or clandestine immigrants from recruitment and exploitation. Subjects may be compensated for their participation, unless they are "minors, adults under guardianship or persons in medical or social institutions" (L.209-15), in which case compensation is prohibited. Volunteers are not allowed to participate in more than one trial at a given time; moreover, after the trial is over there is an "exclusion period, during which they may not register in another trial" (L. 209-17). The total sum of compensation the subjects are allowed to receive during a given year is "limited to a maximum set by the Minister of Health" (L. 209-15) (currently 20,000 francs). The enforcement of such provisions obviously entails strict control. Hence, there is national filing of volunteers, for a period of one year from the time of enrollment in a trial. Decree Number 90-872 (Title VI, Art. R. 2039 through R. 2046) specifies how the data file is to function. The *Commission Nationale de l'Informatique et des Libertes* (CNIL) oversaw the development of the filing procedure, and approved it after several amendments. The same Decree (Title II, Art. R. 2021 through 2028) gives detailed instructions regarding the safety of research facilities and equipment. Facilities, especially devoted to research on healthy volunteers require authorization by the Ministry of Health.

Part IV of the law also established penal sanctions (imprisonment and/or fine) in four cases: violation of clauses pertaining to subjects' consent, violation of clauses pertaining to volunteers not being allowed to accumulate trials, research carried out without the protocol having been submitted to a CCPPRB, and trials pursued after having been suspended or prohibited by the Ministry of Health.

From 1989-1991, researchers have been made aware of the new legislation and regulations. Provisions concerning subject's consent to research were supposed to be implemented in 1990. However, because CCPPRBs have not been operational, the obligation to submit protocols for review has only been theoretical. Some research protocols have been reviewed by the technical section of the National Ethics Committee, others have been reviewed by Disciplinary Ethics Committees such as the ethics committee of the *Societe de Reanimation de Langue Francaise*. Several protocols were reviewed by the older institutional review board, which had spontaneously been created in the 1980s at some hospitals. Nonetheless, during this transitional period, clinical trials were not systematically registered anywhere. Thus, no statistics are available, and no firm predictions can be made regarding the number of protocols which CCPPRBs will be required to review. Although there have been scattered reports on the activity of the aforementioned committees, these imprecise estimates do not allow an analysis of the kind of clinical research currently carried out in France or an estimation of the number of research subjects involved.

A drastic change is expected as a result of the implementation of the new legislation. The French medical profession has had a strong tradition of paternalism and secrecy, including the protection of patients from any "disturbing" knowledge about their condition. The profession has also understood medical research to be in some sense "therapeutic". While in some specialties (such as cardiology or intensive care) there has apparently been little difficulty adjusting to the new situation, in other specialties (such as oncology), lobbies have been formed, claiming that obtaining patients' consent was impossible and/or that such a requirement would spell the end of clinical research. Pediatric oncology, for example, wished to be exempted from the obligation to comply with the rules. Various seminars, meetings, discussion sessions, were held within the profession. The *Societe Francaise d'Oncologie Pediatrique*[2] (SFOP) and a sub-group of it including psychiatrists (SFOP-PSY) met several times in 1990-1991 to study the law and to develop recommendations on how to comply with the new regulations in oncologic pediatrics. At the *Institut Gustave Roussy* near Paris, a group including nurses and trial statisticians discussed how to share information with cancer patients. Also, at the *Institut Curie* in Paris, the former institutional review board made the decision, since the job of reviewing protocols was to be handed over to CCPPRBs, to evolve into a hospital ethics committee, and to concentrate on problems of communication within the institution. In both cases the group's reflection resulted in publication of short education articles in the institution's internal journal[3], which is widely read by staff, patients, and patients' families.

The medical press also reported on local initiatives, raised a number of issues for further reflection, and clarified the making positions of various parties. A daily medical newspaper, *Le Quotidien du Medecin*, ran a series of columns on all aspects of therapeutic trials, with a view to educating as wide an audience as possible. The series was subsequently turned into a small and rather popular paperback [23]. A weekly medical journal devoted a special issue to the functions of institutional review boards in hospitals with the *Assistance Publique de Paris* [14]. Indeed, the third medical ethics congress held by the French Medical Association in March 1991, revealed that professional attitudes had considerably evolved in the last few years (see below, Section IV).

Patients' associations, on the other hand, criticized the 1988 law on three grounds. First, the law accommodated medical paternalism, by allowing doctors to judge what kind of crucial diagnostic information should be concealed from patients. Second, the law legitimized a kind of exploitation of vulnerable persons by permitting (even with restrictions) medical research "without direct-benefit" on categories such as children, adults under guardianship, emergency patients, the aged in retirement homes, and, in general, persons who are hospitalized. An association of psychiatric patients *(Association des Psychotiques Stabilises Autonomes*[4]) argued that such provisions were an insult to the dignity of patients, and that granting the status of human guinea pigs to incompetent patients was in total contradiction with the Nuremberg Code of 1947. The association was backed by a group "*Droit, Ethique et Psychiatrie*", which wrote an informative report on the status of mental patients in various European countries [40], and held a large forum in Paris on psychiatry and human rights in October 1991. Harsh objections also came from lawyers. They focused on the possibility left open by the law to start "cognitive" research in emergency situations without obtaining anyone's consent. Some suggested that the new legislation served the purpose of organizing human experimentation better then the purpose of protecting human rights ([1]; [43]; [54]).

The relative turmoil following adoption of the new legislation by Parliament did not last long, since human experimentation is hardly a popular subject. There were two large public workshops on practical questions organized by a group of pharmacologists who had actively lobbied in favor of legislative action. The interesting phase is yet to come. Hundreds of ordinary citizens have been selected to be members of CCPPRBs. Most of them (including health professionals) have little knowledge of the technicalities of human research, and most researchers have little experience of communicating with lay people. It will be interesting to see how that sort of democratic control over scientific activity will work.

III. REPORT TO THE PRIME MINISTER: A REVIEW OF THE SITUATION IN FRANCE, A DEFINITION OF THE FRENCH APPROACH TO BIOMEDICAL ETHICS, A SET OF PROPOSITIONS

In 1990, Michel Rocard, then Prime Minister, entrusted Noelle Lenoir with the mission of reporting to the government what further changes in legislation were necessary in connection with developments in the field of biomedical ethics. Mrs. Lenoir worked in collaboration with Bruno Sturiese. They consulted with a large number of people, in France and abroad, and they were able to draw from a think-tank of experts gathered by Francois Gros, head of the *Association Descartes*.[5] They handed their Report out to the next Prime Minister, Edith Cresson, in June 1991.

A. *The Report of the Council of State*

The "Lenoir Report" follows two other reports which discussed the legal implications of current trends in bioethics. The first of those was written by the Council of State, at the request of a former Prime Minister. Its working group was chaired by Guy Braibant [9].

The members of the Council of State focused on general principles which should govern ethics and law, and they drew concrete implications from principles to three areas of concern: therapeutic or other interventions, genetics and procreation, and institutions (which correspond to the three main sections of the report). The chief interest of the report, however, resides in it's effort to clarify the philosophical and moral principles underlying the French constitutional and legal tradition. The Report sets out "*une idee juridique de l'homme*" on the following grounds. First, body and spirit are one and indivisible with "relation of identity" between them; our rights are a corollary of our being "embodied". Second, the principle of immunity (*inviolabilite*) of the human body follows from the person's identity with his/her body. Thus, no one is allowed to act upon my body without my consent to the act, and I may only consent to those acts whose purpose is "legitimate". Third, my rights over my own body are limited by the principle of unavailability (*principle d'indisponsibilite du corps*). I may not make my body or any part of my body the object of an illegitimate contract, since my body is not a "thing", but myself. Therefore, I may not engage in actions which might reduce me to slavery. "On ne peut utiliser sa liberte pour se reduire a l'etat de chose at s'asservir" [9]. The Council of State is aware that such principles are currently threatened both by the claims of libertarians to do what they please with their bodies, and by the technical revolution in biomedicine which makes a potential trade in body parts profitable. But the Council firmly maintains traditional rules, putatively derived from the above principles, along with other republican principles such as solidarity. Some of these rules are: body parts may not be bought, sold, or traded for profit; body parts may be donated anonymously to the community; donation must be altruistic, allowing no financial compensation; and directed donations are generally objectionable, although there may be rare exceptions justified by therapeutic necessity.

The "Braibant Report" (named after its chairman) proposed an ambitious body of legislation covering all matters pertaining to "life sciences and human rights" under six headings (*titres*): (1) protection of the integrity of the human body, (2) prenatal diagnosis, (3) ethics committees, (4) organ transplantation, (5) medically assisted procreation, and (6) epidemiology [57]. The project recommended that a number of legal dispositions currently in use should be abrogated or modified in order to fashion a more coherent system. Because the Braibant Report was considered too ambitious and too controversial, it was never examined by Parliament. However, members of Parliament were aware of the need for some degree of legal reform. Hence, two Commissions common to the Senate and the Chamber of Deputies were established. One chaired by Senator Franck Serusclat, the other by Deputy Bernard Biculac.[6] Both Commissions have held public hearings, and a Parliamentary report is scheduled to be published in January 1992.

B. *The Lenoir Report*

The "Lenoir Report", has renounced the ambition to legislate on the whole of bioethics. It seeks to determine those specific areas where legislative action is now imperative, as well as areas where statutory regulation is preferable. The Report is divided into three chapters. Chapter one provides a national and international overview of "law and the practices of biomedicine from the beginning to the end of the human existence". It is subdivided in three sections: A. "Being Born"; B. "Living"; and C. "Dying". Chapter two is entitled "Ethics and Democracy". Chapter three presents the "Conclusions and Propositions" of the Commissioners [84].

In summarizing the contents of the first chapter, I shall set aside international comparisons, and focus on what characterizes the situation in France.

Section I. A. Artificial Insemination in France, especially AID, had been practiced in secrecy until the creation in 1973 by Georges David of the first sperm bank in Bicetre (*CECOS: Centre d'Etude et de Conservation du Sperme*). By maintaining a strict ethics of anonymous donation from fertile couples with children to infertile stable couples, the CECOS have helped to make AID more socially acceptable. By contrast, in vitro fertilization (IVF) was readily accepted in the society of France ever since the first test-tube baby was born in 1982. In recent years, from a total of 770,000 births per year, it is estimated that about

2,000 births resulted from AID and about 4500 from IVF. The accumulative totals are less precise. There may have been 25,000 persons conceived through AID since 1973 and 20,000 persons conceived through IVF since 1982.

In 1988 a modest regulation of IVF-centers by the Ministry of Health began. In principle, 76 centers and 81 laboratories and sperm banks have been licensed through 1991. In fact, however, centers have mushroomed and regulations have not been enforced. Surrogate motherhood has been discouraged. Thus far, however, it has not been prohibited by law and court decisions have been mixed [43]. In summary, artificial procreation techniques are socially well tolerated, currently covered by the national health insurance, with their legal and statutory status unclear. Research on human embryos, while not prohibited by law, and possibly financed by public funds, has operated within the framework of recommendations issued by the National Ethics Committee (see below, Section III). Prenatal diagnosis (*DPN*) is also well accepted and is reimbursed by the national health insurance for women who are medically at risk or more than 38 years old. It is regulated both by the Ministry of Health (15 private and 43 public centers have been authorized), and by professionals (*Association Francaise pour le Depistage et la Prevention des Handicaps de l'Enfant*). Preimplantation diagnosis has been criticized within society because of the dangers of sex selection [102] and has been suspended by the moratorium imposed by the CCNE from 1986 to 1989 (see below, Section III). Mrs. Lenoir, however, expressed a cautious rather than negative conclusion. She wrote that "sex selections, even though in principle debatable, are in our judgment ethically acceptable". Neonatal care has been a subject of concern and of ethical reflection among professionals, more than within society at large.

Section I. B. The Report reviews recent innovative medical research and therapies and discusses the ethical issues they pose.

I. B.1: National and international research programs on the human genome are listed and the Report suggests that an observation group should be established to oversee European developments (*Observatoire Europeen d'Ethique*). Genetic identification of individuals is seen as a potential threat for social peace and individual freedom. Following recommendations already issued by the National Consultative Ethics Committee for Health and Life Sciences (CCNE) (see below) and the High Council for population and family, the Report opposes free access to genetic fingerprinting techniques for ordinary citizens or the police, and concludes the use of such techniques should be ordered by a Court of Justice.

I. B.2: *Genetic Therapy*: Genetic therapy of hereditary monogenic diseases is deemed ethically acceptable if performed on somatic cells. The Report is guarded about germ-life therapy, although it does not call for the legal prohibition of such therapy, and describes as "absurd" a possible "right to inherit an unaltered genetic makeup". The Report recommends that, rather than hurrying to initiate legislation, the government should create a national consultative committee to supervise clinical practices in the field of human fertilization and embryology, and mentioned the British HUFEA as a model. Such a committee would scrutinize all experimental protocols of genetic therapy concerning the beginning of life.

Organ transplantation has been actively promoted in France. Currently, more than 52 transplantations per year and per million inhabitants take place in France, (as compared to 41 in Scandinavian countries and 39 in the United Kingdom). The removal of organs from cadavers for medical purposes (which accounts for 95 percent of the organs transplanted) is made easy in principle by a 1976 bill (*loi Caillavet*), according to which consent is presumed unless the person had made known his/her opposition to transplantation prior to death. In fact, however, medical teams tend not to remove organs unless the family has expressed clear consent. Hence, there is also a shortage of organs in France, in spite of an effective national organization for the procurement and distribution of available organs.

The Report states "Ethical Principles" for organ transplantation. The basic assumption is that, whether alive or dead, the organ donor deserves consideration. Therefore, one should be careful that the bodies of people, even after death, "not be collectively appropriated ... in the name of any sort of social solidarity". (Although such appropriation is implicit in the 1976 Caillavet bill, it is not stated directly.) Three principles to govern organ procurement and distribution are the following. First, the donor should have consented to the removal of his/her organs (although presumed consent, as in the French legislation, is compatible with European instructions). Second, organ donation is free. Donors should not be compensated; transactions should be non-profit; and commercial trafficking of organs should be severely punished. Although there have been no proven cases in France of organ sales, only occasional suspicions, the Report calls for new legislation to sanction all parties possibly involved in organ trafficking. Third, there should be equal access to organ transplantation for all people in need of it. Potential receivers should be selected solely on the basis of medical criteria, regardless of their socioeconomic or other status. Finally, Mrs. Lenoir emphasizes the need to solemnly promote "a new principle". Because medicine and biology tend to promote "an instrumental vision of the human body", the law should "set limits to a utilitarian conception of man" and protect people against any exploitation of their body parts; for "human identity is made of the whole body and soul, and having respect for the human body is having respect for the human dignity" [84]. At the end of this section, there is a discussion of developments in neuroscience, especially of current therapeutic advances in the realm of degenerative diseases. For example, the grafting of fetal cells into the brains of five patients with Parkinson's disease has just been authorized by the CCNE (see below, Section III). Again the Report warns against "biological reductionism", and against the temptation to use neurological treatments for the purpose of suppressing psychological deviations or "alleviating mere existential ill-being". Although there has been no psychosurgery in France for the last ten years, the possible recycling of fetal tissues to treat degenerative disease is viewed with unease [93].

I. B.3: Biomedical Experimentation: Biomedical experimentation on human beings is now under control in France, because of the 1988 legislation (see above, Section II). The "Lenoir Report" includes two amendments: first, to prohibit any non-therapeutic research in emergency situations; and second, to allow CCPPRBs to act as hospital ethics committees. The Report also suggests regulative adjustments so that CCPPRBs may be allowed to review protocols, and may refer them to the CCNE if necessary. The special case of epidemiological research is examined separately, because its legal status in France is notoriously unclear. On the one hand, research which is merely observational is exempted from the obligations stated in the 1988 bill; on the other hand, maintaining epidemiologic registers is incompatible with other French legislation, such as Article 378 of the penal code on medical confidentiality, and the 1978 bill on the protection of citizens against the recording and processing of sensitive personal information. Nonetheless, about fifty epidemiological registers presently function in France. Since 1986, they have been controlled by a national commission (*Comite National des Registres*). The Report urges the legislature to clarify the status of epidemiological research and permit it under specific conditions.

The Report then discusses the general topic of permissible uses of the human body, and asks whether individuals in a liberal society should be free to use their bodies as they please. Its answer is no. Individuals, especially the weakest, should be protected from the temptation to let themselves be exploited. Hence, the principles embedded in the French legal tradition: the human body is both immune and unavailable. My body is immune in that no one else may use it without my consenting to the use. It is unavailable in that there are uses which even my free and informed consent cannot possibly justify; I may only consent to those breaches which are made acceptable by "a legitimate interest and the absence of an excessive risk for my health". Such principles, it seems, apply to the body as a whole (that is, to myself) rather than to its parts. My body as a whole may not be alienated, even by me. What, then, about my body parts? Some laws and regulations in France imply that some body parts may be separated from the whole, and be alienated: milk, hair, and sperm can be bought and sold. Other legislation implies that body parts may not be traded: blood, kidneys, or any organs or tissues for transplantation. The Report urges the legislature to make legal doctrine more coherent, by stating that no part of the human body (even hair) is sheer "material", and that no part of the human body may be traded.

The authors of the Report are aware of the difficulties their position presents. They know that blood transfusion in Europe has entered the industrial era, which makes the French non-profit doctrine obsolete. They know that there have been experimenters from within the drug industry who warn that, as a vicious consequence of the 1988 French law requiring that human subjects of scientific research either be unpaid, or paid very little, it will be easier and cheaper to experiment on human beings than on animals. Nonetheless, the Report recommends that the French tradition of protecting the human body be systematized in the law.

I. *C: Definition of Death*: Defining death is left to the medical profession. In France, there is no legal definition of death. The average duration of life in France in 1987 was 72 years for men and 80.3 years for women, with an increase of 2.5 years within the last ten years. The Report calls for specific diagnostic and prognostic criteria of comatose states, in order to avoid improper use of intensive care units.

Twenty years ago, 30 percent of French people died in hospitals, compared to 70 percent today. Passive euthanasia is tolerated, while active euthanasia remains prohibited. There is a moderately strong movement in favor of the right to die in dignity, supported by the popular figure of L. Schwartzenberg, a medical doctor who publicly declared having helped some of his patients die, and who was sanctioned in 1990 for those actions by the French Medical Association. Living wills have no legal authority at the present time. If a patient has written a living will, doctors have no obligation to follow his/her instructions, and they may even be prosecuted if they deny a patient proper medical care even at his/her request.

A private bill presented by Senator Caillavet on the right to refuse treatment in case of terminal illness was considered and rejected by the Senate in 1980. In April 1991, Doctor Schwartzenberg presented to the European Parliament a proposition denounced as intolerable by virtually all professional circles in medicine and law, and by the CCNE. The "Lenoir Report" excludes the possibility of legalizing euthanasia, because of the "devastating effects" such a measure would have on social solidarity. It calls for better dialogue between doctors and terminal patients. It recommends that patients be granted the right to make a living will known (but with no guarantee that the directive would be systematically followed), and that the obligation for doctors to refrain from unreasonable therapeutic interventions be included in the Code of Medical Ethics. With regard to terminal care, the report merely observes that such care is inadequate in France, as in most European countries except the United Kingdom. An aggravating factor in France has been stubborn resistance to the use of pain killers. The consumption of morphine in France is 50 times lower than in Denmark, 22 times lower than in Canada, 17 times lower than in the United Kingdom, and 10 times lower than in the United States. However, changes have begun, due to the pioneering work of a few individuals such as Dr. Renee Sebag-Lanoe. The Ministry of Health issued recommendations in 1986 and encouraged the creation of the first terminal care unit in 1987 at the *Hopital de la Cite Universitaire* in Paris. In 1990, three medical schools for the first time offered optional qualification in palliative care.

The second chapter of the Lenoir Report, much shorter than the first, is entitled "Ethics and Democracy". It offers a brief survey of the industrial, social, and economic aspects of biomedicine, and of the discrepancy between sophisticated ethical questions raised in developed countries and the very basic sorts of issues facing Third World countries.

II. 1. In France, the right to health is "a constitutional right guaranteed by national health insurance". The drug industry is placed under stringent constraints, both ethical and financial, which results in insufficient money being invested in pharmaceutical research. The patenting of inventions has been the subject of extensive debate. There has been a tendency to extend the length of patent protections; to that effect, a bill was passed in France in 1990, comparable to the 1983 Patent Term Restoration Act in the United States. There has also been much debate around patentability of living organisms [22]. The Report proposes that patenting of the human body or body parts be strictly prohibited, and that greater attention be paid to the potential risks involved in merchandising of genetic information. With 10 percent of its budget devoted to health care, France ranks second behind the United States while the French people pay 20 percent of the expense themselves, compared with 12 percent for the Danes or the British. Because French health care is expensive, the Report, rather timidly suggests methods for curbing the expenses, restricting prescription, disciplining physicians, and educating citizens. Even these modest recommendations are politically sensitive, because physicians have been adamant to retain their freedom to prescribe, and citizens in France take great pride in their putatively excellent health care system.

II. 2. The legal creation of CCPPRBs has led to an imbalance among educational institutions. The Report urges that the CCNE be confirmed by law (thus far, it is only a statutory body created by presidential decree), and that the CCPPRBs be declared competent to oversee not only research, but also other areas. With regard to education in bioethics, the Report recommends that bioethics be taught in all public schools (*lycees*), but opposes, as does the CCNE, teaching ethics as a specific subject. There is strong opposition in France to the training of "ethicists" and/or to the possible teaching of an "official" ethics. Ethics is deemed a concern for all citizens, not a field reserved for specialists. As a consequence, ethics is supposed to be taught in medical schools by medical doctors, in law schools by lawyers, in biology by biologists, and so forth. The Report calls for better information from the media, and for public hearings within research institutions and in Parliament.

II. 3. The first imperative of a "world ethics" should be to reduce disparities between developed countries and Third World countries. In the field of health, an ethical analysis of North-South relations would support greater research on diseases affecting Third World populations, and better supervision (under the aegis of the United Nations and the World Health Organization) of the distribution of drugs to poor countries. The Report advocates the creation of an international committee to oversee western biomedical practices in developing countries.

III. Freedom of Research: The third chapter of the Report concludes that there is little evidence of the need for increased legislation, and that "freedom of research is a benefit of civilization which deserves preservation" [35]. Different countries make different judgments on what should be resolved by law and what may be resolved more flexibly through institutional regulations, professional rules of behavior, discussion within ethics committees, and free agreements between parties. The authors of the Report recommend that the legislature should embrace three general principles to guide decision and action in the realm of biomedicine: (1) respect for the human body, implying its non-commercialization; (2) free and informed consent of individuals to any biomedical intervention; and (3) protection of human genetic heritage. They then urge the legislature to adopt new legislation on three specific matters: (1) to place a ban on, and allow the courts to pursue and sanction, any mercenary use of the human body, including the commercialization of surrogacy and organ transplantation, as well as a ban on the advertising of such practices; (2) to insure both restriction of access and quality control of genetic testing; and (3) to state the conditions under which epidemiologic research is permissible. The authors of the Report conclude that other problems are better handled by other avenues than the law. For example, they recommend that the public debate on artificial procreation continue. And although the restructuring of the CCNE may involve legislation, the balance between Hospital Ethics Committees, CCPPRBs, and the CCNE should be worked out empirically. The Code of Medical Ethics will be revised by the French Medical Association to include the obligation to give patients accurate information, the obligation to refrain from pointless therapeutic interventions, and the obligations to cooperate work in common research objectives. Finally, the authors conclude that a European convention on bioethics, a "European Observatory" or watch-post in ethics, and an international committee to regulate biomedical practices directed at Third World countries should be established.

The "Lenoir Report" was delivered to the government in June 1991. The government has announced that legislative texts are being prepared, which will be submitted to Parliament for discussion during the Spring session of 1992.

IV. NATIONAL CONSULTATIVE ETHICS COMMITTEE FOR LIFE AND HEALTH SCIENCES (CCNE)[7]: RECOMMENDATIONS AND PROBLEMS UNDER STUDY

During 1989-1991, the 35 members of the CCNE have met once a month in plenary session, with President Jean Bernard presiding. The "technical section" (subset of the Committee) has met independently to examine research protocols submitted to the committee. The functions of the technical section will be modified in the future, as a consequence of the Huriet & Seruscilat bill, which requires that research protocols be examined by regional committees (see above, section 1). Working groups, each led by one or two "rapporteurs", and often including experts who are not currently members of the committee, have been in charge of preparing the reports and propositions discussed in the plenary sessions. Since it may take some time, and a significant number of interim drafts before the CCNE reaches a formal consensus, recommendations issued within the period are likely to reflect previous work, and some questions currently under scrutiny, such as that of relations between the CCNE and the newly created CCPPRBs, may not be settled. Eleven reports and recommendations on various questions were published by the CCNE during the period ([10];[26]-[37]), plus one longer document on non-therapeutic, purely cognitive, research [38]. Their contents will be summarized here under seven headings: cognitive research, therapeutic research, reproductive technologies, research on human embryos, genetic diagnosis and/or therapy, euthanasia, and ownership of the human body.

The "Ethics and Knowledge" document is a sequel to the 1984 report and recommendations of the CCNE on clinical trials [38]. The first report only discussed medical research on patients aimed at developing new therapies. The second report discusses biological research on human beings in general aimed at acquiring new knowledge. It investigates the philosophical groundings of such research, its various domains, its funding mechanisms, and its various settings (sports, military, etc). The Report proposes that human beings subjected to investigational procedures be considered "research partners". It sets ethical principles and limits: "While it is of itself a good thing to try and know more about man, scientific knowledge may not be acquired at the price of justice, of safety, or of the autonomy of persons" ([38], p.74).

With regard to the experimental therapy of Parkinson's disease, the CCNE offered two conflicting conclusions (its advice is merely consultative). It first recommended that grafts of nerve cells, and especially autografts of adrenal medulla to the caudate nucleus not be undertaken at present in France, because of serious risks for patients [27]. It later evaluated positively an experimental protocol presented by European collaborative study [31]. The protocol involves a grafting of human fetal mesencephalic cells into the central nervous system of five patients who will be followed over several years. This protocol is currently being carried out. The CCNE justified having changed its mind because the second technique is deemed less hazardous for patients, and in light of encouraging results, reported in February 1990, by O. Lindvall and A. Bjorklund in Sweden. In conformity with earlier recommendations on the use of fetal tissues, the CCNE imposed strict conditions on the way that fetal cells could be collected from aborted fetuses.

Ever since the first French test-tube baby was born in 1982, in vitro fertilization (IVF) techniques have been widely accepted in France, and IVF centers have mushroomed in both public and private settings, despite the CCNE's call for oversight and quality control requirements. The CCNE has called attention to the risks involved in poorly controlled egg or gamete donation, as well as in the proliferation of sperm banks [30]. These risks include the spreading of hereditary or infectious diseases such as AIDS; the commercialization of new reproductive technologies, and the buying and selling of embryos or gametes. It urged legislative action to set limits on what can be done. The CCNE has also called attention to another adverse effect of new reproductive technology, *viz.*, the sharp rise in the number of multiple pregnancies, and as its corollary, the selective termination of some fetuses [36]. The CCNE recommended that women be fully informed of possible consequences, and that professionals be extremely cautious in the use of techniques involving the risk of multiple pregnancy. The Report concluded that selective termination is unjustified, except to alleviate, or as a matter of "therapeutic necessity". Since it is unclear if or how the 1975 abortion law applies to selective termination, legislative action will be needed to specify the conditions under which it is permitted.

In 1986, the CCNE had called for a three-year moratorium on human embryo research aimed at early determination of sex or diagnosis of genetic anomaly. French researchers appear to have complied with the moratorium. After the moratorium expired at the end of 1989, the CCNE reconsidered the problem and reiterated two conclusions: (1) basic research should as far as possible be conducted on animal embryos rather than on human embryos; and (2) with the current state of technology, there is no serious ethical reason for directing fertile couples at risk of transmitting genetic disease to IVF-programs providing services to sterile couples [29]. The Report states that this would involve an effort to discriminate genetically between embryos, and would be more traumatic for fertile couples than having recourse to available prenatal diagnosis followed, if necessary, by termination of pregnancy. The Report stresses once more the conditions under which research on human embryos should be carried out. "Spare" embryos may be used for research if donated freely and anonymously for research by a couple who have successfully achieved pregnancy or abandoned their efforts to do so. Investigations on embryos developing extracorporeally, even beyond the seventh day, are not altogether excluded, but protocols are to be submitted to the CCNE. Finally, "replacement of any embryo after it has been used for experimental research is forbidden" [29].

Human gene therapy is viewed favorably under two conditions: (1) research is aimed at identifying and correcting only monogenic anomalies resulting in medically serious disorders; and (2) therapy applies to somatic cells, not to the germline [33]. Genetic testing for individual, family or population studies of molecular diseases or traits must be conducted under strict conditions of confidentiality, and respect for the autonomy and privacy of patients [35]. Genetic testing in order to identify an individual (DNA fingerprinting) is viewed as potentially threatening for society and to peace of families, since it is estimated that perhaps 10 percent of children in France were sired by someone other than their social father. The technical possibility of identifying an individual or type from a sample of a few body cells endangers the basic freedom and rights of individuals, and since laboratory errors may have disastrous consequences on people's lives, the CCNE has recommended that techniques of DNA fingerprinting be reserved for a few licensed laboratories, and that the tests be performed only at the request of a Court of Justice. Moreover, no information on a person's genotype should be stored, transmitted or used, without the person's written consent [28].

In June 1991, the CCNE responded to the report and recommendations on the care of the dying presented to the European Parliament by the Commission on Environment, Public Health and Consumer Protection ("the so-called Schwartzenberg proposition"). While agreeing with the proposition on the need to develop units of terminal care, it strongly dissented from the suggestion that any form of euthanasia should be authorized by law [34].

In developing its reports and recommendations to various groups, the CCNE also tried to conduct a more fundamental reflection on its own principles, with a view to clarifying its own doctrine. In a secular pluralistic democracy, with a multiplicity of ethnic, political and religious groups, it is problematic to assume any common ethics beyond that embedded in the common constitution and the law. Members of the CCNE may not be representative of society at large, because they are chosen for the "expertise" or "interest" they putatively possess. Nonetheless, their range of opinions is wide, and membership varies since half of the Committee is renewed every other year.

The efforts of the CCNE culminated in a document published in 1987, entitled "Biomedical Research and Respect for Human Persons" [37]. That document, although published prior to the period being reviewed, remains a primary reference document for the CCNE itself. A recommendation issued in 1990, rehearses the steps by which the earlier doctrine was elaborated, applies the 1987 principles to the question of the status of the human body, and draws the following conclusions. First, my body is not my property, but myself. Therefore, I may not alienate it or profit from its sale or the sale of its parts. To do so would be an insult to human dignity; things have a price, while persons are endowed with dignity. Second, society has an obligation to protect people from alienating parts of themselves, and particularly to prevent the most vulnerable populations from being physically exploited. Therefore, selling one's organs or one's blood should be prohibited. Third, when products derived from the human body are marketed, their price legitimately reflects the amount of work invested in manufacturing them rather than the bodily substance itself [32]. The same principles also apply, according to the CCNE, to human genetic information.

Even as the above set of reports and recommendations were made public, working groups have been studying a variety of other issues including ethics and pediatrics, ethics and economics, environmental ethics, the status of animals in biomedical research, the human genome project, eugenics, neurosciences, and the teaching of ethics.

V. ISSUES IN PROFESSIONAL MEDICAL ETHICS

A. *The Ordre des Medecins (OM)*[8]

The French professional association of medical doctors (*OM*) was instituted by a governmental order on September 24, 1945. Since then, professional practice has been conditional on registration with the *OM* and payment of annual dues. The *OM* is charged to proposing to the government the Code of Medical Ethics (*Code de Deontologie Medicale*), and of enforcing it. The current version of the Code was adopted in 1979. Recently, a revision has been deemed necessary, in order to take account of developments such as the 1988 legislation on human research. The *OM*, however, had not been very popular during the 1970s, because of its corporalist and conservative stance. For example, it had tenaciously opposed the decriminalization of abortion which occurred in 1975. A minority of young medical practitioners had therefore engaged in action against mandatory annual payment of dues to the Order. The dismantling of the Order was a stated goal, part of the socialist program when socialists assumed power in 1981. Thus, the 1980s were a crucial test for the capacity of the *OM* to change.

The *OM* evolved, under the presidency of Dr. R. Villey and, during our period, of Dr. L. Rene. Dr. Rene organized an interdisciplinary medical ethics congress where virtually all questions of medical ethics were debated publicly. On the occasion of the Congress, it became clear that the *OM* was committed to change and would remain a vital group. Several difficult issues, however, remain to be addressed. For example, the respective areas of competence among ethics committees will have to be clarified (see above, Section II, Part III). At the local level, professional associations of medical doctors and of pharmacists, nominate candidates for CCPPRBs, but the links between professional committees and hospital ethics committees have not been formalized. In addition, a revision of the Medical Code of Ethics is still in preparation.

1. *The Third International Ethics Conference*

The Third International Ethics Conference was held on March 9-10, 1991. The opening session included a brief welcome by President Louis Rene, a lecture by Marceau Long (Vice-President of the Council of State) analyzing the evolution of medicine and the need for legislative changes, and a friendly address by the (socialist) Minister of Health, Claude Evin. There were 3,000 participants, eleven round tables, and a variety of free communications distributed in four extra sessions. The topics of round tables had been carefully prepared in a series of seven regional meetings of the *OM* held between

October 1989 and January 1991. Those meetings had raised the level of interest among professionals and others by highlighting the need for sustained debate on difficult issues. No revolutionary propositions emerged from the 1991 conference, because its role was mainly educational. But an important aspect of that educational process was that matters of medical ethics were discussed openly and on the basis of equality between professionals and other parties in the community.

The proceedings of the Third Ethics Conference are to be published. The round tables offered an informative overview of current ethical problems as seen by medical practitioners. What follows is a brief summary of the themes and conclusions of the round tables, numbered from 1 to 11 in the order of the program.

1. *Issues of Prenatal Diagnosis and Termination of Pregnancy, Ambiguities Surrounding the Diagnosis and Prognosis of Fetal Anomalies*. These were the subjects of a round table which attracted a large audience and significant press coverage. Its overall tone of prudent toleration is reflected by a subtitle of the medical daily *Le Panorama du Medecin* (Number 3362), "Beethoven: His Mother Had Tuberculosis, His Father Had Syphilis, He Lost Three Brothers, He Himself Was Deaf - And So What?" It was reported that, according to surveys, a large majority of medical practitioners (70 percent) favors offering parents the option of "therapeutic" termination of pregnancy when a monogenic disease such as Huntington's is diagnosed in the fetus. Participants also noted that optional prenatal diagnosis for genetic defects, and optional termination of pregnancy in cases of positive diagnosis, are well accepted by French society at large and are covered by the national health insurance, despite concerns some expressed about the dangers of eugenism ([10]; [16]; [67]; [101]).

2. *Therapeutic Trials and the Protection of Persons*. This round table was honored by the presence of Senator Huriet, one of the "fathers" of the 1988 bill (see above, Section I). The debate focused mainly on issues of consent. It was clear that professional mentalities had become less paternalistic during the past few years, and that the community generally deemed clinical research to be legitimate when properly controlled. Both reflect considerable change. Only ten years ago, medical experimentation on patients could not have been publicly discussed.

3. *Evaluation of Medical Practices and Containment of Health Costs*. This round table, although less popular, was vigilantly attended by hospital administrators, and featured a guest from the Netherlands, who reported on twelve years of experience with a medical audit program. French medicine is notoriously expensive. Tentative measures of control on the part of the national health insurance have been seen as bothersome, and professionals thus far have been reluctant to engage in voluntary cost containment. Such issues will have to be faced clearly in the near future [82].

4. *The Status of Medical Confidentiality*. Is medical confidentiality outdated,

at a time of national filing of research volunteers, epidemiological research, gamete donation, and screening for HIV seropositives? The unanimous answer was that it is not. Medical secrecy has always been highly valued by the profession, and there is a consensus in France favoring the strict protection of personal medical data, even if that is detrimental to public health. The exceptions currently being negotiated, and discussed during the conference, pertained to epidemiological research, in particular to the possibility of researchers having access to nominal information on causes of death (currently covered by strict medical secrecy, in order to protect the memory of the deceased and the peace of families).

5. *The Right to Health, and Equal Access to Health Services*. These issues were discussed in a round table on socio-economic constraints. Overall, equal access to health care is a reality in France, but differences in life expectancy among social categories demonstrate that equal access does not mean equal health. At least for a part, this is probably so because preventive medicine has been insufficiently developed. One of the questions to be faced in the future is that of possible payment of general practitioners for actions of medical prevention [20].

6. *The Implications of Genetic Testing*. The main objectives of this round table were information and education. Most French practioners have had little training in genetics; consequently, their ability to counsel patients or families on genetic problems is usually poor. The profession is aware of this general lack of training, because of inadequate continuing education and because genetics is not yet recognized as a medical specialty. Invited speakers from the Netherlands, Italy, and Canada, contributed to the panel, as did a member of the CCNE who had been influential in the CCNE's reflections on genetic testing (see above, Section III) ([80]; [129]).

7. *Terminal Care and Problems of Incurable Patients, Pain and Dying*. These issues were the focus of a widely attended round table, and received wide media attention. The Minister of Health, in his inaugural address, had declared that "the right to die in dignity is a fundamental right". An expert from the World Health Organization (WHO, Geneva) introduced the subject with a report on the quality of life of terminal cancer patients. Specific issues were then addressed, including the right of patients to information on prognosis, the transition from curative to palliative treatment, palliative care in patients' home, palliative care at the hospital. It was agreed that proper palliative care would obviate the problem of euthanasia. Thus, the main issue facing the profession was how to overcome its traditional resistance to the prescription and use of pain killers (see above, Section II: 1, C). Following the conference, the *OM* engaged in action to persuade practitioners that prescribing pain killers in doses adequate to suppress the pain, and not only mitigate it, is appropriate.

8. *Organ Transplantation and the Status of Body Parts*. These were the subject of a round table during which information was offered about the procurement and distribution of organs, especially by France-Transplant. The

"French" ethics of anonymous donation (see above, sections II and III), allocation based solely on medical diagnosis, and non-commercial use was reasserted.

9. *Medical Responsibility and Problems of Insurance*. This round table pointed to the increase over the last ten years of the number of malpractice actions brought by patients against medical practitioners. The number of actions is, however, still extremely low in France. According to the Mutual insurance company ("Sou Medical") which insures most physicians (around 110,000), especially in the private sector, the number of accidents or incidents reported by practitioners doubled within ten years (1,118 in 1980, and 2,233 in 1989), while the number of actions brought by patients or their families during the same period increased by 18 percent (800 in 1980, and 950 in 1989). French physicians tend to fear a supposed "American trend" leading patients to sue their doctors. Thus far, however, French doctors are very safe, because when a change is made, the onus of proof generally rests with the patient. At that point, the patient only has indirect access to his or her medical file, through a medical doctor of the patient's choice. With the 1988 bill on human experimentation, however, the onus of proof has begun to shift, so there may be a basis for current concern and worry [127].

10. *Artificial Procreation and Social Management of New Reproductive Technologies*. This popular round table considered the sorts of risks for which gamete donors should be screened, *e.g.*, infectious, genetic, etc. A case was reported from the city of Grenoble of a sperm donor with a 50 percent risk of transmitting deafness who was not rejected. There was also ample discussion at the conference of what individual choices society should tolerate. The Minister of Justice expressed his willingness to grant prisoners access to artificial procreation techniques if they so desired, while the Ministry of Health and the *OM* said that artificial procreation techniques were meant to solve problems of sterility, not problems of sexuality in prison (*Le Monde*, March 12, 1991). (see [22];[48];[56];[60];[117].

11. *The Management of Epidemics, Especially of AIDS*. The last round table discussed whether education or legal constraints were the appropriate instruments of policy. The central dilemma considered was whether testing for HIV seropositivity should sometimes be compulsory (*e.g.*, for pregnant women, or patients undergoing surgery), or only optional. In the latter case, patients must be informed of the option, which provides an opportunity to educate them. Medical doctors in France involved in the management of AIDS patients have concluded that systematic testing aimed at protecting the community may result in people at risk shying away from information for fear of being discriminated. There appears to be a slight majority of the public, however, in favor of "systematically offering the option". Medical practitioners are no exception. In the past, they complied with, but never strongly supported, systematic vaccination or mandatory reporting of cases of infectious disease. The current strategy of "systematically proposing" the option (for example, to

pregnant women), and having in each region at least one center where anyone can receive free anonymous testing for HIV seropositivity, is deemed appropriate. It is possible in the future that testing will be made compulsory in specific cases. However, on the whole health professionals tend to favor educational measures over compulsory regulations.[9]

In summary, the 1991 Congress of Medical Ethics was symbolic of changes occurring within the medical profession, and in the relations between medical professionals and the community at large. Its immediate outcome will be a revised version of the "Code de Deontologie Medicale", to be drafted up by the *OM* Commission of Ethics. The issues which emerged as crucial for the profession to face include the following: (a) containment of health costs; (b) legal changes in defining the nature and extent of medical responsibility; (c) legal changes to enforce the obligation to inform patients of potential consequences and risks of medical interventions; and (d) the necessity to articulate professional rules of ethics to determine individual or collective claims for non-therapeutic uses of new medical technologies (such as the recent claim of some prisoners for access to IVF programs).

VI. OTHER ISSUES

A. *Problems in Blood Transfusion*

At the end of the period under review, in the spring of 1991, France was shaken by the disclosure of a scandal involving the transfusion of contaminated blood. Between 1980 and 1985, several thousand persons received blood transfusions, including half of France's hemophiliac population, were contaminated by the AIDs virus. In 1985, officials in charge of the National Blood Transfusion Center (*CNTS*) were aware that their blood products were contaminated, but continued to circulate them for several months. They were motivated, apparently, by economic reasons, because the technology of testing for the virus, or of heating blood products to inactivate the virus, either was not ready (and would have to be imported from abroad) or was deemed too expensive. Until 1991, the facts were deliberately concealed and patients at risk were not warned that they might be seropositive. Even worse, no one appeared to be fully accountable, because the *CNTS* is a public non-profit agency with responsibility divided between politicians and health care professionals. The *CNTS* was therefore unable to compensate patients for prejudice.

Such a disaster should have induced further reflection on the possible links between a national monopolistic non-profit transfusion system and the apparent lack of professionalism or sense of responsibility in this case. Confronted with similar problems in several countries, the European community has instructed member countries that, from 1993 on, blood products should be manufactured under stringent conditions, including quality control and clear lines of accountability. This recommendation, deemed "industrial logic", has been rejected in France. Typical of the French reaction is a CCNE paper appearing in December 1991. This paper characterizes human blood or plasma as gifts of life between persons, not as "materials" for industrial processing or objects of financial transaction. Despite the recent tragedy, the paper assumes that the non-profit rule suffices to guarantee consumer safety. Clearly, further safeguards will be required.

Another aspect of the French system regarding blood transfusion, the so-called "anonymous solidarity", revealed its shortcomings when some officials feared that anonymous blood distributed by transfusion centers might be infected by AIDS (and other) viruses. As was pointed out above (Section II:1, B, 3), one is not allowed in France to receive blood from a known identified donor. From the time when it became clear that supplies released by the blood banks were not safe, surgeons encouraged autologous transfusions when surgery was planned. But in emergencies, one had recourse to the anonymous pool of blood, even if relatives or friends were willing to contribute their own. During 1990, a number of medical doctors claimed the right of parents to donate blood for their young children. The Minister of Health granted such a permission under certain conditions. But what was permitted with parents is not permitted in other cases. The general rule remains: blood banks are not allowed to store other than anonymous blood supplies. That results in society requiring citizens to assume risks in the name of egalitarian solidarity. The CCNE has not questioned the general principle behind such restrictive measures, despite approving the exception for blood transfusions from parents to their young children [97].

B. *Transsexualism*

On May 21, 1990, four verdicts of the Court of Cassation (final court of appeal), denied transsexuals the right to obtain an alteration of their legal sex (in these four cases, from female to male). In recent years, a few medical and surgical teams in France have accepted transsexuals into treatment, rather than sending applicants to Morocco for surgery, as was the case before 1978, at a time of strict adherence of courts to the genetic criterion of sex. The ethics of monitoring individual claims for sex alteration had been worked out between those French teams and the ethics commission of the *OM*. Some (not all) high courts (*tribunaux de grande instance*) agreed to change the sex listed on birth certificates after sex-change surgery. The jurisprudence of the courts of appeal, however, was diverse and conflicting. The Court of Cassation had placed very strict conditions on legal sex alteration, but did not make it impossible ([3];[55];[57];[58]). The Braibant Report included a significant Appendix on transsexualism, which concluded that specific legislation was unnecessary and that relying on the judgment at the court was a more reasonable option [9]. Medical professionals involved in treating transsexuals knew which courts were more flexible than others. These physicians judged recourse to the courts to be acceptable; while reasonably dissuasive, such recourse allowed a "true" transsexual to reconcile his/her legal status with his/her physiological and psychological status after irreversible surgery.

Suddenly, however, the Court of Cassation reversed the recent trend. The Court in its 1991 verdict, declared that: "Transsexualism, even when medically assessed, cannot be analyzed as a real change of sex, since the transsexual, even though losing certain characters of his/her original sex, has not thereby acquired the characters of the opposite sex". The Court did not question the definition of transsexualism given by medical professionals, but argued that a "conviction" to belong to the opposite sex does not amount to an actual change of sex. Thus, the Court asserted its obligation to "stand up for the rules governing social relations and the principles on which society is based". The Court's verdict, therefore, upheld the current state of legislation, although compatible with social tolerance of homosexual unions, does not permit two persons of the same sex to be officially married.

C. *Palliative Care*

A large congress on palliative care was held in Paris from October 17-19, 1990, under the aegis of the European community, and of various national and international associations ([7]). Practical aspects of the clinical management of pain and of terminal conditions were discussed. Other issues were also considered, including the following: problems in communication with patients and in support of families; the comparison of home care vs. institutional care; and the roles and training of nurses, young physicians, physio- and psycho-therapists, clergy, and volunteers. The French medical community learned a great deal from this international exchange. It has been said that in 1990, three medical schools in France began teaching palliative care, a sign that times are changing. Periodicals with local or national circulation have also contributed to educating the general public[10] (see also, [8];[25];[54];[70]).

D. *AIDS*

Until 1988, the management of the AIDS epidemic in France was somewhat disorderly. Three parties were often at odds, making coordination difficult. Research teams were one group (for example, that of Luc Montagnier at the Pasteur Institute), associations of gay patients were a second group with their culture and press, and medical practitioners formed a third group, that tended to underrate the seriousness of the situation. In 1988, a report on AIDS was written for the government by Claude Got [61]. Got is one of a group of five medical doctors and professors of medicine, who in the 1980s publicly proposed reassessing some medical problems in terms of public health and people's welfare. For example, they influenced the Ministry of Health, despite opposing lobbies and indifference from the medical profession, to propose and obtain from Parliament a partial ban on cigarette advertising.

Following the Got Report on AIDS, three structures were created: a National Council of AIDS, a National Agency for AIDS Prevention, and a National Agency for AIDS Research.[11] The latter called for research projects, and currently supports and finances substantial cooperative research in fundamental virology and in the clinical, psycho-social, and anthropological aspects of AIDS. For multicentric clinical trials involving large cohorts of patients, French physicians, now well organized, report the data in "real time" to a unit (SC 10) of INSERM in the city of Villejuif (south of Paris) through the MINITEL system. The ethics of securing confidentiality of that medical data was worked out between INSERM and the CNIL (*Commission Nationale de l'Informatique et des Libertes*). The first trial to be monitored was the Concorde trial (ANRS 002), a "randomized double blind placebo controlled study to determine the efficacy of zidovudine in reducing disease progression among patients with CDC group II and III HIV infection" (presymptomatic seropositives). This trial, organized by the MRC (in the United Kingdom) and INSERM (in France), recruited nearly a thousand patients in each country. In France, 35 centers are involved. The discipline of reporting data in real time facilitates data collection, thus faster detection of the efficacy or possible toxic effects of new treatments. Such a scheme is the modern heir to a structure created in the middle of the 19th century on the French territory, that of "Comites d'Hygiene" and of "Medecins des Epidemies", through which data on infectious diseases were gathered by local doctors and sent to Paris.

Potential vaccines which have already been tried on chimpanzees and rhesus monkeys with apparently encouraging results, have recently become available from the Pasteur-Merieux Institutes. However, at the present time, the problem of how to determine the cohorts of volunteers needed for human experimentation has not been resolved. (See also, [6];[44];[47];[50];[52];[63]; [90];[104];[118]).

VII. CONCLUSION

Only events of national scale have been reported on here. That overview, therefore, does not give a full sense of the growing interest among the populace for questions in biomedical ethics. Three new journals specializing in bioethics have been introduced in France since 1989: *Journal International de Bioethique*, *Bio-ethique*, and *Ethique: La Vie en Question*[12]. In addition, many periodicals in various disciplines have dedicated thematic issues, or parts of issues, to problems of bioethics: Journals of Law (such as *Droits*), Philosophy (such as *Revue de Metaphysique et de Morale)*, Sociology (such as *Cahiers Internationaux de Sociologie*), Theology (such as *Le Supplement*), Biology and Medicine (such as *Medecine/Sciences*), Pharmacy (such as *Lettre du Pharmacologue*), Hospital Management (such as *L'Hopital a Paris*), and of course medical journals, many of which have been cited in the preceding sections (such as the *Journal de*

Medecine Legale). Other journals for a more general audience have also amply contributed to the debate, such as *Agora* (covering topics on society and medicine), *Cahier de Droit et d'Ethique de la Sante* (covering topics on law and ethics of medicine), *Etudes* (covering religious, cultural, social and other topics), and *Esprit* (covering political, cultural and other topics).

Suggestions for further reading on specific themes have been given throughout the course of the paper. It remains to mention a variety of publications on bioethics in general, which have covered several different topics. There were essays by celebrities *e.g.*, [11], by politicians *e.g.*, [105], by prestigious scientists *e.g.*, [23], by theologians *e.g.*, [110], by philosophers *e.g.*, [125]. There were also special volumes, bringing together many authors, of series like "Autrement" [19], and special topical issues of popular journals ([13];[17];[76]).

Finally, a trend noticeable during our period was the broadening of problems in biomedical and environmental ethics and the protection of genetic heritage. The link between the two areas were clearly emphasized at the PGDH symposium held in Paris in October 1989 [102]. In addition, a solemn warning was issued by the Universal Movement for Scientific Responsibility about mankind being threatened by genetic technology already used on other species [98]. Recent issues of certain journals ([106];[109]) and some Reports (*e.g.*, [111]) indicate the likelihood of ongoing discussion about these issues, which will be covered in the next regional yearbook.

Université de Paris-X
Département de Philosophie
Nanterre, FRANCE

NOTES

1 Centre de Documentation et D'Information D'Ethique Pour Les Sciences de La Vie et De La Santé (CDIE), 101 rue de Tolbiac, 75654 Paris Cedex 13, France (Director: Paulette Dostatni), Telephone 33 (1) 44 23 60 92 or 93 or 94. One may also contact information centers for specialized areas: (a) For computer-assisted personalized bibliography: INSRM, Information Medicale Automatisée (IMA), Hôpital de Bicêtre, 94270 Le Kremlin-Bicêtre, France, telecopy 33 (1) 46 58 40, 57; (b) For subscription to a printed selective bibliographic list on "Ethics and Life Sciences" (periodical): INSERM, Service Commun Number 2 (SSM), 44 Chemin de Ronde, 78110 Le Vésinet, France; and (c) For photocopies or microfiches of articles when their reference is known: CNRS-INSIT, 2 Allée du Parc de Brabois, 54514 Vandoeuvre-les-Nancy Cedex, France, telephone 33 83 50 46 64, telecopy 33 83 50 46 66, telex INDIF 961 942 F.

2 Société Française d'Oncologie Pédiatrique, c/o Institut Gustave Roussy, Département de Pédiatrie (Pr Jean Lemerle), Rue Camille Desmoulins,

95805 Villejuif Cedex, France.

3 *IGR.INFO*, Institut Gustave Roussy, 94805 Villejuif Cedex. See the series of issues Numbers 7 (December, 1990), 8 (March, 1991), and 9 (June 1991), following an internal meeting held on March 1, 1990, giving an overview of ethical questions raised within the institution (7: 14-16), reflections on the treatment of pain (8: 4-7), on research protocols (8: 8-15), on terminal care (9: 8-12), on phase 1 trials (8: 13-15, and on the local ethics committee (9: 16-17). See also: *La lettre*, Bulletin de Liaison des Amis de l'Institut Curie (diffusion 320000 copies). *Comprendre et Agir*, Journal de l'Institut Curie, 26 rue d'Ulm, 75005 Paris.

4 Association des Psychotiques Stabilisés Autonomes, BP 603, 75826 Paris Cedex 17, France.

5 Association Descartes (President: Francois Gros), Ministére de la Recherche et de la Technologie, 1 Rue Descartes, 75231 Paris Cedex 05, France. Department, "Science and Technology of the Living" (chief: Gérard Huber), telephone 33 (1) 46 34 39 31, telecopy: 33 (1) 46 34 39 40.

6 Assemblée Nationale & Sénat, Paris. Office Parlementaire d'Evaluation des Choix Scientifiques et Technologiques: Sénateur Franck Sérusclat (telephone 33 (1) 42 34 20 43; telecopy 33 (1) 42 34 21 69). Mission Commune d'Information sur la Bioéthique: Député Bernard Bioulac.

7 The National Consultative Ethics Committee for Life and Health Sciences (CCNE) is located within the National Institute of Health (INSERM: Institut National de la Santé et de la Recherche Médicale), 101 Rue de Tolbiac, 75654 Paris Cedex 13. Its regular publications (since 1984) are an annual report e.g., *Rapport 1989*, publiched in 1990), and a quarterly letter of information (*Lettre d'Information du CCNE*). *Reports* (as other official reports, such as the CNIL's, the Lenoir Report, etc.), are available from: La Documenttion Française, 29-31 Quai Voltaire, 75340 Paris Cedex 07, telephone: 33 (1) 40 15 70 00, telex: 204826 DOCFRAN Paris. Subscriptions to the *Letter of Information*: La Documentation Française, 124 rue Henri Barbusse, 93308 Aubervilliers Cedex, tél: 33 (1) 40 15 70 00.

8 Ordre des Médecins, Conseil National (Président: Louis René), 60 bd Latour-Maubourg, 75340 Paris Cedex 07. Publication: *Bulletin de l'Ordre des Médecins*, monthly (ISSN 0030-4565). Subscriptions: SPPIF, BP 22, 41353 Vineuil, France. Two dittoed documents have been issued on the occasion of the 3rd international ethics congress. One ("Dossier Documentaire") includes the abstracts of all public interventions; the other ("Revue de Presse") is a newspaper review. The proceedings will be published.

9 Ordre des Chirurgiens Dentistes, Conseil National (Président: Eugéne Saint-Eve), 22 rue Emile-Menier, 75116 Paris. Publication: *Bulletin de l'Ordre National des Chirurgiens Dentistes*.

10 *Bulletin de la Société de Thanatologie*, edited by Société de Thanatologie, 17 rue Froment, 75011 Paris, France. *Jusgu'à la Mort Accompagner la Vie* (quarterly), 12 rue Montorge, 38000 Grenpble, France. *Vieillir Ensemble*

Accompagner (quarterly), edited by Dr. Renée Sebag-Lanoë, Hôpital Paul Brousse, 14 avenue Paul-Vaillant-Couturier, 94800 Villejuif, France.

11 Conseil National du Sida (President: Françoise Héritier-Augé), Agence Française de Lutte Contre le Sida (President: Claude Got), Agence Nationale de Recherche sur le Sida (ANRS) President: Jean-Paul Lévy), 66 bis avenue Jean-Moulin, 75014 Paris. Telephone 33 (1) 45 41 12 00. Telecopy: 33 (1) 45 41 14 37.

12 *Journal International de Bioéthique* (quarterly, bilingual: English and French). ISSN: 1145-0762. Postal address: Editions Alexandre Lacassagne, 162 Avenue Lacassagne, 69003 Lyon, France. Scientific director: M. Milani-Comparetti. General editor: Christian Byk.

Bio-Éthique. Réflexions et Informations sur les Sciences de la Vie et de la Santé (bimonthly). ISSN: 1158-0097. Postal address: B.P. Number 9, 78670 Villennes-sur-Seine, France. Telephone: 33 (1) 39 75 61 21. Editor-in-chief: Amar Khadir.

Ethique. La Vie en Question (quarterly). ISSN: pending. Postal address: SFRB, 30 Rue d'Auteuil, 75016 Paris, France. Editor-in-chief: Dominique Folscheid. Other periodicals of interest, besides those mentioned in the list of references, have been:

Génétique et Liberté (irregular), edited by Association Génétique et Liberté (GEL), 45 Rue d'Ulm, 75005 Paris.

Laënnec (bimonthly), edited by Centre Laënnec, 12 Rue d'Assas, 75006 Paris.

Médecine de l'Homme (six-monthly), edited by Centre catholique des médecins français, 5 Avenue de l'Observatoire, 75006 Paris, France.

Ouvertures, edited by Association Médico-Sociale Protestante, 4 Rue des Saules, 94410 Saint-Maurice, France.

Questions de Notre Temps, Questions de Tout Temps, edited by Centre Chrétien des Professions de Santé, 16 Rue Tiphaine, 75015 Paris, France.

BIBLIOGRAPHY

1. Abadie E.: 1990, "Ethique et Essais Cliniques", *L'Encéphale* 16 (4), 275-276.
2. Alméras J.-P.: 1989, "Le Transsexualisme", *Le Concours Médical* 111 (32), 2762-2763.
3. Alméras J.-P.: 1991, "Les Transsexuels et le Droit", *Le Concours Médical* 113 (7), 565.
4. Ambroselli C.: 1990, *Le Comité d'Éthique*, PUF, Paris.
5. Amnesty International, Commission de la Section Française & V. Marange: 1989, *Médecins Tortionnaires, Médecins Résistants. Les Professions de Santé Face aux Violations des Droits de l'Homme*, La Découverte, Paris.
6. Association Didier Seux, Santé Mentale et Sida: 1990, *La Vie au Temps*

du VIH, Editions Fondation Marcel Mérieux, Lyon.

7. Association Européenne de Soins Palliatifs: 1990, "First Congress: Synapsis and Abstracts of Communications", Paris (dittoed). Secretarial offices: Dr. M.H. Salamagne, Unité de Soins Palliatifs, Hôpital Paul Brousse, 14 Avenue Paul-Vaillant-Couturier, 94804 Villejuif, France.
8. "Autour du Suicide": 1990, *Agora, Ethique, Médecine, Sociéte* (Special issues, Nos. 14-15).
9. Avant-Projet de Loi sur les Sciences de la Vie et les Droits de l'Homme ("Rapport Braibant"): 1989, Services du Premier Ministre, 75007 Paris (dittoed).
10. Baulieu E.-E.: 1990, *Génération Pilule*, Editions Odile Jacob, Paris.
11. Bernard J.: 1990, *De la Biologie à l'Éthique*, Buchet-Chastel, Paris.
12. Besanceney J.-C.: 1991, *Initiation à la Bioéthique. Prendre Soin de la Vie*, Coll. "Infirmières d'Aujourd'Hui", Centurion, Paris.
13. "Bioéthique": 1991, *Pauvoirs* (special issue) 56.
14. Beauchamp, T, *et al.*: 1987, "Bioéthique", *Revue de Metaphysique et de Morale* 3, 291-421.
15. "Biologie, Personne et Droit": 1991, *Droits. Revue Française de Theorie Juridique* (thematic issue), 13, 1-122.
16. Bonnet C.: 1990, *Geste d'Amour. L'Accouchement Sous X*, Editions Odile Jacob, Paris.
17. Bouretz, P. *et al.*: 1989, "La Bioéthique en Panne?", *Esprit* 11, 49-123.
18. Brentano, F. *et al.*: 1990, "Questions d'Éthique", *Revue de Metaphysique et de Morale* 1, 3-129.
19. Brisset-Vigneau, F. (ed.): 1990, *Le Défi Bioéthique. La Médecine entre l'Espoir et la Crainte*, Autrement, Série Mutations Number 120, Paris.
20. Brucker G. & Fassin D. (eds.): 1989, *Santé Publique*, Ellipses, Paris.
21. Buxton, M. *et al.*: 1991, "Quantifier la Qualite, Mesurer les Soins? Défis et Dilemmes de l'Economie de Santé": 1991, *Projections. La Santé au Futur* (4), 2-148.
22. Byk C. ed.: 1989, *Procréation Artificielle: Où en Sont l'Éthique et le Droit? Une Analyse Internationale et Pluridisciplinaire*, Editions Alexandre Lacassagne, Lyon.
23. Changeux J.-P. and Connes A.: 1989, *Matière à Pensée*, Editions Odile Jacob, Paris.
24. Chapouthier G.: 1990, *Au Bon Vouloir de l'Homme, l'Animal*, Denoël, Paris.
25. Chardot, C., Rene, L., and Schaerer, R.: 1991, "Ethique et Concertation Pluridisciplinaire en Cancérologie Clinígue", *Bulletin du Cancer* 78, Supplement 1.
26. Comité Consultatif National d'Éthique: 1989 (October 16), "Avis sur le Dépistage des Toxicomanies dans l'Enterprise", in [39].
27. Comité Consultatif National d'Éthique: 1989 (October 16), "Avis sur les Greffes de Cellules Nerveuses dans le Traitement de la Maladie de

Parkinson", in [39].
28. Comité Consultatif National d'Éthique: 1989 (December 15), "Avis Relatif a la Diffusion des Techniques d'Identification par Analyse de l'ADN (Technique des Empreintes Génétiques)", in [39].
29. Comité Consultatif National d'Éthique: 1990 (July 18), "Avis sur les Recherches sur l'Embryon Soumises à Moratoire Depuis 1986 et Qui Visent à Permettre la Réalisation d'un Diagnostic Génétique Avant Transplanation", in [40].
30. Comité Consultatif National d'Éthique: 1990 (July 18), "Avis sur l'Organisation Actuelle de Don de Gamètes et Ses Conséquences", in [40].
31. Comité Consultatif National d'Éthique: 1990 (December 13), "Avis Concernant les Greffes Intracérébrales de Tissus Mésencéphaliques d'Embryons Humains Chez Cinq Malades Parkinsoniens dans un But d'Expérimentation Thérapeutique", in [40].
32. Comité Consultatif National d'Éthique: 1990 (December 13), "Avis sur la Non-Commercialisation du Corps Humanin", in [40].
33. Comité Consultatif National d'Éthique: 1990 (December 13), "Avis sur la Therapie Génique", in [40].
34. Comité Consultatif National d'Éthique: 1991 (June 24), "Avis Concernant la Proposition de Résolution sur l'Assistance aux Mourants, Adoptee le 25 Avril 1991 au Parlement Européen par la Commission de l'Environement, de la Santé Publique, et de la Protection des Consommateurs", in [41].
35. Comité Consultatif National d'Éthique: 1991 (June 24), "Avis sur l'Application des Tests Génétiques aux Études Individuelles, Études Familiales, et Études de Populations (Problemes des 'Banques d'ADN', des 'Banques de Cellules', et de l'Informatisation des Donnees)", in [41].
36. Comité Consultatif National d'Éthique: 1991 (June 24), "Avis sur les Réductions Embryonnaires et Foetales", in [41].
37. Comité Consultatif National d'Ethique (CCNE): 1987, *Recherche Biomédicale et Respect de la Personne Humaine*, La Documentation Française, 29 quai Voltaire, 75007 Paris.
38. Comité Consultatif National d'Ethique (CCNE): 1990, *Ethique et Connaissance. Une Réflexion sur l'Éthique de la Recherche Biomédicale*, La Documentation Française, Paris.
39. Commission Nationale de l'Informatique et des Libertés 1989: *Rapport 1989*, La Documentation Française, Paris.
40. Commission Nationale de l'Informatique et des Libertés: 1990, *Rapport 1990*, La Documentation Française, Paris.
41. Commission Nationale de l'Informatique et des Libertés: 1991, *Rapport 1991*, La Documentation Française, Paris.
42. Conseil d'Etat: 1988, *Sciences de la Vie. De l'Éthique au Droit*, La Documentation Française, Paris.

43. Cour de Cassation, Assemblée Plénière, 31 Mai 1991: "L'Illicéité de l'Adoption Plénière de l'Enfant d'une Mère Porteuse", Rapport et Note par Yves Chartier et Dominique Thouvenin, *Recueil Dalloz Sirey* 30 (September 19, 1991): 417-428.
44. D'Agostino M.: 1990, *Sida, Enfant, Famille: Les Implications de l'Infection à VIH Pour l'Enfant et la Famille*, Centre International de l'Enfance (CIE), Paris.
45. Doray B.: 1991, *Ethique et Psychiatrie. Document de Travail. La Formation Professionnelle*, Ministère de la Santé, MIRE, Paris.
46. "Droit et Humanite": 1989, *Actes* (special issues, Nos. 67-68.
47. "Droits de l'Homme, Éthique et Médecine": 1990, *Cahier du Droit et d'Ethique de la Santé* (special issue, No. 2.
48. Ducret R, *et al.*: 1990, "Procréations Humaines et Société Techno-Scientifique", *Le Supplément, Revue d'Ethique et de Théologie Morale* 174, 1-108.
49. Edelman B.: 1989, "Le Droit et le Vivant", *La Recherche* 20 (212), 966-976.
50. Enel P., Charrel J., Larher M.P., Reviron D., Manuel C., Sanmarco J.L.: 1991, "Ethical Problems Raised by Anti-HIV Vaccination", *European Journal of Epidemiology* 7 (2), 147-153.
51. Enel P., Manuel C., Charrel J., Larher M.P., Reviron D., Sanmarco J.L.: 1991, "Aids, a Social Dilemma: Detection of Seropositives", *European Journal of Epidemiology* 7 (2), 139-146.
52. Espinoza P. *et al.*: 1990, "Les Années Sida", *Revue Française des Affaires Sociales* (thematic issue) 10.
53. Eschwège E., Bouvenot G., Doyon F., Lacroux A. (eds.): 1990, *Essais Thérapeutiques, Mode d'Emploi*, Le Quotidien du Médecin & INSERM, Paris.
54. Fagot-Largeault A.: 1989, *Les Causes de la Mort. Histoire Naturelle et Facteurs de Risque*, Vrin, Paris.
55. Fauré G.M.: 1989, "Transsexualisme et Indisponibilité de l'État des Personnes", *Revue de Droit Sanitaire et Social* (January-March), 1-13.
56. Frydman R., Letur-Konirsch H., de Ziegler D., Bydlowski M., Raoul-Duval A., Selva J.: 1990, "A Protocol for Satisfying the Ethical Issues Raised by Oocyte Donation: The Free, Anonymous, and Fertile Donors", *Fertility and Sterility* 53 (4), 666-672.
57. Gobert M.: 1988, "Le Transsexualisme, Fin ou Commencement?", *La Semaine Juridique*, Edition G, Number 47, I, 43: 3361.
58. Gobert M.: 1990, "Le Transsexualisme ou de la Difficulté d'Exister", *La Semaine Juridique*, Edition G, Number 49, I, 45: 3475.
59. Gobert M.: 1991, "La Maternité de Substitution: Réflexions à Propos d'une Décision Rassurante", *Les Petites Affiches. La Loi* 127 (October 23): 4-25.
60. Gobert M.: 1991, "Procréations Médicalement Assistées: le Problème des

Sources du Droit", *Actes. Les Cahiers d'Action Juridique*, 77 (October): 30-36.

61. Got C.: 1988, *Rapport sur le Sida*, Paris, Flammarion.
62. Gouttard-Dojat V.: 1989, "Réflexions sur la Mort en Milieu Hospitalier à Base d'Entretiens avec le Personnel Soignant, Réalisés en Région Oarisienne, Master's Thesis, University of Paris-X, Sociology & Philosophy.
63. Grmek M.D.: 1990, *Histoire du Sida. Début et Origine d'une Pandémie Actuelle*, Nouvelle Édition, Payot, Paris.
64. Gros F., Huber G., Kahn F., Augé M., Briand P., Peschanski M.: 1990, "La Biologie, l'Homme et la Société', *Médecine/Sciences* 6 (2), 125-151.
65. Gros F.: 1989, *La Civilisation du Gène*, Hachette, Paris.
66. Gros F.: 1990, *L'Ingénierie du Vivant*, Editions Odile Jacob, Paris.
67. Haut Comité Médical de la Sécurité Sociale (HCMSS): 1990, *Rapport Concernant la Médecine Prédictive*, HCMSS Number R/90.466.1. AP 152, Paris.
68. Hirsch E.: 1990, *Médecine et Éthique. Le Devoir d'Humanité*, Editions du Cerf, Paris.
69. Huber G. & Legrand C.: 1991, *Présentation Thématique des Avis et Publications du Comité Consultatif National d'Ethique, de Mai 1984 à Juin 1991*, Association Descartes, Paris (dittoed).
70. Huguenard P.: 1991, "L'Accompagnement à la Mort", *Urgences Médicales*, 10 (3): 113-115.
71. Isambert F.-A.: 1989, "Ethics Committees in France", *The Journal of Medicine and Philosophy* 14 (4), 445-456.
72. Jacquard A. *et al.*: 1990, "Dossier 'Bio-Éthique': Raison et Déraison", 2, 7-60.
73. Jonas H. *et al.*: 1991, "Ecologie, Bioéthique, Démographié, Quelles Responsabilitiés?", *Esprit* 5, 5-79.
74. Klarsfeld A. & Granger B.: 1991, "Faut-Il Dépister les Prédispositions aux Troubles Psychiatriques?", *Médecine/Sciences* 7 (1), 58-61.
75. "La Bioéthique": 1990, *Le Bulletin* (special issue) 23.
76. "La Demande d'Ethique": 1990, *Cahiers Internationaux de Sociologie* (special issue), No. 88.
77. "La Loi du 20 Décembre 1988. Deux Aspects Sensibles: Les Phases IV, Le Volontaire Malade": 1990, *Lettre du Pharmacologue* (special issue 4 (3), Supplement.
78. "La Loi du 20 Décembre 1988. Statut et Responsabilités de l'Investigateur": *Lettre du Pharmacologue* (special issue) 5 (4), Supplement.
79. "L'Annonce de la Séropositivité": 1990, *Cahier de Droit et d'Ethique de la Santé* (speical issues, Nos. 3-4.
80. Laberge C.M. *et al.*: 1991, "La Génétique", 2 (1), 5-32.
81. Langlois A.: 1991, "L'Altruisme et le Donneur", *The Canadian Nurse* 87

(7): 40-43.

82. Lazar P.: 1989, *Les Explorateurs de la Santé. Voyage au Centre de la Recherche Médicale*, Editions Odile Jacob, Paris.
83. "L'Ethique au Quotidien": 1990-91, *Revue du Practicien. Médecine Générale* (special issue), excerpts from Vos. 95, 96, 97, 98, 99, 117, 118, 127.
84. Lenoir N. & Sturlese B.: 1991, *Aux Frontières de la Vie. Pour une Démarche Française en Matière d'Éthique Biomédicale*, La Documentation Française, Paris.
85. "Loi No. 88-1138 du 20 Dec 1988 Relative à la Protection des Personnes Qui se Prêtent à des Recherches Biomédicales": 1988, *Journal Officiel de la République Française JO* (December 22), 16032-16035; Modifiée: "Loi No. 90-86 du 23 Jan 1990, Art 35-49": 1990, *JO* (January 25), 1013. Complétant le *Code de la Santé Publique*, Livre II bis, Art L. 209-1 à L. 209-23. Published with English translation: Bulletin Officiel No. 90-4 bis, Direction des Journaux Officiels, 26 Rue Desaix, 75727 Paris Cedex 15. Published with statutory and regulative dispositions: Bulletin Officiel No. 91-12 bis and 91-13 bis, same address. Statutory and regulative dispositions: "Décret No. 90-872 du 27 Sep 1990 Portant Application de la Loi No. 88-1138 du 20 Dec 1988 Modifiée Relative à la Protection des Personnes qui se Prêtent à des Recherches Biomédicales et Modifiant le Code de la Santé Publique (Deuxième Partie: Décrets en Conseil d'Etat)": 1990, *JO* (September 29): 11862-11868. "Décret No. 91-440 du 14 Mai 1991 Définissant les Conditions de l'Assurance que les Promoteurs de Recherches Biomédicales Sont Tenus de Souscrire en Application de l'Article L. 209-7 du Code de la Santé Publique": 1991, *JO* (May 16): 6478. Arrêtés: December 27, 1990 (Droit Fixe Versé par les Promoteurs des Recherches Biomédicales), December 28, 1990 (Indemnité Maximale des Participants), February 14, 1991 (Déclaration d'Intention de Recherche), May 7, 1991 (Perception du Droit Fixe). Circulaires de la Direction de la Pharmacie et du Médicament: DPHM/01/08 No. 90-3 du October 1, 1990 (Mise en Place des Comités de Protection des Personnes), DPHM/01/08 No. 90-4 du October 24, 1990.
86. Louzoun C. *et al.*, (eds.): 1990, *Législations de Santé Mentale en Europe* (Angleterre, Ecosse, Espagne, France, Italie), Comité Européen "Droit, Ethique et Psychiatrie", 153 Rue de Charonne, 75013 Paris.
87. Lucas M.: 1991, *Transfusion Sanguine et SIDA en 1985. Chronologie des Faits et des Décisions pour ce qui Cnceme les Hmophiles*, Inspection Gnérale des Afaires Sciales, 14 Aenue Duquesne, 75350 Paris 07 RP (dittoed).
88. Lucas Ph.: 1990, *Dire l'Éthique*, Actes-Sud & INSERM, Paris.
89. Lycée International de Saint-Germain-en-Laye: 1989-90 & 1990-91, "Initiation à la Bioéthique", Rapport d'Activité du Groupe de Réflexion du Lycée.

90. Manuel C., Enel P., Charrel J., Reviron D., Larher M.P., Thirion X., Sanmarco J.L.: 1990, "The Ethical Approach to AIDS: A Bibliographical Review", *Journal of Medical Ethics* 16 (1), 14-27.
91. "Medicine, Droit, Éthique": 1989, *Journal de Médecine Legale - Droit Médical* (special issue) 32 (2).
92. "Médecine, Santé, Information": 1990, *Agora. Ethique, Médecine*, Société (special issue), No. 16.
93. "Medicament: dela Réalite au Mythe": 1989, *Agora, Ethigue, Médecine, Société* (special issues), Nos. 10-11.
94. Mémeteau G.: 1990, "De Quelques Droits sur l'Homme: Commentaire de la Loi du 20 Décembre 1988 Relative à la Protection des Personnes Qui se Prêtent à des Recherches Biomédicales", *Recueil Dalloz Sirey*, Cahier No. 25: 165-178.
95. Michaud J., Pierson M., Thouvenin D.: 1991, "Richesse et Complexité d'une Loi: Loi Huriet-Sérusclat", *Revue Française des Affaires Sociales* 91 (1): 13-24.
96. Ministère de La Solidarité, de La Santé et de La Protection Sociale: 1990, *Guide des Textes Relatifs à la Protection des Personnes Qui Se Prêtent à des Recherches Biomédicales*, Direction de la Pharmacie et du Médicament, Bureau des Affaires Juridiques, 1 Place de Fontenoy, 75350 Paris Cedex 07 SP, Telecopy 33 (1) 40 56 53 55 (dittoed).
97. Moulin A.-M.: 1991, *Le Dernier Langage de La Médecine. Histoire de l'Immunologie de Pasteur au Sida*, PUF, Paris.
98. Mouvement Universel de la Responsabilité Scientifique (MURS): 1989, *Science et Devenir de l'Homme. Alerte à l'Homme*, Cahier No. 18 (Automne 89).
99. Novaes S. (ed.): 1991, *Biomédecine et Devenir de la Personne*, Seuil, Paris.
100. Osborn J. *et al.*: 1990, "Procréation Artificielle: Oùen Sont l'Éthique et Le Droit", *Journal International de Bioéthique* 1 (1 & 2), 4-66, 77-110.
101. Papiernik E.: 1990, "Diagnostic Anténatal et Problèmes Éthiques", *Chirurgie* 116 (6-7), 541-544, discussion 544-545.
102. Patrimoine Génétique et Droits de l'Humanité (PGDH Colloquium, Paris, Oct 1989): 1990, *Livre Blanc des Recommandations*, Osiris, Paris. Proceedings to appear in 1992: Editions Odile Jacob, Paris.
103. Poirier-Littré M.-F.: 1990, "Consentement Éclairé du Malade en Psychiatrie: Conséquences sur Les Recherches Biomédicales", *L'Encéphale* 16 (4), 277-279.
104. Pollak M.: 1990, "Aids Policy in France: Biomedical Leadership and Preventive Impotence", in: Misztal & Moss (eds.), *Action on Aids: National Policies in Comparative Perspective*, Greenwood Press, New York, 79-100.
105. Pompidou A.: 1990, *Souviens-toi de l'homme. L'éthique, la vie, la mort*, Payot, Paris.
106. "Pour une Nouvelle Genération de Droits et de Devoirs. Bicentenaire,

Nouvelle Citoyenneté et Environnement. Le Fait deDisposer d'un Environnement Preservé est-il un Droit de l'Homme?": 1990, *La Baleine* (special issue), 78.

107. "Pratiques Médicales et Libertés": 1990 *L'Hôpital à Paris* (special issue), 116.
108. "Procréations, Droits, et Droits de l'Homme": 1991, *Actes* (special issue, No. 77.
109. "Quel Environement Nous Prépare le Génie Génétique?": 1990, *Diplômees* (special issue) (September), 1-252.
110. Quéré F.: 1991, *L'éthique et la Vie*, Editions Odile Jacob, Paris.
111. Rapports d'Ambassades: 1990, *Biotechnologies. Période* 1990-91, ADITECH, Paris.
112. "Réflexions sur l'Europe de la Santé et de la Protection Sociale": 1989, *Agora. Ethique, Médecine, Société* (special issue), No. 12.
113. "Regards sur les Inégalités Sociales de Santé": 1990, *Agora, Ethique, Médecine, Société* (special issue), No. 13.
114. Roche L.: 1991, "Actualités en Bioéthique", *Journal de Médecine Légale - Droit Médical*, 34 (1): 51-75.
115. Saint-Sernin B.: 1990, "Dimensions de l'Éthique", *Etudes* 373 (6), 625-634.
116. "Secret Médica, Experience des Comités d'Ethique Hospitaliers: Troisieme Journée des Comités d'Ethique des Hôpitaux de l'Assistance Publique de Paris": 1989, *Concours Médical* (special issue), January 31.
117. Secrétariat d'État Chargé des Droits des Femmes: 1991, *Procréations Médicalement Assistées. Auditions, Réflexions*, 31 Rue Le Pelletier, 75009 Paris.
118. "SIDA et Sociéte": 1991, *Agora, Ethique, Médecine, Sociéte* (special issues), Nos. 18-19.
119. Sinding C.: 1991, *Le Clinicien et Le Chercheur. Des Grandes Maladies de Crence à La Médecine Moléculaire (1880-1980)*, PUF, Paris.
120. Souché A., Zekri J.R., Thermoz P.: 1990, "Essais Thérapeutiques de Médicaments. Un Protocole Clinique Établi a priori, Pourquoi Faire?", *Psychologie Médicale* 22 (14), 1483-1484.
121. Soutoul E.: 1991, *Les Essais Cliniques à l'Hôpital. Recherche, Industrie et Centres Hospitaliers Universitaires*, Mémoire Hospitalier (Thèse), Ecole Nationale de la Santé Publique (ENSP), Rennes.
122. Terrenoire G.: 1991, *Information en Éthique Biomédicale. Sources et Ressources Françaises*, Centre de Sociologie de l'Éthique, 59-61 Rue Pouchet, 75849 Paris Cedex 17.
123. Testart J.: 1990, *Le Magasin des Enfants*, Editions F. Bourin, Paris.
124. Thévenot J.P., Constant J., Douche F., Wiener M., Bourcier G.: 1989, "Quelques Aspects Actuels de l'Éthique du Soin en Psychiatrie Infanto-Juvénile", *Annales Médico-Psychologiques*, 147 (2), 190-193.
125. Thomas J.-P.: 1990, *Misère de la Bioéthique. Pour une Morale Contre les*

Apprentis Sorciers, Albin Michel, Paris.

126. Thouvenin D.: 1989, "Commentaire Législatif. La Loi du 20 Déc 1988: Loi Visant à Protéger Les Individus ou Loi Organisant Les Expérimentations sur l'Homme?", *Actualités Législatives Dalloz*, 10: 89-104, 11: 105-16, 12: 117-128.
127. Thouvenin D.: 1991, "De la Qualification de l'Acte Thérapeutique à Son Contrôle", *Recueil Dalloz Sirey* 33, 221-228.
128. "Une Mémoire Essentielle": 1990, (special issue), December.
129. Valabrègue-Wurzburger O.: 1990, "Introduction Chez l'Homme de Matériel Génétique Modifié. La Thérapie Génique", *Jal Inter Bioéth/ Inter Jal of Bioeth* 1 (4), 245-263.
130. Véreecke L. *et al.*: 1991, "La FIVETE, du Vertige à la Responsabilité", *Le Supplement, Revue d'Ethique et de Theologie Morale* 177, 1-149.

MAURICE A.M. DE WACHTER, G.M.W.R. DE WERT,
R.H.J. TER MEULEN, R.L.P. BERGHMANS,
H.A.E. ZWART, I. RAVENSCHLAG,
AND A.K. SIMONS-COMBECHER

BIOETHICS IN THE NETHERLANDS: 1989-1991

I. INTRODUCTION: DUTCH HEALTH POLICY MAKING

Policies express a country's mind and arise from structures that may be particular to a given country. We think that summaries and reports on major legislation, court rulings, regulatory changes, and policy announcements by the Dutch government and professional associations in the Netherlands will be better understood and appreciated if one also understands how the system develops its health policy. Therefore, this introduction outlines the major institutions that contribute to the preparation of health care policy. Study groups and councils, health care professionals, and patient organizations play roles in Dutch policy making. The actual implementation of the policy itself then becomes the prime responsibility of politicians and government.

Many of these institutions, such as the Health Council or the National Council for Public Health, have been appointed by the government to oversee the field of health care. Provided that they maintain their independence in choosing topics for study and in stating their advice, these institutions contribute to the development of health care regulations.

There are six major advisory organizations that assist the departments, the government, and Parliament. The Health Council offers information and advice, either when requested or on its own initiative, concerning "the state of the science in matters of public health". The National Council for Public Health is charged to facilitate relations between government and particular organizations and institutions of health care. The Sickness Fund Council consists of employers, insurers, unions, consumers, and patients organizations, and serves as the liaison with international conventions on matters of social security. The Central Organization for Fees in Health Care determines or approves fees for services. The College for Hospital Provisions, since the early 1970's, has controlled all hospital construction and increases in hospital beds. And recently, the Netherlands' Organization for Technology Assessment (NOTA) was created at the request of Parliament in order to study the ethical issues raised by new technologies.

In the Fall of 1991, the government made plans to suspend the Central

B. Andrew Lustig (Sr. Ed.), Bioethics Yearbook: Volume 2, 191-210.

Organization for Fees in Health Care and the College for Hospital Provisions. It also intends to replace the Sickness Fund Council by a Council for Health Care Insurance. Major changes may occur if these institutions lose their freedom to offer advice without being asked.

Professional organizations have, no doubt, significantly contributed to policymaking. Dutch lawmakers of the 1980's adhere to the tenet that self-regulation by the professions requires legislators to intervene less and less. All categories of medical, nursing, and paramedical organizations are, therefore, invited to have their say in fashioning policy.

Patient organizations, whether general in nature or organized for specific concerns, are well developed in the Netherlands. The ministries and the government are creating channels to have these organizations participate in the health care policy development.

II. NEW REPRODUCTIVE TECHNOLOGIES

A. *Agreement of Coalition Government (1989)*

When Christian Democrats and Socialists drafted their mutual agreement in 1989 to direct the coalition government, it included the following points concerning artificial fertilization:

(1) The principle that the mother who gives birth is also the legal mother remains unchanged.

(2) A legal prohibition on commercial surrogate motherhood will be decreed.

(3) In principle, artificial insemination with donor sperm (AID) should be available to couples in all types of relationships.

(4) A study will be made about the consequences of AID for children who lack access to information about the identity of their biological/genetic parents. Pending the findings from this research, donors retain their right to have their anonymity respected.

(5) Embryo experimentation will be legally regulated. The regulations should stipulate that no embryos will be created solely for research purposes. The law should limit the decision-making power of medical ethics committees. A national commission will judge whether medical ethical norms are being met [1].

B. *The Royal Dutch Medical Association (RDMA): The Physician's Responsibility in Artificial Procreation (1989)*

In its report, the RDMA stresses that physicians who participate in artificial reproduction share co-responsibility for the creation of human life. With regard to in vitro fertilization (IVF), this means that, in addition to the presence of medical indications, the future situation of the child to be born should be evaluated from a psychosocial viewpoint. However, in line with the advice given by the Health Council, the RDMA considers extensive psychosocial evaluation impossible and undesirable. The Association, therefore, recommends, only that level of assessment necessary to assure that the child will be able to grow in a situation characterized by continuity and stability in order to offer the child safety and security and to allow the child to bond emotionally with one or more adults. Only in the case where the physician is convinced that procreation would pose serious risks for the psychosocial development of the child is there a psychosocial counterindication [2].

C. *The Council of the Health Insurance: Advice on IVF* (March 23, 1989) presented to the Secretary of Welfare, Public Health, and Culture

In its Advice [3], the Council considered the question of whether or not IVF should be covered by the Health Insurance Fund [4]. The Council decided not to preempt the broader social debate about the composition of a basic benefits package or about future expansion of health insurance coverage. Instead, the Council opted for a pragmatic approach, and decided to continue a previously introduced coverage of IVF until January 1, 1991. The most important conditions agreed upon at the end of 1989 were:

(1) All medical indications qualify for coverage (no longer will coverage be limited to women with tubal pathologies);

(2) All patients of the Union (to which the Council belongs) who have a medical indication qualify for coverage; and

(3) A maximum of three treatments per patient will be reimbursed.

The Secretary of Health has agreed with these extensions in coverage through late 1991. This policy has remained unchanged.

D. *Jurisprudence and Legal Initiative: A.I.D. and Surrogacy*

On the basis of Article Eight of the European Declaration of Human Rights (right to family life), a sperm donor requested his right to visit the child conceived with his sperm. A Court in the city of Utrecht judged that the donor, by offering semen anonymously, indicated that he did not want to be perceived as a father [5]. Donorship does not create family life in the sense of the Declaration. Given the resistance of the social parents, the Court deemed that it would not be in the child's interest to allow social visitations with the donor. The judge declined the request.

A Bill of Law to ban commercial surrogacy was tabled during 1987 (Number 21 968). Commercial surrogacy will be banned by adding stipulations to Articles 151b and 151c of the Criminal Code. The highest punishment will be either one year imprisonment or a fine of the fourth category (25,000 florins). The intention of these stipulations is to discourage not only commercial surrogacy, but any and all forms of the practice. Among the reasons mentioned for discouraging surrogacy are the following: (1) surrogacy may cause emotional and psychological problems for surrogate mothers and the children born of the arrangement; (2) surrogacy may involve new forms of exploitation of women; and (3) surrogacy may undermine the place of marriage in society. However, the partners of the current government coalition have not reached agreement on whether all forms of surrogacy should be punished.

III. MATERNAL FETAL CONFLICTS

The Health Council issued a report on the unborn child as a patient. Its particular focus was on invasive diagnostics and on fetal therapy [6]. The Council concluded that invasive diagnostics and therapy, whether experimental or not, should be practiced only in rare cases, in limited numbers, and exclusively in specialized centers. For such interventions it is mandatory for the pregnant woman to give informed and free consent. The Council rejected the idea that enforceable rights should be granted to the unborn, because such adjudication would create the possibility of compulsory treatment, which conflicts with the pregnant woman's right to bodily integrity.

IV. CARE OF SEVERELY ILL NEWBORN BABIES

This issue was explored in an interim report of the Royal Dutch Medical Association [7]. The Commission intended to highlight medical and moral aspects of life-prolonging treatment for severely disabled newborn children. When discussing life termination, the Commission meant all actions performed by physicians aimed at the death of the incompetent patient, regardless of whether death occurs through the administration of lethal drugs, pain medication, or withholding or withdrawing treatment. The Commission concluded that the main reasons for life terminating-interventions in today's clinical practice are either a prognosis that the newborn has virtually no chance of survival, or that it will not be able to sustain a "liveable" life. The criteria invoked to support such conclusions, however, seem to be very diverse.

The Commission concluded that a medically meaningless treatment should be withheld or withdrawn. "Medically meaningless" are all medical actions that do not meet the primary goals of medicine, *viz.*, the preservation of life and the alleviation or suppression of somatic and/or psychic suffering. To insist on the singular goal of preserving life is not always ethical. Similarly, it may be ethical, to withdraw treatment in order to alleviate pain, even if this leads to the patient's death. The administration of lethal drugs is allowed only in exceptional cases where treatment has been withdrawn but the infant, rather than expiring, continues to suffer.

The Commission discussed four general guidelines to direct neonatal decision-making:

(1) In all cases it is imperative that, according to prevailing clinical insight there be sufficient certainty about the diagnosis underlying the prognosis. In the absence of such certainty, the Commission thought that, despite the rule, "*in dubio abstine*" ("when in doubt abstain"), it is permissible to maintain the newborn alive with artificial means until a diagnosis/prognosis can be made with sufficient certainty. If the latter diagnosis can be made, the treating physician must be willing to stop treatment and possibly even to terminate the life, should the diagnosis warrant that action.

(2) With regard to the prognosis of "no real chance for survival", the Commission called for a consensus to be reached among professionals on the criteria for such a judgment.

(3) The Commission found prognosis of "no liveable life" to be excessively subjective, and recommended that it be made more precise. Professionals should develop a frame of reference that allows them to chart nature, severity, and scope of the (expected) defects of the newborn. However, setting limits within which the notion "unliveable life" can be responsibly invoked is, according to the Commission, not only a question for professionals, but for society as a whole.

(4) In order to make responsible decisions about life termination, careful attention to substantive standards of decision-making is required. The opinion o parents should receive its proper weight. And certainly, in cases where the judgments of parents differ from those of physicians, consultation with an experienced and independent colleague by the treating physician is essential.

(5) Finally, the Commission considered it essential that the physician extensively document the deliberations which lead to the eventual decision.

A year before the RDMA Commission's report, the Supreme Court of the Netherlands decided to grant a petition about a life-saving intervention on a severely disabled newborn baby. As part of the petition for protection against unjust prosecution by the District Attorney, the decision did not concern the content of the petition but only its reasonableness. The Supreme Court apparently decided that treatments of disabled newborn babies can be withdrawn, provided that (1) there is the likelihood of very serious suffering, and (2) treatment would no longer serve any medically useful purpose but only prolong the dying process and cause more suffering [8].

V. CONSENT TO TREATMENT

Two documents were issued concerning informed consent in therapy. The first was a model developed by The Royal Dutch Medical Association in collaboration with the National Platform for Consumers and Patients. The second was a Bill concerning the medical contract between doctor and patient.

Because the model procedure was accepted by the General Assembly of the RDMA, its formulation of a number of rights and duties may be read as reflecting a professional standard. The model contains stipulations on the right to be informed, the requirements of consent, the right of patients to have access to their medical files, and the right to privacy [9]. The model procedure anticipates in fact what will be said in the Bill on the medical contract.

The so called "Patients' Bill", tabled on May 29, 1990, was intended to complement the Civil Code [10]. To the rights already mentioned in the RDMA model, this Bill adds a section on the medical secrets and confidentiality.

VI. HUMAN EXPERIMENTATION: CONSENT AND IRB's

Although the Bill on Human Experimentation (tabled since 1987 in its latest version) had not, as of late 1991, been made law, guidelines on the matter issued in 1984, have greatly contributed to the growth of ethics committees. The core of the 1984 guidelines was the statement that no human experimentation may take place in hospitals unless a specially instituted committee has approved it.

The Bill has been favorably reviewed by the State Council [11]. It contains five sections on general rules, informed consent, liability, insurance, and evaluation by the Institutional Review Board. Of these, the sections on informed consent and on IRB's will be presented here.

A. *Informed Consent*

The Bill on human experimentation requires that research subjects be informed about several matters: (1) the purpose, nature, and duration of the experiment; (2) the risks of participation; (3) the risks of interrupting the experiment; (4) the level of discomfort; and (5) the assessment by the IRB. Information may be oral or in writing. Consent, however, should always be in writing.

Excluded from participation are prisoners and civilly committed psychiatric patients. Experiments with incompetent subjects and with subjects under 12 years old cannot be done without the consent of their parents, guardian; or legal representative. For adolescents (12 to 18 years of age), both their own consent and the legal representative's consent is required.

B. *Independent Review Boards*

Since 1984, numerous research ethics committees have been established at the local or regional level. The Central (now National) Council for Public Health made recommendations in 1982 on the mandate, composition, and objectives of such committees.

The Bill on human experimentation (1987) proposed the establishment of a Central Committee which, in terms of composition and method of procedure, would also be a model for all other committees. Members of the Central

Committee are nominated by the Minister. The Central Committee accredits other committees, compiles all applications for research from all committees, and receives annual reports from them. The Central Committee can issue guidelines for all other committees to follow.

While the Bill awaits passage into law, an interim Central Committee called the "Core Committee Medical Research Ethics", was established by the government in 1989. Its sole purpose is to offer expert advice on new medical developments with societal repercussions; *e.g.*, embryo research and research with incompetent subjects. The Core Committee can be solicited for its advice by local IRBs, directors of clinics and research institutes, granting agencies, and the Health Council. The Committee can also offer advice on its own initiative.

VII. COST CONTAINMENT

Health care in the Netherlands, as in most Western countries, struggles with problems of cost containment. In 1980, the Netherlands devoted 9.8 percent of its GNP of health care; that figure decreased to 9.4 percent in 1990. Of the 17 OESO countries, the Netherlands rank fourth in health care expenditures, after the United States, Sweden, and France. Thus far, several legislative efforts have been made to contain costs: a law on Hospital Provisions (1971), a law on Fees and Fares in Health Care (1980), and the regulation of budgeting in hospitals (1983). Often, however, these measures have generated red tape and bureaucracy, thereby decreasing the sense of responsibility among health care providers, insurers and the insured.

A. *The Simons Plan: A Reform of the System (1990)*

Drawing on a report from a previous State Commission [12], the Secretary of Health Simons, proposed a set of dramatic changes in May, 1990. Simons wished to combine regulation with elements of the free market. He also called for greater restraint in the demands of health care consumers [13].

The central piece of this reform was the introduction of a universal compulsory health care insurance. This system would replace the previous one where people with a yearly income below 52,000 Florins (25,000 U.S. dollars) were insured by the Sickness Fund, while those with incomes above 52,000 Florins bought private insurance. The new system rests on two pillars: a social one meant secure solidarity among the insured, and an economic one meant to bring efficiency and cost control. Thus, the government recedes and allows for self-regulation.

The reform plan contains four major components:

(1) The government lessens its involvement, and is now responsible only for ensuring the conditions necessary in order to maintain quality of care, access, and cost.

(2) Health care providers (hospitals, nursing homes, etc.), are allowed to

plan in accordance with local needs and with the wishes of insurers.

(3) Health care insurers (including the present Sickness Funds), are free to market the most cost efficient care.

(4) Insured individuals may choose their insurer, and, to some extent, their benefits package and premium.

As of late 1991, the Simons Reform Plan has been tabled as a bill. A gradual introduction of the reform plan should begin by January, 1992, and be finished by 1995.

B. *The Drug Compensation System*

Another initiative for controlling cost intends to slow the annual increase of drug expenditures, which are presently at 9 percent. As of July 1, 1991, the Secretary of Health introduced the Drug Compensation System. This system applies only to those insured under the Sickness Fund, which is compulsory for those with an income below 52,000 Florins.

While ensuring that all necessary drugs will be available in a patient's benefits package, the following mechanisms are designed to control costs:

(1) Drugs are divided into groups that are interchangeable. This is done according to the ATC (Anatomical- Therapeutical-Chemical classification). In this classification, Level Four determines the place of a drug in the group of those with similar chemical structure. Level Three, which uses similarity in indication and effect as the sole criterion for inclusion, is used only in exceptional cases.

(2) For each group, an upper limit for reimbursement is established, determined by the average cost of all drugs belonging to any given group.

(3) For drugs belonging to the same pharmacotherapeutical group, but whose price exceeds the upper limit, the patient pays the difference, unless the industry adjusts its price to the upper limit.

The Drug Compensation System does not result in only the least expensive drugs being reimbursed. Nor does it mean that new drugs are unavailable to patients. In fact, the full price of new drugs will be reimbursed.

VIII. COMATOSE AND THE INCOMPETENT ADULTS

A. *The National Hospital Council: A Manual for Practice*

In 1988 the National Hospital Council published a manual for treating comatose patients [14]. The manual discussed the following five concerns:

(1) There may be conflicts generated by the emotional involvement of family and caretakers, on the one hand, and medical or scientific judgments about the efficacy of continued treatment, on the other.

(2) The line between medical treatment and care may be hard to draw, with artificial nutrition as an obvious example of this problem. According to the

report, recent jurisprudence still considers artificial nutrition and hydration to be normal care.

(3) Although a coma longer than three months reduces the patient's chance of full recovery to almost nil, there is never complete certainty about the prognosis.

(4) Opportune moments for decision-making may vary. For the patient in an acute coma, there may be numerous moments for decision-making. It is important that such moments be recognized by the medical staff and that family members be included in drawing up an explicit plan of care.

(5) Voluntary euthanasia, in the strict sense, is not applicable to comatose patients, because they are not able to request it. Moreover, there is still great uncertainty about whether or not comatose patients feel pain. However, foregoing medically futile treatment will not be punishable, because in such cases, the patient is considered to die a natural death.

B. *Long Term Comatose Patients (1991)*

Another report by the RDMA discusses long term comatose patients who are in a persistent vegetative state [15]. At stake are two issues: (1) whether or not, and if so, under what conditions it is acceptable to terminate the lives of long-term comatose patients; and (2) the extent to which one should treat such patients with all available medical and technological means, once they have become unable to express their own choice. Initially, with an uncertain prognosis all means are used. But as the prognosis of irreversible coma becomes more certain, fewer urgent medical decisions are involved. The report repeats the considerations of the National Health Council, but criticizes them on the following grounds.

First, the small chance that those patients do suffer cannot be excluded. Second, family members and caretakers may find continued care to be a great burden. Third, if present and future quality of life have been deemed unacceptable, treatment is no longer needed. According to the Commission, artificial nutrition and hydration are forms of medical treatment. Without an explicit and attainable medical goal, no treatment is justified. The Commission calls it a doctor's duty to withdraw treatment which affronts human dignity. The mere biological residue of life is an insufficient basis for human existence. Once treatment is stopped and the patient's death has been accepted, the Commission concludes that administering lethal drugs may be indicated. Physicians who cannot accept this choice should refer or transfer their patients to another doctor.

The role of nurses is essential in providing good care for the dying. Nurses are often better informed and may contribute to careful decision-making about treatment. Physicians, therefore, should consult with nurses as well as with colleagues. Nurses should participate in all team discussions about patients, especially in order to avoid bonding that would impede medical decisions about discontinuing treatment. Decisions, nevertheless, remain the sole responsibility of the treating physician. A conscience clause is included to protect nurses in case of conflict.

Finally, the death of a comatose patient after treatment has been withdrawn is for legal purposes, a natural death.

C. *Non-Treatment Decisions About The Incompetent Adult* (1991)

The Health Council published a background study on nontreatment decisions for adult incompetent patients including the mentally retarded, psychiatric, psychogeriatric patients, comatose patients, and patients in a persistent vegetative state [16].

This background paper emphasizes the need for further study by a commission to be installed by the Health Council. It also suggests a number of issues which that committee should address. These are: (1) decisions about the substantive criteria for incompetence, and about the procedure for establishing incompetence; (2) an evaluation of current categories that are used, especially "medical futility" and "quality of life"; (3) a procedure for careful and deliberate decision-making; (4) a clearer understanding of the health care proxy and his/her powers; and (5) the status of written wills and oral statements.

IX. ACTIVE EUTHANASIA AND ASSISTED SUICIDE

As of late 1991, no changes in the criminal status of active euthanasia or assisted suicide had occurred in the Netherlands. The law, which remains unchanged, states that "he who takes another person's life, even at his explicit and serious request, will be punished by imprisonment of approximately twelve years or a fine of the fifth category" [17]. In practice, however, and at the level of jurisprudence and professional regulation, matters are different. In the what follows, we will describe initiatives for legislation, guidelines of professional organizations, the latest report on medical practice concerning euthanasia, and the government's position on the issue.

The latest initiative for legislation on euthanasia dates back to 1987, when the then ruling coalition of the Christian Democrats and Liberals tabled a proposal [18]. The initiative states that the legal prohibition of euthanasia should not be lifted, but that a physician may invoke "emergency" or overwhelming power (*force majeure*). It does not guarantee that physicians will not be prosecuted, nor that an "emergency" will be accepted by courts. On the other hand, the Bill offers physicians the legal certainty that some cessation of treatment will not be viewed as termination of life. Moreover, the bill lists a set of requirements which will guarantee careful and responsible decision-making about euthanasia.

The intention of this proposal is to steer developments in jurisprudence on euthanasia. Indeed, case law has grown ever since the first case in 1973, peaking in 1984 when the Supreme Court emphasized the judge's duty to examine the physician's conflict of duties. This conflict should be understood as both a subjective conflict of conscience and as an objective conflict of duties. The Court of Appeal, to which the Supreme Court had referred a case, acquitted the physician because he had resolved his conflict of duties by reasonably and responsibly making an objective and justified decision [19]. Although the Supreme Court's decision was influential, it did not change the law. In principle, therefore, every case has to be prosecuted.

Professional organizations, in particular the Royal Dutch Medical Association (RDMA), have recently contributed several discussion papers. The paper on comatose patients was discussed in Section VIII above. On February 23, 1991, the RDMA, in collaboration with the Ministry of Justice and with the Attorneys General, also issued a procedure on notification of euthanasia to be followed by the physician who has performed euthanasia [20].

During the same period, a government commission was preparing its final report. In January 1990, the Remmelink Commission (named after its president) was appointed to study physicians' practices with regard to medical decisions and patient care at the end of life. On September 10, 1991, the Commission issued its report [21]. Research done for the Commission showed that about 9,000 patients annually requested euthanasia, of whom 2,300 received it. The research also revealed that in an additional 400 cases, physicians assisted patients to commit suicide. Finally, it showed that in about 1,000 cases, active euthanasia without request occurred. In several of these latter cases, however, patients had expressed such a desire at an earlier time. The Commission, though wishing to consider such interventions as ways to help patients to die, recommends that future eases of termination without request follow the requirements of careful medical practice.

The Remmelink Commission also recommended not only that cases of euthanasia should be reported, but also cases in which the patient's life is terminated without his or her express request. It further urged that general practitioners consult specialists with regard to end-of-life decisions and vice versa, and that all physicians follow strictly the rules set out for careful implementation of euthanasia. Finally, the report recommended that people in various professional groups and sectors of health care be trained in making at the end of life, and that postgraduate programs devote greater attention to these issues.

The current coalition government of Christian Democrats and Socialists agrees with the Commission's proposal that a physician who performs euthanasia remains punishable. "The maintenance of this norm allows the government services to check any active intervention intended to end a patient's life" [22]. Therefore, the physician should not simply presume that he will not be prosecuted because he acted according to requirements of careful medical practice. The government concludes that there is no absolute contradiction between, on the one hand, pleas for legal recognition that, under certain circumstances, euthanasia will not be punished and, on the other hand, the opinion that such practice would erode respect for and protection of life. While the government keeps euthanasia under the criminal code, it also proposes to change the Law on disposing of the deceased body in such a way that a declaration of unnatural death does not necessarily entail legal prosecution.

X. ORGAN TRANSPLANTATION

A. *Health Council: Advice on Heart Transplantation* (Feb 27, 1989)

In this report the commission of the Health Council concluded that heart transplantation is a valid treatment. Although the Commission recognized that a great shortage of donor hearts is likely, this should not to be a reason for using non-medical standards (such as a strict age limit) in selecting patients. Moreover, the shortage of donor hearts may be reduced in part by the government's active promotion of people's willingness to donate their organs [23].

B. *Health Council: Advice Concerning the Protocol for Heart Transplantation* (June 22, 1990)

According to the protocol developed by the Dutch Heart Transplant Centers, patients over 55 years of age do not qualify for heart transplants. According to the Health Council, however, a strict age limit contradicts the principles of just distribution and equal opportunity to receive health care. Criteria for selection are acceptable only if they are medically relevant; *i.e.,* criteria which allow one to predict the outcome of the intervention in terms of its success. The Commission does not object to using age as a criterion when fitness and bodily condition are relevant indices, but does object to using calendar age as an absolute limit [24].

C. *Bill on Organ Transplantation* (Number 22 358)

This future law is meant to accomplish several objectives: (1) to protect the donor, once he/she has decided to donate an organ; (2) to establish strict requirements governing procurement of organs; (3) to encourage organ donations; (4) to increase equity in the distribution of organs; and (5) to discourage commercialization.

The law would adopt a system of "opting in". Persons must decide to donate; their consent cannot be presumed. In order to provide individuals with the opportunity to donate, all Dutch citizens at least 18 years of age will receive a codicil at home, and a central registration of all individual choices will be established. For those who have not expressed their will during their lifetimes, family members will be entitled to decide.

The future law also considers donation among living persons. Such donation, which may have permanent consequences for the donor's health, is allowed only if the life of the potential recipient is in danger. Young children may donate organs only if no permanent damage to their health occurs.

XI. CLINICAL GENETICS

A. *Health Council, Heredity: Society and Science* (December 29, 1989)

This report offers advice on the normative aspects of genetic testing and gene therapy. The most important conclusions and recommendations of the Health Council are the following:

(1) The Health Council recommends that equal opportunities for access to genetic diagnosis and counseling be made available, without privacy and personal freedom of choice being violated.

(2) The Health Council calls for further research and reflection by genetics professionals regarding the conditions for which prenatal diagnosis is indicated.

(3) The Health Council does not support those who wish to issue special legal regulations about the rights of counselees and their relatives. Rules that are generally valid in health care apply here as well. The Council emphasizes both the right to be informed and the right to be uninformed.

(4) With regard to genetic registries, the Council states that registered persons have rights to be established by future legislation. Of major importance is the consent requirement for any introduction of data into the registry. Moreover, the Council deems essential a person's right to delete information and to render it anonymous. The Council recommends that, prior to the introduction of data, the counselee be asked, in specific terms, to authorize what use may be made of his data in order to help relatives or for scientific research.

(5) Cell banks should require a code of behavior which, on the one hand, does justice to the donor's rights and, on the other hand, does not impede unnecessarily the use of material for the benefit of others or for scientific research.

(6) With regard to genetic screening programs, the Council recommends strict application of the rules that have been developed for mass health screening. The Council opposes the screening of newborn babies for untreatable later onset affections.

(7) The Council recommends that targeted genetic diagnosis of people who apply for any type of insurance be forbidden. Moreover, the right of insurers to ask for extant genetic information should be limited; *i.e.*, there must not be a "duty to inform" about the results of a previous genetic diagnosis of the person to be insured or of his family. The Council recommends legislation on this matter.

(8) The Council rejects diagnostics about hereditary dispositions for

employment examinations. However, should future tests prove to be reliable, their application may be considered, in particular, in situations where it can be shown that the health of the employee or of other persons will benefit.

(9) Pre-implantation diagnosis and research with human pre-embryos aiming at such diagnosis is for the majority of the Council acceptable under certain circumstances.

(10) With regard to somatic gene therapy, the Council believes that, once the experimental phase is over, this treatment will prove to be no different than other methods of medical therapy, such as organ transplantation. In the meantime, the Council calls for a national committee to evaluate this practice.

(11) Because germ line therapy is still fraught with so many uncertainties about its safety for humans, the Council recommends that genetics professionals should propose a moratorium [25].

B. *Government: Positions and Policy Intentions Concerning Genetic Testing and Gene Therapy (Reaction to Health Council's Advice)* (November 30, 1990)

The Government agreed with most of the recommendations made by the Health Council. With regard to the use of genetic information by insurers, the politicians note that the insurers have shown their intent to exempt candidates from their obligation to inform about data resulting from a previous genetic diagnosis. (In the case of life insurance this exemption applies to a limit of 200,000 Florins). The insurers are willing to try this policy for a period of five years. They will not ask for additional genetic investigation. The government deems this measure sufficient to meet the recommendations of the Health Council.

With regard to the development of pre-implantation diagnostics, the government considers research on embryos to be permissible only for strictly limited purposes. Permissible experiments must be directed to research on a given embryo aimed at improving its healthy development and its being brought to term. These exceptions do not justify the production of embryos for special research purposes.

The government intends to prohibit germ line therapy [26].

XII. MENTAL HEALTH

A. *Bill of Law on Special Commitments in Psychiatric Hospitals*

Intended to replace an 1884 law, this bill was tabled in 1971 and accepted by the Lower House but consequently heavily criticized in the Upper House. Because of the criticism, a new bill to change the law on special commitments to psychiatric hospitals was tabled on July 20, 1989.

The new text tries to accommodate the criticism of the extant law's scope.

The law's intent is to regulate involuntary commitments. Its basic criteria are a suspicion of mental disturbance with resultant danger an individual poses to himself or others, and (2) and that person's unwillingness to be hospitalized [27]. The latter criterion is, in fact, essentially different from the current law, where resistance/objection to hospitalization is the criterion. The new criterion of "unwillingness" suggests that certain mentally handicapped and psychogeriatric patients would now come within the law's scope. For them, too, the extensive procedural legal safeguards vis-a-vis coerced commitments would be applicable. This extended ambit would, in turn, lead to a considerable increase in cases about (40,000 decisions annually) for the judiciary apparatus, compared with 5,600 cases under the existing law.

The government has opted for a limited scope to the law. Mentally handicapped psychogeriatric patients who are unable to express their own will, but do not resist hospitalization, are excluded from the court's authority. These individuals may be hospitalized only by the decision of a specially appointed committee that will judge their cases on medical grounds.

B. *Health Council: Advice Concerning Neurosurgical Treatment of Patients With Severe Psychiatric Diseases* (February 4, 1990)

Already in 1980, the Secretary for Public Health and Environmental Hygiene had asked the Health Council for advice concerning psychosurgery. Five questions were raised:

(1) What is the state of the art regarding so-called stereotactic operations?

(2) Are these interventions acceptable from the viewpoint of medicine, ethics and law?

(3) Are there any areas where these interventions are indicated?

(4) Who is competent to perform such interventions and under what conditions?

(5) Is it desirable that further legal steps be taken in the matter?

The following conclusions and recommendations of the report were the most important ones:

(1) Neurosurgery, on psychiatric indication from a medical viewpoint, is acceptable in very rare cases and under the strictest conditions; *viz.*, (a) previous well implemented treatments were of insufficient help to the patient; (b) the patient's suffering is serious and unbearable; and (c) the intervention offers a reasonable chance for a considerable improvement in the patient's condition, without causing serious side effects.

(2) The neurosurgical intervention requires the approval of the Neurosurgery Workshop. The intervention may be performed only on patients who freely and knowingly have chosen it. For an incompetent patient, his or her legal representative decides. The Commission rejects neurosurgery as a coercive form of treatment. The latter is defined as therapy against the will of

either a competent or an incompetent patient.

(3) The following should not be considered indications for neurosurgery: schizophrenia, obesity, anorexia nervosa, aggression, "deviant" sexual behavior, and psychopathic personality. The Commission also recommends restraint with regard to the behaviorally, disturbed, or mentally retarded (who may exhibit automutilation), because there are no other ways of either preventing or limiting such behavior.

(4) The neurosurgical intervention should to be performed by an experienced neurosurgeon [28].

C. *Assisted Suicide in Psychiatric Patients* (A Pronouncement by the Central Medical Court of Discipline)

On March 29, 1990, the Central Medical Court pronounced in a case against a psychiatrist who had helped a patient to commit suicide [29]. Having consulted two colleagues, and in close consultation with the family, this psychiatrist had accompanied the patient to his home and assisted him in taking a lethal potion. Consequently, the psychiatrist informed police, the district attorney, and the Inspection of mental health.

The patient's earlier history was well known. During his hospitalization in 1983, he was chronically depressed with psychotic phases. The patient repeatedly requested assistance in suicide. After various relevant therapies had been tried without any improvement, it was decided that no further treatment possibilities were available.

The district attorney dismissed the case because the psychiatrist had acted according to accepted requirements of careful practice in euthanasia and assisted suicide. The Inspection of Mental Health, however, sued the psychiatrist at the Medical Court of Discipline.

The Inspection's position was that the psychiatrist should not have assisted, because it was doubtful, in view of the patient's psychiatric problems, that he was competent to freely decide in a matter of such utter and irreversible consequence as termination of life. Moreover, it is nearly impossible to make long-term predictions for chronic psychiatric patients, and the possibility of improvement or cure can never be excluded.

The Medical Disciplinary Court judged the complaint to be without foundation because experts had shown that: (1) the patient was sufficiently competent to decide about his wish to commit suicide; (2) his chance for improvement was extremely small, and genuine alternatives were unavailable; (3) a patient who is able to judge autonomously may not be denied the right to self-determination in matters of life and death; and (4) the psychiatrist acted with sufficient care by conforming to the rules of euthanasia and assisted suicide. The Court judged the patient's pain to be unbearable and permanent, and concluded that the patient had unmistakably expressed his wish to end his life.

The Inspection, however, appealed the case, and the Central Medical Disciplinary Court considered the accusation valid. For them, the psychiatrist had failed in the following respects. First, about two weeks before the suicide occurred, civil commitment was changed to "voluntary" commitment at the instigation of the psychiatrist. In his request to end civil commitment, the psychiatrist had failed to mention that he was considering assisting the patient to commit suicide. Second, it was known that the nursing staff, on the basis of their own experience, did not consider the suicide to be indicated and did not wish to be part of it. Moreover, the nursing staff had ethical objections. The Central Medical Disciplinary Court, therefore, found it reprehensible that the nursing staff had been involved only at a very late stage of the psychiatrist's original plan to assist the patient on the ward. Third, two consulted colleagues, who were also linked with the same hospital, had given their opinion only orally; and fourth, no external psychiatrist had been consulted.

The Central Disciplinary Court emphasized that it is quite difficult to establish the extent to which the patient's will is being influenced by his psychiatric condition: "The question of whether the desire to die and the concomitant depth of suffering comes from, is connected with, or is part of the depression, is hardly ever to be answered in a final way".

These various considerations led the Central Medical Disciplinary Court to its final conclusion, *viz.*, that the psychiatrist in this case should not have assisted the patient's suicide. However, no action was taken by the Court.

D. *Assisted Suicide in Psychiatric Patients* (April 1991)

The Chief Inspectors of Public Health and of Mental Health sent a letter to all physicians in the Netherlands instructing them that the procedure concerning the notification of euthanasia and assisted suicide does not apply to psychiatric patients or the mentally handicapped, but applies only to somatically and seriously ill patients. "A mental handicap or a psychiatric condition cannot as such be a reason for a physician to administer euthanasia or assist in suicide".

XIII. AIDS

HIV seropositivity and AIDS are afflictions which, because of their sensitivity and implications for privacy, have been excluded from the law on tracing and combatting causes of disease. For the same reasons, there has been no large scale public screening for these conditions. HIV testing has been required of those who apply for life insurance exceeding 200,000 Florins (about 100,000 U.S. dollars). Testing is also compulsory for disability insurance exceeding 40,000 Florins (about 20,000 U.S. dollars). Until now, no significant jurisprudence on AIDS has been developed.

XIV. ANIMAL EXPERIMENTATION

Of the three levels where animal experimentation occurs – fundamental, applied, and for teaching purposes – only applied research is still prevalent in the Netherlands. In practice, this means that such experiments must deal directly with either the health or nutrition of humans or other animals in terms of prevention or treatment.

A law on animal experimentation, introduced on January 12, 1977, regulates the qualifications and licenses for those who perform animal experimentation [31]. In 1980, a bill on Animal Health was introduced to regulate the relationship between humans and animals [32]. The Upper House is currently considering a revised version of this bill.

Article 18 of the Law of 1977 states that an advisory committee should assist in developing the Order in Council by which the Minister further applies the Law and its execution. Members of this committee should be experts in the fields of animal experimentation and of animal protection. In addition, there now exists a provisional committee, the "Committee on Ethics and Biotechnology in Animals", which was established by the Ministry of Agriculture and Fisheries on April 21, 1989.

Animal experimentation is controlled by the granting of licenses which must be obtained from the Minister of Welfare, Public Health and Culture. The Veterinary Chief Inspection of the Department Animal Experimentation registers all experiments and reports on an annual basis.

Institute for Bioethics
Maastricht, THE NETHERLANDS

BIBLIOGRAPHY

1. Regeerakkoord mei 1989.
2. Koninklijke Nederlandse Maatschappij ter bevordering van de Geneeskunst, De verantwoordelijkheid van de arts bij kunstmatige voortplanting. Utrecht, 22 December 1989.
3. Ziekenfondsraad, Advies inzake in vitro fertilisatie, 's-Gravenhage, 23 maart 1989.
4. The Health Insurance Fund covers health care for every Dutch citizen whose yearly income is less than 50.000 Florins.
5. NJ 1989, No. 237.
6. Gezondheidsraad, Het ongeboren kind als patient: invasieve diagnostiek en behandeling van de foetus. 's-Gravenhage, 11 juni 1990.
7. Koninklijke Nederlandse Maatschappij ter bevordering der Geneeskunst, Levensbeeindigend handelen bij wilsonbekwame patienten. Deel 1: zwaar defecte pasgeborenen. Interimrapport. 's-Gravenhage, April 27th, 1990.

8. Beschikking van de Hoge Raad d.d. 28 April 1989: Niet uitvoeren van een levensreddende operatie bij een zeer ernstig gehandicapte pasgeborene. Zie: Nederlands Tijdschrift voor Geneeskunde 133, 1281-1283.
9. Koninklijke Nederlandse Maatschappij ter bevordering der Geneeskunst, Modelregeling arts-patient relatie. Utrecht, 199O.
10. Wijziging van het Burgerlijk Wetboek en enige andere wetten in verband met de opneming van bepalingen omtrent de overeenkomst tot het verrichten van handelingen op het gebied van de geneeskunst, Tweede Kamer, vergaderjaar 1989-1990, 21561, Nos. 1-3.
11. Ministerie voor Welzijn, Volksgezondheid en Cultuur, Regelen inzake medische experimenten, 's-Gravenhage, 1987.
12. Commissie Structuur en Financiering Gezondheidszorg, Bereidsheid tot verandering, 's-Gravenhage, 1987.
13. Wet Stelselwijziging Ziektekostenverzekering, Kamerstuk 21 592.
14. Nederlandse Ziekenhuisraad, Omgaan met coma. Een handreiking voor de praktijk. Utrecht, 1988.
15. Koninklijkke Nederlandse Maatschappij ter bevordering van de Geneeskunst, Commissie aanvaardbaarheid levensbeeindigend handelen. Dlscussienota levensbeeindigend handelen. Deel 2: langdurig comateuze patienten. Utrecht, 1991.
16. J.J.M. van Delden, Het nalaten van medische behandelingen bij meerderjarige wilsonbekwame mensen. Den Haag, 1991.
17. Penal Code art. 293.
18. Voorstel van wetswiiziging met betrekking tot de hulpverlening door een geneeskundige die zich beroept op overmacht bij levensbeeindiging op uitdrukkelijk en ernstig verlangen van de patient. No. 20 383.
19. Nederlandse Jurisprudentie 1984: 106.
20. "Aandachtspunten", zie Medisch Contract No. 2, Januari 1991.
21. Medische beslissingen rond het levenseinde. Rapport van de Commissie onderzoek medische praktijk inzake euthanas Gravenhage, 1991.
22. Standpunt van het kabinet inzake medische beslissingen rond het levenseinde, 8 November, 1991.
23. Gezondheidsraad, Advies inzake harttransplantatie. 's-Gravenhage, 27 Februari 1989.
24. Gezondheidsraad, Advies inzake protocol harttranspalantatie. 's-Gravenhage, 22 Juni 1990.
25. Gezondheidsraad, Erfelijkheid: maatschappij en wetenschap. 's-Gravenhage, 29 December 1989.
26. Standpunten en beleidsvoornemens van het kabinet ten aanzien van erfelijkheidsdiagnostiek en gentherapie. 's-Gravenhage, 30 November 1990.
27. Wetsontwerp bijzondere opnemingen in psychiatrische ziekenhuizen. 's-Gravenhage, 20 Juli 1989.
28. Gezondheidsraad, Advies inzake neurochirurgische behandelingen van

patienten met zeer ernstige psychiatrische aandoeningen. 's-Gravenhage, 4 Februari 1990.

29. Centraal Medisch Tuchtcollege. Hulp bij zelfdoding bij psychiatrische patienten, 29 Maart 1990.
30. Hoofdinspecties voor de Gezondheidszorg en de Geestelijke Volksgezondheid. Hulp bij zelfdoding bij psychiatrische patienten, April 1990.
31. Wet op de dierproeven. Staatsblad van het Koninkrijk der Nederlanden, 1977, No. 67.
32. Ontwerp Gezondheids- en welzijnswet voor dieren, 1990-1991, No. 16 447.

HANS-MARTIN SASS

BIOETHICS IN GERMAN-SPEAKING WESTERN EUROPEAN COUNTRIES: AUSTRIA, GERMANY, AND SWITZERLAND

I. PRIORITIES IN PUBLIC DEBATE AND LEGISLATION

Reporting on developments in three quite different German speaking countries – Austria, Germany, and Switzerland[1] – allows one to compare similarities and differences in regulation, legislation, and public discourse on issues in bioethics.

In all three countries, bioethical debate and legislation on the new reproductive technologies plays a prime role. Germany has passed an Embryo Protection Law [28]; in Switzerland the Academy of Medical Sciences has proposed guidelines [75]; and in Austria legislation is pending ([54];[58]). In Switzerland, issues of patient self-determination in refusing intensive treatment at the end of life have prompted local societies of physicians to introduce forms for patients to specify their wishes regarding treatment (*Patiententestament*) ([68];[74], pp. 12ff.), while in Austria issues of research ethics have been legislatively addressed in animal experimentation [56] and in drug development [57]. Animal protection is a prime issue in all three countries ([19];[22];[56]). In Germany, a dormant abortion debate has intensified because the federal legislation is being drafted to supersede two different ethical and legal approaches to abortion in the former East and West Germany [32;17]. Moreover, following the Council of Europe's initiative of the Human Genome Project and Peter Singer's lectures on euthanasia and the withholding of treatment from severely ill newborns, ethical issues surrounding genetics and euthanasia have caused strong emotional debates, in part related to Germany's Nazi past ([65];[67];[69]).

As in many countries with strong professional organizations and a tradition of professional self-regulation, there is controversy in the German-speaking countries over the role of governmental regulation and legislation versus professional self-regulation, individual responsibility and civil liberty. Major

[1]I gratefully acknowledge the contribution of Jean-Marie Thevoz, General Secretary of the Schweizerische Gesellschaft für biomedizinische Ethik (Director: Dr. Alberto Bondolfi), Zürich, Dr. Gabriele Mras, Mag. Beatrix Varga, and Dr. Alfred Choueki of the Wissenschaftliche Landesakademie für Niederösterreich (Director: Prof. Dr. Peter Kampits), Krems, Brigitte Heerklotz of Bundesaerztekammer, Köln, and Carmen Kaminsky MA and Susanne Eschen of the Zentrum für Medizinische Ethik, Bochum, in the collection of legal and regulatory documents for this report.

B. Andrew Lustig (Sr. Ed.), Bioethics Yearbook: Volume 2, 211-231.

proponents of self-regulation are the Swiss Academy of Medical Sciences in Switzerland ([74];[75]) and the Federal Chamber of Physicians (*Bundesärztekammer*) in Germany ([14];[15];[16]).

There are also controversies about the most efficient way to deal with bioethical challenges in professional and political settings. The German debate on bioethics may serve as a case study for three different approaches in public culture and administrative attitude towards new challenges in biomedical technology, approaches which might be called "political correctness", "defensive ethics", and "differentiated ethics".

The approach of "political correctness" is most prominently represented in the public debate on issues of reproduction and human genetics ([65];[66]), and has had major influence on regulatory and legislative bodies. The impact of this approach on legislative action has shaped the Embryo Protection Law in Germany ([24];[28]), the debate there on the Human Genome Project and on genetic engineering ([62], pp. 3-15; [77]), and the discussion of animal research regulation and legislation in all three countries. The forces of political correctness have been successful in silencing Peter Singer in the German lecture circuit ([67];[69];[70]) and in shifting the topic of the bioethical evaluation of euthanasia and infanticide to the question of whether or not there should be ethical limits to academic bioethical discourse. E. Seidler, President of the *Akademie für Ethik in der Medizin*, suggests a discourse "about the limits to discourse", which might be the true "product of a new freedom to think" in Germany after the horrible experiences of Naziism 50 years ago ([69], p. 174). One author associated with the *Gen-Archiv* in Essen and other groups, defines bioethics as an "ethics of technocrats", a new "form of research in acceptability", a "service ethics", which provides "acceptance for risky technologies, in order to promote easy introduction of a health policy ready to accept dead persons and the realization of cost-benefit-based calculations of killing utopias". The author predicts that bioethics will serve as a tool of "further suppression of critical assessment of [genetic] and reproductive technologies or of a health policy exclusively oriented according to economic criteria" ([66], pp. 35ff.). The politically correct thing to do therefore was to reschedule the meeting of the European Society for Philosophy of Medicine and Health Care from Bochum, Germany to Maastricht in the Netherlands, because, as this author said: "It is mandatory to prevent, ...under the cloak of tolerance, democracy, and liberalism, strategies of annihilation [from being discussed]. This is the reason that we will prevent the Bochum congress" ([66], p. 43). In the field of party politics, one of the more recent cases of poorly informed political correctness was a request by Social Democratic members of Baden-Württemberg State Parliament that the executive branch of that State formally acknowledge that the teaching and the foundations of bioethics, as developed "by Anglo-Saxon institutes such as the Kennedy Institute of Ethics (Washington), the Hastings Center (New York) or the Center for Human Bioethics (Peter Singer, Clayton, Australia)...is incompatible with the norms of the Constitution (*Grundgesetz*)" [65].

A more moderate form of critical response towards the bioethical challenges of modern technology can be called "defensive ethics", most prominently represented by the strong influence of Has Jonas's lectures and his book *Ethics of Responsibility* [50]. Jonas's influence reaches from the ultra-conservatives to the Green Party. The targets of defensive ethics are new technologies such as nuclear power, information processing, and gene technologies. Jonas argues that, because these technologies might be misused, their development and application should not be supported by the government. Rather, the government should move to ban the further development of these technologies because ethics will not be able to keep pace with developing technology. The Embryo Protection Law [28] and the Gene Law [27] in Germany, adopted under severe technophobic public pressure, represent an attitude of defensive ethics because they issue a general ban on several applications of new technologies without differentiated ethical or legal argument.

"Differential ethics" and regulations, which distinguish between different applications and goals, understand the new technologies as value-free instruments for achieving either moral or immoral goals. This approach is represented by the self-regulation of the Federal Chamber of Physicians (*Bundesärztekammer*) ([9];[11];[12];[13];[14];[15];[16]) and the Society for Medical Law (*Deutsche Gesellschaft für Medizinrecht*) [32], and is reflected in many new laws. A differentiated approach has always been successful in Germany in creating frameworks within which citizens and professionals are called upon to identify their specific obligations. A good example of prudent and differentiated policy is the establishment and continuing reform of the statutory health care system and other elements of the German welfare system [25]. The concept of the "order of politics" (*Ordnungspolitik*), first developed by the Freiburg school in economics [46], had a particularly strong impact on the formation of networks of welfare and health insurance. Drug research regulation is not different from the legal and regulatory framework of the United States ([23];[74], p. 6ff;[55]).

II. NEW REPRODUCTIVE TECHNOLOGIES AND EMBRYO RESEARCH

A. *Germany*

In Germany, the ethical and legal aspects of medically assisted reproduction and related research were first discussed by the Central Commission of the Federal Chamber of Physicians. The Central Commission issued guidelines for infertility treatment using in-vitro fertilization (IVF); the guidelines limited these procedures to married women using only the spouse's sperm. Moreover, surrogate motherhood was not approved [10]. The Central Commission also drafted guidelines for research on improving methods of infertility treatment.

The 1985 guidelines, which governed research on the early embryo, defined as ethically acceptable only those research protocols which seek to improve methods of infertility treatment. In addition, only the use of spare embryos was accepted. The generation of embryos exclusively for research purposes was rejected. Finally, in order to proceed with a research protocol, the recommendation of the local Ethics Committee and a positive vote by the Central Commission was required [8]. A dispute then arose concerning the use of spare embryos for research purposes.

A national association of judges (*Deutscher Richterbund*) in 1986 concluded that IVF procedures would be illegal if they were not used exclusively for the purpose of giving life, *i.e.*, for implantation, because the production of embryos solely for research would violate the principle of human dignity protected by constitutional law [38]. Other legal experts disputed the conclusions of the *Richterbund* [47]. Nonetheless, the clinical practice of IVF changed, and all embryos generated by IVF were implanted. This new clinical approach, however, in turn produced the ethical conflicts posed by selective feticide. Guidelines by the Central Commission [9] recognize the risk that a greater number of fetuses pose to the mother and the increased risk of early pregnancy termination. The Commission guidelines recommend the careful use of hormonal stimulation in order to avoid the maturation of too many eggs, but also recognize the duty to reduce the number of implanted fetuses, if harm to them or to the mother can be prognosed [11]. In the latter cases, the moral principle to save life is paramount to accepting the loss of all lives. Feticide, therefore, should be discussed with the parents, should not be performed without their consent, and should be non-selective, *i.e.*, the fetus easiest to reach should be terminated.

Since the end of 1990, self-regulation of new reproductive technologies has been replaced by governmental legislation, the *Embryo Protection Law* (*Embryonenschutzgesetz (ESchG)*) [28]. The *EschG* primarily uses Penal Code legislation to enforce its provisions. Imprisonment for up to three years and monetary fines are the penalties for those who fertilize oocytes without the permission of the person from whom they are taken, those who implant an embryo into a woman without her consent, or those who use gametes of a deceased man for fertilization. Germ-line manipulation carries a penalty of up to five years, except for: (a) research on oocytes whose use for fertilization is excluded; and (b) research on other germ-line or fetal cells if it is assured the manipulated cells will not be transferred to embryos, fetuses or humans, and that other germ-line cells will not be generated from them. Cloning, *i.e.*, the artificial generation of an embryo with identical genetic information to another embryo, fetus, or person also carries a penalty of five years. The same penalty applies to the creation of chimeras or hybrids using human embryos, for the implantation of those chimeras into women or animals, and for the implantation of human embryos into animals. The term "embryo", as this law uses it, includes totipotent cells derived from embryos. An earlier version of the law contained a paragraph expressly excluding from protection embryos prevented by hormonal or technical devices from implantation [24]. The final version of the *ESchG* does not address the contraceptive and contra-implantative issue, but does not infringe on the use of any contraceptive or anti-implantation devices. The "protection" of early embryos from research and not from non-implantation might be criticized as hypocritical or as "research-bashing". In addition, as more and more DNA-based diagnosis on severe hereditary abnormalities becomes available, the law will preclude pre-implantation diagnosis. At the same time, other laws will allow abortion for medical or other reasons at later stages of fetal gestation, thus causing ethical double standards and complex medical and ethical risks for pregnant mothers and their physicians.

The *ESchG* does not directly address the use of fetal tissue. A policy statement of the Central Commission of the Federal Chamber of Physicians, however, discusses the use of fetal tissue from living and from cadaver fetuses. The statement concludes that the use of fetal tissue from a living fetus has to be restricted to the direct and carefully assessed benefit to this particular fetus. The use of tissue from a dead fetus will only be acceptable for clinical research after an ethics committee has concluded that the research cannot be performed without using fetal tissue and that the protocol is well designed. Permission of the mother for the use of fetal tissue must be obtained. If the dead fetus has been aborted, the probable use of its tissue for research may not be mentioned to the mother prior to abortion, because this possibility may influence her ethical judgment regarding the abortion [13].

B. *Austria*

In Austria, the Ministry for Science and Research put forward a policy statement for the legal and ethical regulation of new reproductive technologies and gene technologies in 1986 [54]. In July 1991, a federal Reproductive Medical Law (*Fortpflanzungsmedizingesetz, FMedG*) was introduced, whose ratification is pending. It includes changes in the Civil Code and the Code of Family Law [58]. The new law will allow IVF and artificial insemination only for couples and those partners who have lived together for at least three years. Medically assisted reproduction may only be performed by physicians, but physicians are not required to perform those services. A consulting session is required prior to assisted reproduction and both partners must agree in writing to the procedure(s). No research on gametes or embryos is legal, nor is germ-line manipulation. Heterological insemination is legal, if performed in a selected number of specially certified hospitals, but may only be performed on married couples or those living in marriage-like companionship. The offspring of heterological insemination, after reaching maturity, has the right to information about his or her father ([58], pp. 3-4). Conservation of oocytes and embryos for more than a year is illegal. Files on assisted reproduction have to be kept for thirty years. Finally, the sperm from one donor may only be used for three treatments ([58], pp. 3-4). The Austrian Conference of University Rectors in 1986 recommended that it "be carefully assessed whether eminent medical research promising progress in prevention or treatment of diseases should be allowed to use healthy spare embryos" and urged ethics committees to discuss the issue further. A heated debate arose about whether or not the morula was protected by law; in the meantime, a consensus has seemed to develop against using spare embryos for any research [53].

C. *Switzerland*

In Switzerland, two initiatives on the "Prevention of Abuse of Reproductive and Gene Technology in Humans" were introduced in 1991 for ratification by the Swiss people, one an initiative by the people, the other by the Federal Assembly [76]. The people's initiative has since been withdrawn, but the initiative by the Federal Assembly will be submitted for a vote in 1992. Its main proposals for a new article 24 of the Federal Constitution are: (1) manipulation of germ-line cells and embryos is illegal; (2) production of chimeras is illegal; (3) IVF is legal only as an infertility treatment or to prevent the transmission of a disease to the mother and only according to a law to be passed; thus, no spare embryos may be produced; (4) embryo donation and all forms of surrogate motherhood are illegal; (5) no commerce is allowed in human oocytes and embryo products; (6) genetic material may only be diagnosed, registered, or made public with the consent of the carrier; and (7) a right to hereditary data must be established ([77], p. 2). The new article of the Federal Constitution will supersede various cantonal laws in the cantons Aargau, Basel-Land, Geneva, Glarus, Neuchatel, St. Gallen, Ticino, Vaud, and the directives by the Health Council on assisted human reproduction from 1986 [72]. Most of these regulations allow IVF or artificial insemination by the husband and prohibit all other forms of intervention, but some simply refer to the 1985 guidelines of the Swiss Academy of Medical Sciences.

III. ABORTION

A. *Germany*

When Germany, after the political, economic, and moral bankruptcy of the Marxist-Leninist system of the German Democratic Republic, became unified in 1990, most laws and regulations of the West were applied in the East, but not Penal Code paragraph 218 addressing abortion. Until 1970, all abortion, with certain exceptions for medical indications, were crimes and the individuals involved were prosecuted. Paragraph 218 was changed in 1974 to allow the pregnant woman to choose abortion until the end of the first trimester [17]. However, the German Constitutional Court (*Bundesverfassungsgericht*) found the new law unconstitutional because it did not protect the rights of the unborn, which according to the Court were guaranteed by constitutional law (*Grundgesetz*). In interpreting the *Grundgesetz*, the court drew the following conclusions:

> The state's duty to protect life does not only prohibit direct governmental interference with developing life, it requests that the state protects and supports [developing] life. The responsibility of the state to protect developing life exists also against the mother. The protection of life of the fruit of the womb (Leibesfrucht), as the legal term says, during the entire period of pregnancy generally has higher value than the self-determination of the mother and may not be put at her disposal for a certain period of time ([18], p. 1).

This court decision and its legal reasoning has often been heavily criticized by legal experts [47]. A reformulation of the Penal Code then allowed abortion until the end of the first trimester, if certain social or medical indications were present. This reformulation, which formally satisfies the court's legal requirements, has been known since that time as the "indication model" (*Indikationsmodell*).

The German Democratic Republic in 1972 enacted a special Law on Interruption of Pregnancy which stated that the: "[e]quality of women in education, occupation, marriage, and family requires that women [be] free to make decisions about pregnancy and delivery" [30]. Since the law allowed abortive decisions within the first trimester, the regulation was called the "term model" (*Fristenmodell*). While the East German law gave full legal recognition and support to the principle of the mother's self-determination for the first 12 weeks, the West German model balanced the right of the fetus to live not against the right of the mother to exercise self-determination, but against social or medical hardship. Nonetheless, under the West German law, "social indications" (including "too many children") or medical indications (including a threat to commit suicide!) were easily met; thus, a double standard developed between the legal concept and its practical implementation. Legal and medical practices, however, were not consistent throughout different states of the Federal Republic. The more conservative states made abortion difficult or impossible, while the more liberal ones made it very easy to obtain the necessary documentation for abortion. The Unification Law in 1990 did not resolve the abortion issue, because citizens of the Marxist-Leninist state emphasized that they had the more liberal, less intrusive law which allowed a pregnant woman to choose. However, statistics show that the number of abortions is only slightly higher in the former East Germany.

The Federal Parliament (*Bundestag*) is charged to resolve the legal debate soon by ratifying a new law. Until that time, two laws on abortion exist in Germany, whose application is governed by residency. The requirement to ratify a new law opened a heated debate, with severe pressure from the Roman Catholic Church to pass a conservative law. Most citizens residing in the former East Germany want the "term model" to prevail. In the former West, the dispute takes place along party lines. The Liberals (*Freie Demokraten*) introduced a law which, very similar to the "term model", allows free choice to the mother, offers free but mandatory counseling prior to abortion, and promotes a "child-friendly" society [35]. Proposals include free availability of contraceptives, free kindergarten, and three years of paid leave. The Social Democrats also favor the more liberal "term model", and intend to improve it and to enhance respect for unborn life by introducing a large package containing sex education, family counseling, workplace and other benefits for pregnant women and mothers [36]. The Christian Democrats (*CDU*) are divided between those who favor an even more conservative law than the existing paragraph 218 and those who recognize the double standards in existing law and the inherent limits to enforcing reproductive behavior by law. A CDU proposal distinguishes medical indications from psycho-social indications; but both must be made by a physician, who must explain his judgment in writing [37]. Rita Süssmuth, a member of the Christian Democrats and the Speaker of the Bundestag proposed a "third way" to deal with the abortion issue. She urges the introduction of a new paragraph into the *Grundgesetz*: "[U]nborn, handicapped, and dying life requires special protection by means of appropriate help" ([73], p. 213). She also recommends a regulatory environment in which the pregnant woman must undergo consultation on reproductive choices, but is allowed to make the final decision within the first 12 weeks. Other proposals, such as the passing of a Life Protection Act, which would protect human life from the beginning of brain life (*i.e.*, synapsis formation at the end of the 10th week) until brain death have not found much support [64]. The Roman Catholic Church strictly opposes abortion, while the Protestant churches are divided. Some Protestant theologians and lay people criticize the traditional model, where physicians make the final decision about indications, for failing to respect the dignity and self-determination of the pregnant woman. The issue will have to be settled by the end of 1992. Because there is wide consensus on the need to improve social conditions which lead to the prevalence of abortion, any solution will include features which will make it easier for singles and families to decide to have and to raise children.

IV. HUMAN GENETICS

A. *Germany*

In Germany, two key aspects of the Human Genome Project have been discussed: the social acceptability of gene technology in general and of its eugenic aspects in particular. Given the historical background of Nazi experimentation and the existence of strong so-called "alternative groups", which are well connected to the media, public concerns have been voiced about the moral risks associated with genetic projects and discussed under "worst case scenarios" ([62], pp. 3-15). Amidst a basically technophobic attitude towards new forms of high technology, special fears have been expressed regarding future discrimination toward the retarded and the handicapped – discrimination in the work place, in health care and insurance systems, and perhaps reflected in an increased number of selective abortions following predictive prenatal screening. Prudent science policy has recognize the historical and cultural parameters of the ongoing debate, and has emphasized the need for the formation of public consensus on these issues. In 1983, Dr. Riesenhuber, the Minister of the *BMFT* [Federal Ministry for Research and Development], invited a diverse group of experts to discuss the ethical and legal issues raised by applying gene technology to humans. The German Research Foundation [*DFG*] and the Max Planck Society [*MPG*] introduced programs for the social and moral assessment of genetic research and sponsored conferences and publications. The new approach by Minister Riesenhuber indicates that the executive and legislative branches, as well as the research funding institutions in the Federal Republic, have learned a valuable lesson from the political and moral mismanagement of the nuclear debate in Germany; *viz*, of the need to encourage public ethical debate before introducing technology.

In 1987, a special commission of the Federal Parliament issued a report on "Chances and Risks of Gene Technology" [21]. The report discussed the social and moral acceptability of gene manipulation, including human genome screening in a variety of settings. It recognized the diagnostic benefits of genome analysis and recommended discussing prenatal screening and work place screening separately. Its recommendations included non-directive counseling, non-eugenic use of new technology, cooperation with the medical profession to develop a catalogue of treatable diseases for prenatal screening, self-regulation in order to disclose only non-treatable severe anomalies or diseases prior to the twelfth week of pregnancy, and the development of binding legal guidelines governing employer-employee cooperation on genetic screening in the work place.

A joint Bund-Laender commission of senior executives from state and federal governments issued guidelines, in May 1990, which rejected a general ban on genetic screening and instead favored oversight of specific areas where abuse is likely to be greater: prenatal screening, screening in occupational health, and the protection of data and privacy in civil and penal court cases. The Bund-Laender group offered the following recommendations: (1) legal and voluntary access to prenatal screening; (2) screening of totipotent cells and screening of non-totipotent trophoblast cells only after clinical safety standards have been developed; (3) non-directive and non-acute counseling; (4) special licensing for screening physicians; (5) limited screening for severe and untreatable anomalies and diseases; (6) restricted information to parents prior to the twelfth week of pregnancy, particularly regarding the sex of the fetus; (7) costs to be paid by statutory or private insurance; (8) no official large scale screening for non-treatable diseases; (9) screening in the work place which does not include general DNA diagnosis but only diagnosis for diseases or risks related to the particular work place; (10) informed consent of the employee; (11) no general screening to be allowed for those applying for health care or life insurance; and (12) no screening to classify groups of persons, although other specific screening in forensic medicine (such as DNA fingerprinting) may be done.

These guidelines mark an important consensus in German public policy and will determine the future course of regulation, self-regulation, legislation and employer-employee cooperation. The "right to know or not to know", according to the Working Group, is an essential aspect of self-determination and should be respected accordingly; thus the optimal framework for prenatal genetic testing is the sequence of "counseling – testing – counseling" (p. 141; [26]). The findings and recommendations of the Bund-Laender group are widely supported, except by the Green party. A 1986 public statement by the Green party described its position in this way:

> Green's politics is directed against the concepts of biology and medicine which apply genetic engineering in an attempt to merely repair social and environmental problems. Genetic engineering techniques are the product of a relationship to Nature which is based on exploitation and domination rather than sustainability. This is true not only for applied research but for basic research as well; with the advent of genetic engineering, both are becoming increasingly directed by the objectives of accelerating industrial utilization of Nature...<u>We are demanding an immediate halt to all genetic engineering research and to all related applications and forms of production</u> ([45], p. 2).

In July 1990, the German Genetic Engineering Law (*GenGesetz*), subtitled "The Law for the Protection of Man and Environment" (and including five related regulations) was implemented [25]. The law supplements the recommendations of the Bund-Laender group in ensuring protection against the risks and dangers arising from recombinant nucleic acids constructed *in vivo* ([62], pp. 81-96; [62], pp. 125-139).

A study group of Christian Democratic legal experts issued a 1987 policy statement which emphasized the benefits of prenatal testing for severe disorders but recommended the prohibition of testing for sex or minor disorders [6]. The Federal Chamber of Physicians, in a similar statement, took the same position, but also stressed the importance of protecting data and privacy, of respecting a woman's free choice, and the need for post-abortion counseling [9]. The Chamber also issued guidelines for human genome therapy which supported the use of somatic cell therapy but rejected germ-line intervention ([12];[62], pp. 331-334).

The German federal government (*Bundesregierung*) and the federal parliament (*Bundestag*) support the European Genome Project. However, the first draft of the EC Predictive Medicine Project (COM (88) 424 final - SYN 146, July 20, 1988) caused a stormy debate in the federal parliament concerning the possible eugenic consequences of the project. In particular, the phrase "predictive medicine" was interpreted as introducing a eugenic public health policy for all of Europe. All parties, except the Greens, were in favor of possible diagnostic benefits from the project. But all participants, including the federal government, criticized possible or actual eugenic ideology and intentions of the program. As a result, the wording and focus of the project were subsequently changed. One participant asked whether the program would later call for a "eugenic oriented European medical and health policy and mass screening, ... a European abortion program motivated by eugenics". Another questioned whether "we Germans, in the light of experiences during the years 1933 through 1945, should be sensible, even supersensible" with regard to the protection of human rights and human dignity. Minister Riesenhuber pointed out that germ-line intervention should not be accepted, that genome analysis should not be used for a health policy based on eugenics, and that voluntariness and data protection should be essential [65].

V. WITHHOLDING TREATMENT, EUTHANASIA AND PATIENT SELF-DETERMINATION

Euthanasia is illegal in Austria, Germany, and Switzerland and will be prosecuted according to the national Penal Codes. The Swiss Guidelines concerning Assistance in Dying state: "The duty of the physician and of the nursing staff, which embraces the noble aim of healing, aiding, and alleviating, also includes assisting the dying patient until his death. This assistance comprises medical treatment, personal support and nursing care" ([74], p. 12). The wishes of patients who are competent and suitably informed, however, must be respected, even if honoring these wishes does not appear to be the best medical treatment. Humane support must be given as long as any contact is possible with the dying patient. Withholding treatment, and the withdrawal of drugs or various technical measures including artificial feeding, is justified from the medical point of view "if a postponement of death would represent an intolerable prolongation of suffering, and if the basic disease has taken an irreversible turn with a fatal prognosis" ([74], p. 15). In cases of severe cerebral disorders when, after a long period of observation the physician has decided that the process is irreversible, "he may renounce special measures to prolong life, even when spontaneous respiration and swallowing are still present. In these cases, treatment may be limited to nursing care" ([74], p. 15). In making decisions about reducing technical levels of medical treatment, the physician must take into account "the personality of the patient and his syndrome or presumed will, his tolerance of pain and mutilation, the acceptability of medical interventions by the patient, the availability of certain therapeutic measures, [and] the attitude of the patient's family and other persons around him" ([74], p. 16). German Guidelines for Assisting the Dying, adopted by the Federal Chamber of Physicians, mirror the Swiss guidelines ([61], pp. 378-383).

With regard to withholding treatment from severely defected newborns, the Swiss guidelines state the following:

"If a new-born baby or an infant has severe deformities or has suffered perinatal damage to the central nervous system which would result in irreparable disturbances of development, and if, furthermore, life is possible only with the permanent help of special technical measures, it is permissible to refrain from initiating or continuing their use" ([61], p. 15).

The German Society for Medical Law in 1986 has adopted guidelines which state that the medical duty to treat must be based not only on the "possibilities of medicine", but also on "human-ethical assessment criteria and the physician's obligation to care" ([61], p. 376). The Society differentiates between two sorts of handicaps: (1) cases of severe abnormalities or multiple severe handicaps where traditionally physicians did not prolong life or intervene with heroic measures, such as severe microcephaly or a combination of severe abnormalities which will result in imminent death; and (2) cases involving handicaps which did not and do not allow for withholding treatment, such as Down syndrome. The Society did not make recommendations, but asked the physician to decide each case individually by taking the traditional ethics of withholding or intervention into account.

Given the increased technical capacities of intensive care and the illegality of active euthanasia, discussion has begun about changing the legal framework and allowing patients to determine the circumstances and the time of their dying. Clinical experiences with longtime comatose patients and patients in persistent vegetative state (PVS) have prompted physicians and legal experts to discuss the ethical issues of actively assisting in the shortening and easing of the dying process. Some have proposed reformulating Paragraph 216 of the German Penal Code to allow for euthanasia "when, because of a severe illness, the continuation of life is against the interest of the person concerned and the killing is performed by a physician" ([49], p. 295). It is against the background of these proposals and the already mentioned publicity of Singer's invited and aborted lectures that *very* emotional anti-euthanasia activities have increased [*e.g.*, [71].

Living wills are not routinely recognized as valid expressions of the patient's will in Germany. A Society for Dying in Dignity (*Gesellschaft für Humanes Sterben*) distributes forms which allow individuals to express treatment preferences and to make advanced directives regarding resuscitation or other forms of intensive treatment at the end of life. In Switzerland, the Academy of Medical Sciences advises physicians to take advance directives into account in clinical decision-making, but not to base decisions exclusively upon them. According to the Academy,

"What is decisive...is...[the patient's] presumed will at the present time, which can be determined only by careful appraisal of all circumstances. Because a previous declaration can be withdrawn at any time, it does not put the physician under obligation. One always has to consider, therefore, if the patient would reasonably at the present time have revoked the decision or not" ([74], p. 18).

The legal situation in Austria is similar. Since the binding character of a living will is legally uncertain ([51], p. 74), physicians are not obligated to respect advanced directives.

A prominent German group, made up of two surgeons, a theologian and a law professor, has recently issued a policy statement paper emphasizing that the law does not require intensive surgical therapy at all times for all patients. Rather, intensive and surgical therapy should be intended to restore the health of the patient or to ease his pain. Thus, surgeons, together with their patients, should evaluate high technology interventions on an individual basis. Moreover, decisions about intervention or non-intervention should be made solely for medical reasons, not from economic considerations. The paper concludes by discussing different scenarios of withholding, reducing, and terminating therapy ([5], p. 2462).

VI. ORGAN DONATION

Germany is a member of the Netherlands-based Eurotransplant, an organization which computerizes information about organs and tissues and organizes the distribution of cadaver organs. Organ information and distribution centers in Switzerland and Austria are loosely affiliated with Eurotransplant. In all three countries, the use of cadaver organs is not legal without the documented permission of the donor or his or her family's consent, except in urgent cases where the presumed will of the donor or the consent of the family cannot be obtained and no objection can be presumed ([32], pp. 199-210). Several organizations are working to increase the number of donor card holders.

In Austria, the legality of "encapsuled organ transplantation", *i.e.*, the transplantation of an organ from an unborn fetus to the mother, has been discussed [40]. The procedure of encapsuled transplantation would arrange for the non-birth of the fetus and the removal of its head and other organs and tissues so that, in cases of kidney transplantation, the fetal kidneys would become part of the mother's body. Since the Austrian law does not distinguish between abortion and the killing of a fetus and allows for abortion until the end of the first trimester, it appears to be legally uncertain whether the "operation" which results in the liquidation of the fetus but the survival of an organ-mass is legal or not. Eder-Riedel suggests that encapsuled transplantation, which has not yet been expressly addressed by law, should be evaluated by ethics committees and that an embryo protection law should be passed later [40].

VII. ANIMALS IN RESEARCH

Bioethical concern for the protection of animals in research is very strong all over Europe. A special Animal Protection Law (*Tierschutzgesetz*) was passed in Germany in 1986 [19]. Various regulatory interventions followed. General administrative regulations implementing the Law on Animal Protection were passed by the Federal Ministry for Agriculture and Forestry in 1988, with detailed requirements for the type of experiment performed, selection of animals, supervision by qualified veterinarians, and regulations on the treatment of animals in agriculture and as pets [22]. Education and information have been stressed as essential elements in improving the ethical treatment of animals within and outside the research setting [79]. In addition, public authorities must be notified of the election of animal welfare commissioners who have academic degrees in veterinary medicine, medicine or zoology, or additional technical skills and training. Finally, animal welfare committees have been established on the Laender level, *i.e.*, in all states of the Federal Republic. One-third of the members are chosen from lists submitted by animal welfare groups, and two-thirds must have academic degrees and professional experience as prerequisites ([79], pp. 180-181).

Swiss guidelines, initiated by animal rights activists years ago, have been exemplars for new regulations in other countries. Austria passed its Animal Research Law (*Tierversuchsgesetz*) in 1989 [56]. The law requires that research be reviewed and performed according to the recent scientific standards, that the number of animals involved and the number of experiments be reduced whenever possible, and that all persons involved exercise ethical and scientific responsibility ([56], p. 3480). Violations of the law will result in criminal charges ([56], p. 3483).

VIII. ETHICS COMMITTEES IN CLINICAL RESEARCH

The Swiss Academy of Medical Sciences has had a central Ethics Committee since 1979. Among its members are a law professor and two delegates of the Swiss Nurses Association. The yearly reports of the Committee to the General Secretary of the Academy are published in the annual report of the Academy ([74], pp. 3-5).

In Germany, ethics committees are required by law to review clinical research [20]. All Chambers of Physicians on the state level and all Medical Faculties have research ethics committees. These are mostly peer review boards. Some have legal experts, but only a few ethicists or philosophers are members. Hospital ethics committees are not widely spread. The study group of the Ethics Committees meets once a year, according to their statutes [19].

The working group has passed guidelines to regulate ethics committees ([77], pp. 163-164) and a checklist for the review of clinical research protocols [3]. The guidelines require that ethics committees have at least five members, among them at least four physicians and one legal expert. For prospective ethical review of new medical technologies, the Federal Chamber of Physicians has established the already mentioned central ethics committee, the Central Commission for the Protection of Ethical Principles in Reproductive Medicine, Embryo Research, and Gene Therapy. The Central Commission publishes annual reports ([14]; [15]; [16]), and also serves, in addition to local ethics committees, as a review board for clinical research in IVF treatment.

The Austrian *Wiener Krankenhausgesetz* requires that all research hospitals have at least one commission charged with reviewing the ethical aspects of clinical protocols (57], p. 16). The commission must include two physicians who are not involved in the research under review, one person representing the clinical staff, one legal representative of the hospital, one representative of the hospital chaplains, one pharmacist, and one person representing hospital workers.

IX. CARE AND COST CONTAINMENT IN STATUTORY HEALTH CARE SYSTEMS

The discussion of the structures of the health care systems of Germany, Austria, and Switzerland is ongoing. All three countries have statutory health care systems of different kinds, with the general structure of the German system over one hundred years old [60]. Health care systems are based on the principles of solidarity and social justice, allowing for limited market competition among institutional and individual providers of health care. Advances in costly medical technologies and the increasing costs of acute care have led to discussions about reducing costs by making minor changes in the system. Welfare attitudes of citizens have been criticized ([60], pp. 79-90; [60], pp. 151-168). More emphasis has been placed on individual health literacy and responsibility ([60], pp. 133-134). In addition, the segregation of general health insurance into a basic and mandatory and a supplementary and voluntary system has been proposed ([60], pp. 93-112).

In Germany since 1987, a federal working group, the Council of Experts for Concerted Action in Health Care, composed of high ranking bureaucrats, representatives of insurance companies, physicians, hospital organizations, health economists, and organized labor and business has been addressing the issues in routine meetings and annual reports ([60], pp. 13-33). The most recent discussion focused on the issue of nursing home care costs ([1];[2]). One side has argued for improving the existing system based on private insurance and market competition, while the other side has called for including nursing home care costs in the statutory system [31].

Center for Bioethics
Kennedy Institute of Ethics
Georgetown University
Washington, DC, U.S.A.

BIBLIOGRAPHY

1. Anonymous: 1991, "F.D.P. Für Private Pflegeversicherung", *Deutsches* Ärzteblatt 88, D-4082.
2. Anonymous: 1991, "Solidarische Pflegeversicherung, *Gesellschafts-politische Kommentare*, 32, No. 2 [Sonderausgabe: Modelle und Vorschläge zur Absicherung des Pflegerisikos].
3. Arbeitskreis Medizinischer Ethikkommissionen in der Bundesrepublik Deutschland: 1990, "Checkliste zur Überprüfung der Vollständigkeit von Anträgen an die Ehtikkommission vor der Durchführung Klinischer Versuche order Epidemologischer Forschung am Menschen", in: Toellner, R. (ed): 1990, *Die Ethikkommission in der Medizin*, Fischer, Stuttgart, 170-171.
4. Arbeitskreis Medizinischer Ethikkommissionen in der Bundesrepublik Deutschland: 1990, "Verfahrensgrundsätze", in: Toellner, R. (ed.): 1990, *Die Ethikkommission in der Medizin*, Fischer, Stuttgart, 163-165.
5. Berger, H. G., Oettinger, W., Rössler, D., Schreiber, H. L.: 1991, "Grenzen der Intensivtherapie in der Chirurgie", *Deutsches Ärzteblatt*, 88, C-2461-2465.
6. Bundesarbeitskreis Christlich-Demokratischer Juristen: 1987, "Leitsätze zur Genomanalyse", in: Ruth Faden, *Die Genomanalyse, Zentrum für Medizinische Ethik*, Bochum [Medizinethische Materialien, No. 9].
7. Bundesärztekammer (Wissenschaftlicher Beirat): 1984, "*Pränatale Diagnostik*, Empfehlungen des Wissenschaftlichen Beirates, Deutsches Ärzteblatt" 84, B-434-436.
8. Bundesärztekammer: 1985, "Richtlinien zur Forschung an Frühen Menschlichen Embryonen", *Deutsches Ärzteblatt* 82, B-3757-3764.
9. Bundesärztekammer: 1988, "Prädiktive Medizin: Analyse des Mmenschlichen Genoms", *Deutsches Ärzteblatt*, 85, 3595.
10. Bundesärztekammer: 1988, "Richtlinien zur Durchführung der In-vitro-Fertilisation mit Embryotransfer und des Intratubaren Gamenten- und Embryotransfers als Behandlungsmethode der Menschlichen Sterilität",

Deutsches Ärzteblatt 85.

11. Bundesärztekammer: 1989, "Mehrlingsreduktion Mittels Fetozid", *Deutsches Ärzteblatt*, 86, A-2218-2222.
12. Bundesärztekammer: 1989, "Richtlinien zur Gentherapie beim Menschen", *Deutsches Ärzteblatt*, 86, A-2957-2960.
13. Bundesärztekammer: 1991, "Richtlinien zur Verwendung Fetaler Zellen und Fetaler Gewebe", *Deutsches Ärzteblatt*, 88, D-4432-4436.
14. Bundesärztekammer (BÄK) Zentrale Kommission zur Wahrung ethischer Grundsätze in der Reproduktionsmedizin, Forschung an Menschlichen Embryonen und Gentherapie: 1989, *Arbeits- und Erfahrungsbericht 1988*, BÄK, Köln.
15. Bundesärztekammer (BÄK), Zentrale Kommission zur Wahrung Ethischer Grundsätze in der Reproduktionsmedizin, Forschung an Menschlichen Embryonen und Gentherapie: 1990, *Arbeits- und Erfahrungsbericht 1989*, BÄK, Köln.
16. Bundesärztekammer (BÄK), Zentrale Kommission zur Wahrung ethischer Grundsätze in der Reproduktionsmedizin, Forschung an Menschlichen Embryonen und Gentherapie: 1991, *Arbeits- und Erfahrungsbericht 1990/91*, BÄK, Köln.
17. Bundesrepublik Deutschland: 1974, "Fünftes Gesetz zur Reform des Strafrechts", *Bundesgesetzblatt*, I, 1297.
18. Bundesrepublik Deutschland, Bundesverfassungsgericht: 1975, "Urteil des Ersten Senats vom 25. Februar 1975 [zur Neufassung des Paragraphen 218 und 219]", *Bundesverfassungsgerichtsentscheide [BVerfGE]*, 39, 1.
19. Bundesrepublik Deutschland: 1986, "Tierschutzgesetz", *Bundesgesetzblatt*, I, 1319 (18. August 1986).
20. Bundesrepublik Deutschland, "Bundesregierung: 1987, Bekanntmachung von Grundsätzen für die Ordnungsgemäße Durchführung der Klinischen Prüfung von Arzneimitteln", *Bundesanzeiger*, 243, 16617 [30. 12. 1987].
21. Bundesrepublik Deutschland, Deutscher Bundestag: 1987, "Bericht der Enquete-Kommission: 'Chancen und Risiken der Gentechnologie'", *Drucksache 10/6775*, Deutscher Bundestag, 10. Wahlperiode.
22. Bundesrepublik Deutschland, "Bundesminister für Landwirtschaft und Forsten: 1988, Verordnung vom 20. Mai 1988 über die Behandlung von Tieren in der Forschung", *Bundesgesetzblatt*, Vol. I, No. 21, 31. Mai 1988 [English version, in: *Federal Journal* 40, (July 29, 1988) ISSN 0720-6100].
23. Bundesrepublik Deutschland: 1988, "Drittes Gesetz zur Änderung des Arzneimittelgesetzes vom 20. Juli 1988", *Bundesgesetzblatt*, I, 1050.

24. Bundesrepublik Deutschland, Der Bundesminister der Justiz: 1989, *Entwurf eines Gesetzes zum Schutz von Embryonen (IIAI-4000727-230.577/89)*, BMJ, Bonn.
25. Bundesrepublik Deutschland, Bundesminister für Arbeit und Soziales: 1989, *Social Security*, Bundesminister für Arbeit und Soziales, Bonn.
26. Bundesrepublik Deutschland, Bund-Länder Arbeitsgruppe: 1990, *Abschlußbericht der der Bund-Länder Kommission "Genomanalyse"*, Bundesministerium der Justiz, Bonn.
27. Bundesrepublik Deutschland: 1990, "Gesetz zur Regelung von Fragen der Gentechnik", *Schutz von Mensch und Umwelt. Das Gentechnikgesetz*, Presse- und Informationsamt der Bundesregierung, Bonn, pp. 21-40.
28. Bundesrepublik Deutschland: 1991, *Gesetz zum Schutz von Embryonen* [ESchG], Presse und Informations der der Bundesregierung, Bonn [also Deutscher Bundestag Drucksache 11/5460, 24. Okt. 1990 and Bundesrat Drucksache 745/90, 26. 10. 1990; English version, translated by R. Nicholson, in *Bulletin of Medical Ethics* 64, (Dec. 1990) 9-11] .
29. Bundesrepublik Deutschland, Bundesminister für Forschung und Technologie: 1991, *Ethische und Sozial Aspekte der Erforschung des Menschlichen Genoms*, Campus, Frankfurt.
30. Bundesversammlung der Schweizerischen Eidgenossenschaft: 1991, Vorlage der Redaktionskommission für die Schlussabstimmung: Bundesbeschluss über die Volksinitiative "Gegen Missbräuche der Fortpflanzungs- und Gentechnologie" vom 21. Juni 1991, *Bundesversammlung, 89.067*, pp. 1-2.
31. Clade, H.: 1992, "Keine Wende in der Gesundheitspolitik", *Deutsches Ärzteblatt*, 88, C-9-10.
32. Deutsch, E.: 1991, *Arztrecht und Arzneimittelrecht*, Springer, Heidelberg.
33. Deutsche Demokratische Republik: 1972, "Gesetz über die Unterbrechung der Schwangerschaft", *Gesetzblatt I, 89ff*.
34. Deutsche Gesellschaft für Medizinrecht: 1986, "Grenzen der Ärztlichen Behandlungspflicht bei Schwerstbehinderten Neugeborenen, *Medizinrecht*, 1, 281.
35. Deutscher Bundestag, F. D. P. Fraktion: 1990, *Vorschlag für den Entwurf eines Gesetzes zum Schutz des Werdenden Lebens, der Förderung einer Kinderfreundlichen Gesellschaft, für Hilfen im Schwangerechaftskonflikt und zur Regelung des Schwangerschaftsabbruchs (Schwangerenhilfsgesetz)*, Deutscher Bundestag, Bonn.
36. Deutscher Bundestag, SPD Fraktion: 1991, *Entwurf eines Gesetzes zur Beratung in Fragen der Sexualität und Familienplanung, über Hilfen mit Rechtsanspruch für Schwangere und Mütter sowie Neuordnung des*

Schwangerenhilfsrechts (Schwangerenhilfsgesetz-SchHG), Deutscher Bundestag, Bonn.

37. Deutscher Bundestag, CDU Fraktion: 1991, *Gesetzentwurf zur Novellierung des des Paragraphen 218*, Deutscher Bundestag, Bonn.
38. Deutscher Richterbund: 1986, *Thesen zur Fortpflanzungsmedizin und zur Humangenetik*, Deutscher Richterbund, Bundesvertreterversammlung 17 (April 1986).
39. Eder-Riedel, M. A.: 1988, 4,32 "Die In-vitro-fertilization (IVF) in Rechtlicher Sicht", *Gynäkologische Rundschau*, 203ff.
40. Eder-Riedel, M. A.: 1990, "Strafrechtliche Aspekte der 'Kapselgeschützten' Organtransplantation", *Österr. Jur. Zeitung*, 627.
41. Eser, A.: 1989, *Neuartige Bedrohungen ungeborenen Lebens. Embryonenforschung und "Fetozid" in Rechtsvergleichender Sicht*, C.F. Müller, Karlsruhe [Juristische Studiengesellschaft Karlsruhe, No 187]
42. Evangelische Kirche in Deutschland, Deutsche Bisschofkonferenz: 1989, *Gott ist ein Freund des Lebens*, Mohn, Gütersloh.
43. Federation of Swiss Physicians [FMH]: 1990, "Patientenverfügung / Dispositions de Fin de Vie", *Schweizerische Ärztezeitung* 71 [33] 1338-1339.
44. Gesellschaft für Humangenetik, München: 1989 "Anwort der Gesellschaft für Humangenetik auf die vom BMFT Gestellten Fragen", in Bundesärztekammer (BÄK), Zentrale Kommission zur Wahrung ethischer Grundsätze in der Reproduktionsmedizin, Forschung an Menschlichen Embryonen und Gentherapie: 1990, *Arbeits- und Erfahrungsbericht 1989*, BÄK, Köln, section No. 9.5.
45. Grünen, Die: 1986, "Public Statement on Gene Technology and on the Human Applications of Reproductive and Genetic Engineering" (Adopted by the 8th National Assembly of the Green Party, Hagen, February 15-16, 1986), *Pressedienst*, Die Grünen, Bonn.
46. Herder-Dorneich, P.: 1983, *Orndnungstheorie des Sozialstaates*, Mohr, Tübingen.
47. Hoerster, N.: 1989, "Forum: Ein Lebensrecht für die Menschliche Leibesfrucht", *Juristische Schulung* 3, 173-178.
48. Hoerster, N.: 1989, "Tötungsverbot und Sterbehilfe", in H. M. Sass (ed.), *Medizin und Ethik*, Reclam, Stuttgart, pp. 287-294.
49. Hülsmann, C., Koch, H. G.: 1990, "Regelungen in der Bundesrepublik Deutschland", in A. Eser (ed.), *Regelungen der Fortpflanzungsmedizin und Humangenetik*, Vol. I, Enke, Frankfurt, 29-156.
50. Jonas, H.: 1980, *Das Prinzip Verantwortung*, Insel, Frankfurt.
51. Lachmann, J.: 1991, "Zur Bindungswirkung des 'Patiententestaments'", *Anwalts Blatt*, 74.

52. Landtag von Baden Württemberg: 1991, "Antrag des Abg. Ulrich Lang (SPD) u.a.: 'Bioethik' und Grundgesetz", *Landtag von Baden-Württemberg, 10. Wahlperiode*, Drucksache 10/6122.
53. Lewisch, P.: 1990, "Leben Order Sterben Lassen. Zur Frage der Verbrauchenden Experimente an Embryonen", *Österr. Jurist. Zeitung* 45, 133-142.
54. Republik Österreich Bundesministerium für Wissenschaft und Forschung: 1986, *Zu grundsätzlichen Aspekten der Gentechnologie und der humanen Reproduktionstechnologie*, BMWF, Wien.
55. Republik Österreich: 1989, "Arzmeimittelgesetz", *Österreichisches Recht*, December 12, Lieferungen 909 und 910.
56. Republik Österreich: 1989, "Bundesgesetz vom 27. 9. 1989 über Versuche an Lebenden Tieren", *Bundesgesetzblatt*, 207. Stück, 3479-3483.
57. Republik Österreich: 1990, "Wiener Krankenanstaltengesetzgesetz [WrKAG], *Rechtsvorschrift S 740-000*, Seite 1-39" [Anlage zur Wiederverlautbarungskundmachung der Wiener Landesregierung vom 24. 3. 1987].
58. Republik Österreich: 1991, "Regierungsvorlage: Bundesgesetz, mit dem Regelungen über die Medizinisch Unterstützte Fortpflanzung Getroffen Werden" (Fortpflanzungsmedizingesetz – FMedG), *Stenographgische Protkolle des Nationalrates XVIII.GP*, 216 der Beilagen [ausgedruckt am 30. 7. 1991].
59. Sass, H. M.: 1988, "Biomedical Ethics in the Federal Republic of Germany", *Theoretical Medicine* 9, 287-297.
60. Sass, H. M. (ed.): 1988, *Ethik und öffentliches Gesundheitswesen*, Springer, Heidelberg.
61. Sass, H. M. (ed): 1989, *Medizin und Ethik*, Reclam, Stuttgart.
62. Sass, H. M. (ed): 1990, *Genomanalyse und Gentherapie. Ethische Herausforderungen in der Humanmedizin*, Springer, Heidelberg.
63. Sass, H. M. (ed): 1991, *Güterabwägung in der Medizin*, Springer, Heidelberg.
64. Sass, H. M.: 1991, *Will there ever be a Consensus in the Abortion Debate?*, Zentrum für Medizinische Ethik, Bochum [Medizinethische Materialien, Nr. 72].
65. Sass, H. M.: 1992, "National and Religious Concerns and Input: A View from Germany", in D.S. Grisolia (ed.), *The Human Genome Project*, Fundacion BBV, Madrid.
66. Sass, H. M., Viefhues, H.: 1990, *Vierte Jahrestagung der European Society for Philosophy of Medicine and Health Care*, Zentrum für Medizinische Ethik, Bochum [Medizinethische Materialien, No. 59].

67. Schöne-Seifert, B, Rippe, K. P.: 1991, "Silencing the Singer: Antibioethics in Germany", *Hastings Center Report* 21 (6), 20-27.
68. Schweizerische Ärztegesellschaft: 1990, "Patientenverfügung. Dispositions de Fin de Vie", *Schweizerische Ärztegesellschaft*, 71, 1338-1339.
69. Seidler, E.: 1990, "Editorial", *Ethik in der Medizin* 2, 173-174.
70. Singer, P.: 1991, "On Being Silenced in Germany", *The New York Review of Books* (August 15), 36-42.
71. Spämann, R. *et al.*: 1991, *Kinsauer Manifest*, Sophienstiftung [Herzogstraße 5], Kinsau.
72. Stepan, J (ed.): 1990, "Canton Laws on Assisted Procreation in Switzerland", *International Survey of Laws on Assisted Procreation*, Schulthess, Zürich [Publications of the Swiss Institute of Comparative Law, No. 15], 173-184.
73. Süssmuth, R.: 1990, Schutz des Ungeborenen Lebens im Geeinten Deutschland. Ein Dritter Weg, *Ethik in der Medizin* 2, 211-218.
74. Swiss Academy of Medical Sciences: 1983, *Guidelines and Recommendations on Medical Ethics*, Schwabe, Basel.
75. Swiss Academy of Medical Sciences: 1990, "Medizinisch-Ethische Richtlinien für die Ärztlich Assistierte Fortpflanzung", *Schweizerische Ärztezeitung* 72 [10], 374-376.
76. Switzerland, Bundesversammlung: 1991, "Bundesbeschluss über die Volksinitiative 'Gegen Missbräuche der Fortpflanzungs- und Gentechnologie beim Menschen'", *BBl, 89.067*, 1-2.
77. Toellner, R. (ed.): 1990, *Die Ethikkommission in der Medizin*, Fischer, Stuttgart.
78. Wessels, U.: 1991, *Genetic Engineering and Ethics in Germany*, Universität des Saarlandes, Saarbrücken [Diskussionsbeiträge zur Ethik, Nr. 5].
79. Wiemer, U.: 1989, "Animal Experimentation: Legislation and Education in the Federal Republic of Germany", in L.F.M. Van Zutphen (ed.), *Animal Experimentation and Legislation*, Veterinary Public Health Inspectorate, Rijswijk, Utrecht.

RICHARD NICHOLSON[1]

BIOETHICS DEVELOPMENTS WILL COME LATER IN THE NEW EASTERN EUROPE: 1989-1991

I. INTRODUCTION

It should come as no surprise to readers that in eastern Europe there was little legislative or regulatory activity relevant to the time period and concerns of this yearbook. From 1989 to 1991, there were immense changes throughout the region as authoritarian communist regimes gave way to mainly democratic societies. Even the countries existing in the region changed substantially. The German Democratic Republic was reunited with the Federal Republic, while constituent republics of the Union of Soviet Socialist Republics (USSR) began to declare their independence. The latter process has continued since June 1991, the end of the period under consideration, as further independent states have been formed in both the former USSR and Yugoslavia.

These changes have put immense pressure on national legislatures as political and economic structures have been completely reformed. Ministries of health have had to concentrate on maintaining health care systems while preparing for changes such as the introduction of health insurance and private practice. Bioethical problems were not high on the national agenda in these countries during communist rule, and there has yet been little opportunity either to increase public awareness and discussion of them, or to address them legislatively. However, the opportunities that have been available have often been used imaginatively. On World AIDS Day 1991, for instance, Dr. Martin Bojar, Minister of Health for the Czech Republic, ascended by balloon over one of the old town squares in Prague and showered those below with condoms!

What follows is an incomplete picture of how the bioethics topics that are the subject of this yearbook have been handled in eastern Europe. The principal reason for the incompleteness is that there has been little relevant activity by legislatures or professional associations, and virtually none in the courts. Another reason, however, is that it is not much easier to obtain precise information now than it was under communist rule. A third consideration is

[1]The author gratefully acknowledges the help of the following correspondents in eastern Europe in producing this chapter: Petko Dontschev, Eugenius Gefenas, Jiri Haderka, Dagmar Pohunkova, Juait Sandor, Zbigniew Szawarski.

B. Andrew Lustig (Sr. Ed.), Bioethics Yearbook: Volume 2, 233-250.

that under communist rule, the difference between what really happened and the official line on a particular problem was often considerable.

When teaching in Yugoslavia a few years ago, for instance, I was told very firmly that nothing that could remotely be viewed as euthanasia was permitted. Thus, if the heart of a 90-year old with widely disseminated cancer stopped beating, it would be the duty of the health professionals present to try to resuscitate her. After the session, however, a medical student quietly pointed out that what happened in practice was very different; the view expressed during the session was the official line to which the students had to adhere since a senior public health official was present. This student suggested that I should visit mountain villages in the south of the country. When old people there became a burden on their families and felt that they were no longer making a worthwhile contribution, they knew that they could go to the local general practitioner, who could provide the wherewithal for a gentle death.

Since there is not enough material available from each country in eastern Europe to provide a comprehensive picture of their approaches to bioethics, the material has instead been divided according to topic.

II. NEW REPRODUCTIVE TECHNIQUES

Artificial insemination is the reproductive technique most regulated in Europe at present. The regulations tend to be extremely detailed, although Hungary has produced a succinct definition of who is entitled to receive such treatment:

> Artificial insemination may be performed at the request of any woman living in a state of marriage, who is not yet 45 years of age, is in full possession of her physical and intellectual capacities, resides permanently on Hungarian territory, and according to medical opinion, is in all probability unable to conceive a healthy child by natural means ([8], p. 76).

A decree published in the USSR gave considerable detail about which establishments were allowed to provide artificial insemination by donor; how sperm banks were to be organized; what examinations of donor and recipient were to be performed; and how much a donor was to be paid *viz.,* 20 roubles per ejaculate ([17], p. 597).

The Bulgarian regulations are also extremely detailed, as shown by the following extracts:

> 6. Artificial insemination shall be authorized if the following conditions are fulfilled:
>   [a]. The woman is in a satisfactory state of health ...; and
>   [b]. the person responsible for the consultation center on sterility has established the presence of ovulation and potency of the fallopian tubes in the woman.
>
> 7. Artificial insemination with genetic material from the spouse shall be performed after verification of the following:
>   [a]. A cervical secretion in the woman prevents the penetration of her husband's

spermatozoa;

[b]. the husband has a mild degree of reduced fertility; or

[c]. one of the spouses is suffering from a disease or pathological condition that renders normal sexual relations or natural insemination impossible.

8. Artificial insemination with genetic material from a third party may be performed, on the following medical indications:

[a]. There are permanent disorders of fertility or deterioration of basic indicators that control the possibility of fertilization by the husband's sperm;

[b]. treatment provided to the husband in order to restore his fertility has proved ineffective;

[c]. the husband is suffering from a hereditary disease that jeopardizes the viability of his progeny;

[d]. iso-immunization is present, in cases where there is a serious risk that the fetus would suffer from a hemolytic disease; or

[e]. there is a history of habitual abortions, that have been determined to be due to morphological or genetic anomalies in the husband's spermatozoa.

9. [With regard to gamete donors:]

[a]. Only persons of Bulgarian nationality between 18 and 40 years of age, in good physical and mental health, free from hereditary diseases, and presenting normal results in para-clinical investigations and as regards the indicators of the spermogram, may donate genetic material.

[b]. The persons referred to in subsection [a] shall have their state of health determined in a consultation center on sterility for the purposes of item [a] of Section 6 and shall submit the conclusions of the district neuropsychological clinic and of the center for genetic counselling of the place of residence of the person in question, which shall certify that such persons are free from mental or hereditary disease and that the results of tests for the causative virus of AIDS, syphilis, and viral hepatitis B are negative.

[c]. Each of the persons referred to in subsection [a] shall sign a declaration that the genetic material is to be donated only to a health care establishment and that no information relating to health has been concealed.

10. [With regard to the timing of genetic donation:]

[a]. The donation of genetic material shall be made within a period of 10 days following receipt of the results of the tests referred to in subsection 2 of Section 9. During this period, the persons concerned may donate genetic material up to three times, over a period of several days.

[b]. Any subsequent donation of genetic material shall be authorized after an interval of at least three months, after repetition of the serological tests referred to in subsection 2 of Section 9.

11. [The husband as gamete donor:]

[a]. In the case of artificial insemination with the genetic material of the husband, he shall not be required to undergo tests.

[b]. Genetic material from a husband intended for the artificial insemination of his wife may not be used in another woman.

12. Genetic material from persons referred to in Section 9 may normally be used for up to three inseminations over the total period between the ages of 18 and 40.

13. [The medical records of gamete donors:]

[a]. Any person donating genetic material shall be registered in the consultation center, unit, or department for the control of sterility, and the following medical records shall be kept for each such person: (i) the results of serological tests; (ii) the results of the examination of genetic material, accompanied by relevant findings as to its appropriateness; (iii) a medical certificate drawn up by, and the findings of, the psychoneurological clinic and the center for genetic counselling; (iv) the number of donated ejaculates and the mode of their preservation in the bank of genetic material; (v) the results of each insemination; and (vi) the number of inseminations.

[b]. The state of health of the persons referred to in subsection [a] shall be verified once a year by means of an examination performed by the local specialist in medical therapeutics as well as, if deemed necessary, another specialist ([7], pp. 73-74).

On December 29, 1989, the Lithuanian Ministry of Public Health approved an order on artificial insemination by donor. It requires that the recipient must be younger than 40 years old. Spouses must pledge themselves "not to sue the doctor in the case of an ineffective procedure or fetal abnormality, or in the case of the child having severe phenotypic differences from their nationality". At the same time, however, "spouses are entitled to choose the donor according to their wishes regarding the donor's national features (height, color of eyes and hair, form of face and nose)". Spouses must also pledge themselves "not to try to establish the donor's identity and to keep secret that conception of the child was the result of artificial insemination by donor sperm". Likewise, a donor must commit himself "not to seek to establish the identity of the recipient or of a child born as a result of artificial insemination by his sperm". If unmarried women request artificial insemination, each case should be considered individually by a special committee chaired by the chief doctor of the hospital.

Other reproductive techniques have yet to be subject to significant legislation. In Czechoslovakia, however, surrogacy is prevented by a clause in the "Family Act" which declares that the legal mother of a child is the woman who carried the child through pregnancy. In vitro fertilization and embryo transfer have thus far been controlled only by internal health service regulations, since all doctors have been state employees.

In Poland, situation regarding new reproductive techniques changed greatly during 1991. Research into, and provision of, in vitro fertilization virtually came to a halt. Towards the end of the year, a new draft code of practice for doctors appeared that was heavily influenced by the Roman Catholic Church. It includes, at clauses 35 and 36: "The doctor has no obligation to carry out artificial fertilization procedures as they are beyond the scope of therapeutic activity. Only if the patient so wishes should the doctor advise him concerning fertilization and contraception".

III. ABORTION

Attitudes of the former communist regimes toward abortion were generally, although not always, quite liberal. Abortion on demand in the first trimester has been generally available during the last three decades. In earlier times, however, the extent to which abortion was permitted sometimes depended on perceived manpower and population needs. From 1936 to 1955, for instance, abortion was totally banned in the USSR ([4], p. 20). Recently, efforts have begun to alter abortion laws to make abortion less available – most notably in Poland, but also in Lithuania and Czechoslovakia.

In Poland, attempts to try to alter the abortion law have been led by Roman Catholic doctors, politicians and clergymen. There was a desire in 1990 and early 1991 to present Pope John Paul II with a new anti-abortion law as a special present when he visited his homeland in May 1991. So far, however, the law has not been changed. This may reflect the increasing public support for abortion to continue to be permitted, even though the vast majority of the population are Catholics ([1], p. 23; [3], pp. 3-4). In late 1991, an attempt was made to achieve the same result by altering the Doctor's Code so as to prevent doctors from performing abortions, but it is unclear whether this move will succeed.

A ministerial ordinance issued on April 30, 1990 has already, however, made it more difficult to obtain an abortion, even though the relevant law remains unaltered. The following are some extracts from the new ordinance:

Section 2:

[a]. A pregnant woman who intends to terminate her pregnancy must request a physician holding a specialty at first or second degree level in gynecology and obstetrics to issue a ruling permitting an operation for termination of pregnancy.

[b]. If a pregnant woman intends to terminate her pregnancy because of difficult personal circumstances, she must present to the physician a written declaration indicating the circumstances justifying the termination of pregnancy.

Section 3:

A physician who has received a request from a pregnant woman to issue a ruling permitting a termination of pregnancy on medical grounds shall, after carrying out a gynecological examination and confirming the pregnancy, seek the opinion of a competent specialist if he does not have the specialized knowledge needed in the particular case and the pregnant woman has not submitted a suitable certificate from a specialist.

Section 4:

[a]. A physician who has received a request from a pregnant woman to issue a ruling permitting a termination of pregnancy on the grounds of difficult personal circumstances shall, after carrying out a gynecological examination and confirming the pregnancy, inform the woman of the adverse effects on health of an operation for termination of pregnancy, especially a first pregnancy, and of the possible complications after the operation, and shall consider and discuss in detail the motives that in the opinion of the woman making the declaration are inducing her to terminate the pregnancy, and shall also attempt to dissuade the woman from her intention.

[b]. If a woman intends to terminate her pregnancy and the circumstances she describes justify the issuance of a ruling permitting the termination of pregnancy, the physician shall: (i) seek the opinion of another physician holding a specialty at first or second degree level in gynecology and obstetrics, as well as the opinion of a physician in general practice, as to

whether there are medical contraindications to performing an operation for termination of pregnancy; and (ii) ask the woman to submit a certificate attesting that she has spoken to a psychologist about her intention, the circumstances, and the motives for the termination of pregnancy on the grounds of difficult personal circumstances...

Section 5:

[a]. In issuing a ruling permitting an operation for termination of pregnancy, the physician shall conform to the model shown in Annex 1 to this Order.

[b]. The grounds for the ruling shall be entered by the physician on the medical card, including any medical opinions obtained and other documents.

[c]. If the grounds for the ruling permitting an operation for termination of pregnancy are the difficult personal circumstances of the pregnant woman, the physician shall in addition enter the following declaration on the medical card, and give it to the woman to sign: "I have been informed about the adverse effects, dangers, and consequences of an operation for termination of pregnancy but still wish a ruling to be issued permitting termination of pregnancy".

Section 6:

A physician who gives a woman a ruling permitting an operation for termination of pregnancy shall:

[a]. instruct her on methods of preventing unwanted pregnancies;

[b]. inform her of the need for a check-up after the operation; and

[c]. direct her to a hospital and instruct her on the dangers to health of a termination of pregnancy outside hospital.

Section 7:

If there are no medical indications for performing an operation for termination of pregnancy but an operation for termination of pregnancy is to be undertaken on the grounds of difficult personal circumstances, the physician shall issue a ruling prohibiting termination of pregnancy ... if he finds medical contraindications associated with the woman's state of health or because the pregnancy is of more than 12 weeks duration, or if he does not consider that the difficult personal circumstances described by the woman are sufficient grounds for termination of pregnancy.

Section 8:

[a]. A woman to whom a physician has issued a ruling prohibiting termination of pregnancy may submit to the director of a public health service establishment competent for her place of residence an appeal for the matter to be considered before a medical commission.

[b]. The woman shall attach to the appeal the ruling prohibiting termination of pregnancy.

Section 9:

[a]. If an appeal as referred to in Section 8 is submitted, the director of the public health service establishment shall immediately, and in any case not later than three days after the date of submission of the appeal, appoint a commission made up of three physicians, of whom at least two shall hold specialties at first or second degree level in gynecology and obstetrics.

[b]. The provisions of Sections 5-7 shall be applicable mutatis mutandis.

Section 10:

In the issuance of rulings concerning the termination of a pregnancy, all actions relating thereto shall be performed within sufficient time to ensure that the development of the pregnancy does not give rise to contraindications to its termination or dangers to the woman's life...

Section 14:
A physician may refuse to issue a ruling permitting termination of pregnancy and refuse to perform an operation for termination of pregnancy, except in those cases where failure to perform an operation for termination of pregnancy would directly threaten the woman's life ([13], pp. 463-466).

The important changes introduced by this ministerial decree lie in paragraphs 2(2), 4(1) and 4(2)2. The requirements that a woman seeking an abortion for social reasons must submit a written explanation of her circumstances, and that a doctor must try to dissuade her, may be of more nuisance value than of positive effect in reducing abortions. But the requirement that all such women should have spoken to a psychologist before obtaining an abortion is much more restrictive. When the requirement was made, only one gynecological ward in Warsaw had a staff psychologist.

Lithuania, until 1990, had an abortion law similar to the rest of the USSR. Abortion was permitted: a) on medical grounds; b) on demand in the first twelve weeks; and c) for certain non-medical reasons, up to 28 weeks. Those reasons, specified by Order 1342 of the USSR Ministry of Public Health and issued on December 31, 1987, were the following:

- the woman's husband has died during the pregnancy;
- the woman or her husband have been confined in places of detention;
- the number of children in the family is already five or more;
- the woman has been deprived of maternal rights (e.g. for alcoholism, drug addiction or prostitution);
- divorce is occurring during the pregnancy;
- the pregnancy is the result of rape; or
- a child in the family is already handicapped ([4], pp. 20-21).

These non-medical reasons were abolished by a special decree of the Lithuanian Ministry of Health in early 1990, although abortion remains lawful on medical grounds, and lawful on demand during the first 12 weeks of pregnancy. Doctors, members of the Lithuanian Parliament, and Catholic clergy made an "Appeal to the Nation" in 1990 that read in part:

Abortion has been legal already for 34 years in the USSR. This law imposed on the Lithuanian people is an encroachment on the human right to life, and legalization of violence against the most innocent human being, who is unable to protect himself. Let us seek and insist on an unequivocal condemnation of abortion by the law, considering it a crime against man ([2], p. 24).

In Czechoslovakia, abortion has been available on demand in the first 12 weeks of pregnancy – free during the first eight weeks, and for a nominal charge thereafter. Free abortions are available until the 24th week for medical or genetic indications, or until term if the mother's life is in danger. It is now proposed, however, to charge the woman with the full cost of abortions on demand up to 12 weeks. In addition, parental consent will be required for any abortion on a minor, and the medical and genetic grounds for later abortions will be stricter.

Other countries in the region have issued decrees confirming their liberal approach to abortion. Hungary issued an ordinance that includes the following these provisions. Section 1 of this Ordinance lays down that an abortion may be performed at the written request of the pregnant woman when the required conditions are fulfilled. Section 2 reads as follows:

A pregnancy may be terminated when:
(a) the pregnant woman's state of health justifies an abortion;
(b) a medical opinion indicates that the conceptus has sustained a lesion liable to result in a severe physical or mental handicap at birth or to render it unviable;
(c) the pregnant woman is unmarried or has been separated from her husband for a continuous period of at least six months;
(d) the pregnancy results from a criminal act;
(e) neither the pregnant woman nor her husband owns vacant accommodation or rents independent accommodation;
(f) the pregnant woman is more than 35 years of age;
(g) the pregnant woman has two or more surviving children;
(h) the pregnant woman or her husband is serving a sentence of imprisonment;
(i) the pregnant woman's husband is performing his military service and will not be discharged until more than six months after the date of submission of the application for an abortion; or
(j) other social indications justify an abortion.

Under Section 4, a medical specialist is responsible for determining whether the conditions referred to in items (a)-(i) are fulfilled, whereas the existence of social indications covered by item (j) is to be determined by a family and women's protection counsellor [9].

In 1990, Bulgaria also issued a decree defining a woman's rights to abortion ([5], pp. 624-625). It provides for abortion to be available on request up to 12 weeks of pregnancy, provided that it does not endanger the woman's health. Such abortion is to be performed in obstetric/gynecological units or departments within three days of an authorization for abortion being presented. Abortion may be performed on medical grounds in the first 20 weeks of pregnancy if the woman has been proven to be suffering from a disease that could endanger the life or health of herself or the child. Later abortions may be performed if the mother's life is in danger, or if evidence is available of a severe malformation or genetic abnormality of the fetus. Abortions on medical grounds must be approved by a Special Medical Commission.

IV. CARE OF SEVERELY ILL NEWBORNS

There are a few guidelines or laws everywhere in eastern Europe that specifically govern the care of newborn babies. One potentially relevant decree has been issued in Lithuania. The decree moves the border between miscarriage and premature birth from 28 weeks of gestation to 22 weeks, as of January 1, 1991.

One of the few studies of how ethical problems in the care of handicapped neonates are handled by physicians in eastern Europe was published by a Polish philosopher and a pediatrician ([16], pp. 11-17). The authors received 74 responses to a questionnaire sent to pediatricians and intensive care doctors in major hospitals and clinics in Warsaw. The authors' conclusions included the following:

In the Polish medical community surveyed, unconditional respect for life is a more dominant attitude. If life is a sacred value, it must not be shortened deliberately or purposefully, and therefore half of the Polish doctors would be willing to preserve the lives of severely defective newborn infants at all costs. Our study has revealed a deeply-entrenched paternalistic attitude among Polish doctors, a strong unwillingness to distinguish between "ordinary" and "extraordinary" means of prolonging life, as well as an ambivalent attitude towards legal regulations binding in Poland ([16], p. 17).

V. CONSENT

The doctrine of informed consent, as understood in most Western democracies, has largely been ignored in eastern Europe. The situation in Hungary was, until recently, typical of attitudes in the region as a whole. As one author observes, "There are still pockets of medical paternalism where [informed consent] is considered dangerous, uncomfortable, and impractical" ([15], p. 19).

The doctor's duty to provide information to patients is laid down in the Hungarian Health Act of 1972. Section 45(1) requires general information to be given about a patient's illness:

The doctor has to inform adequately the patient, his next of kin, or – if the interest of medical treatment requires – the person who is responsible for the care of the patient, about the illness and the status of the patient. In some cases when the interest of the patient requires, doctors are entitled to withhold medical information.

Section 47 states more specifically that information must be given before medical treatment or a surgical operation. The rule, therefore, should have been that information is ordinarily given, and only withheld in exceptional cases. In practice, however, the reverse was true. Doctors have behaved, for the most part, as though they had to provide information only in exceptional cases.

In March 1990, an apparently unrelated law came into force. The Modification of Acts Act provided amendments to a wide range of laws. Section 35(1) stated that the second sentence of section 45(1) of the Health Act was repealed. Thus, there can no longer be any discretionary exceptions to the legal duty that patients must be provided with information. It appears, however, that many doctors remain ignorant of this duty, and that the gap between the law and clinical practice has become even larger.

VI. CONFIDENTIALITY

Just as there appears to be little legislation or regulation in eastern Europe about giving information or obtaining consent, so the issue of confidentiality does not appear to be high among health concerns. One recent instruction with interesting implications for confidentiality also illustrates a medical procedure now uncommon in developed countries.

Bulgaria has issued an instruction on the procedures for prenuptial medical examinations ([6], pp. 265-266). Persons wishing to marry must submit a medical certificate indicating that they do not suffer from certain categories of disease. A prenuptial examination is carried out by the physician responsible for a person's work place or place of residence, and requires very detailed data to be recorded. If various categories of disease are found or suspected, the physician must seek a specialist's opinion.

Annexes to the instruction contain lists of diseases that may be an impediment to marriage. Thus, Annex 1 includes diseases, such as congenital or acquired mental retardation, psychoses, and serious personality disorders that constitute absolute impediment to marriage. Marriage may also be prohibited if there is a serious risk to the life or health of any potential progeny arising from either an hereditary disease or from syphilis. A third category of diseases, including other sexually transmitted diseases, tuberculosis, leprosy and AIDS, are those that might pose a serious threat to the health of the prospective spouse. Marriage can only be permitted if the prospective spouse has been informed of the risk.

VII. ACCESS TO HEALTH CARE

Under the communist regimes, there was universal, but tightly controlled, access to health care. Usually there was no choice, one's doctor was determined by one's place of work or residence. Health care workers were all state employees. Thus far, there has been little change, but the first signs of changes in the system may be seen in Czechoslovakia and Hungary. In Hungary, in particular, an ordinance has been issued permitting the private practice of medicine. Its main provisions are as follows:

Section 2:

[a]. A license shall be required in order to engage in the private practice of medicine.

[b]. The private practice of other professions shall be open to physicians and dentists (referred to hereinafter as "physicians") who: (i) are registered in the National Register of Physicians; (ii) have not been, as a consequence of a disciplinary sanction, prohibited to practice their profession by a judicial order, prohibited to practice medicine by a legal decision of the health authorities, or prohibited to engage in medical or private practice by a disciplinary legal decision; (iii) own or rent a private consulting room whose premises, outfitting, equipment, and sitting meet the exigencies of public health and the necessary professional requirements; and (iv) hold a specialist diploma (referred to hereinafter as "diploma") that corresponds to the specialty which they intend to practice, subject to the condition that they are working in this specialty at the time of submission of their application, or have not terminated their practice of this specialty more than three years previously.

[c]. Whenever a physician has not practiced a medical activity within the meaning of subsection

2(d) for more than three years, he may be authorized to practice a private medical activity only if he can show evidence, in the form of a certificate, of having undergone continuing education that concludes with an examination and is dispensed in an Institute of Postgraduate Medical Training.

Section 3:
[a]. The private practice of medicine may be authorized only for the specialties laid down in the regulations on specialist qualifications...

Section 13:
[a]. The regulations on the continuing education of physicians employed in the state health service shall apply to physicians in private practice.
[b]. Any physician in private practice who fails to participate in continuing education programs shall be required to suspend his private activities until he has fulfilled this obligation.

Section 14:
[a]. The only activities that may be performed by a physician in private practice shall be those specified in his license.
[b]. First-aid and life-saving interventions must be performed by a physician engaging in private practice. Such interventions may not be remunerated...

Section 17:
A physician may issue prescriptions, within his private practice, for medicaments, narcotics, and therapeutic devices in accordance with the provisions in force. Contraceptive medicaments may be prescribed only by physicians specializing in gynecology and obstetrics...

Section 21:
Any physician engaging in private practice shall be subject to obligations relating to the supply of data and notifications laid down in the regulations in force ([10], pp. 592-593).

VIII. EUTHANASIA

In most of eastern Europe, euthanasia is absolutely forbidden. As a consequence, discussion of euthanasia in any form, or of withdrawing or withholding treatment, is most unusual. That is not to say that decisions about withdrawing or withholding treatment are not made every day, because health care resources are finite, as in other regions. Discussion, however, is rare. The decisions are taken by physicians without consultation with patients, families or nurses ([12], p. 16; [16], pp. 12-14). Whether discussion of these issues will become more open as democracy is established remains a moot point. The desire to prevent all discussion of euthanasia has been the focus of the "antibioethics" movement in German-speaking central Europe. There have been some early signs that similar attitudes may be developing in Poland, whence they could spread readily to Lithuania and Slovakia. It will be very interesting to see what has developed by the time the next regional bioethics yearbook is produced.

IX. TRANSPLANTS AND THE DEFINITION OF DEATH

Several of the countries in eastern Europe appear to have no specific laws on organ transplantation, and none appear to have been enacted in the period under consideration. Harvesting of organs is generally undertaken on the principle of presumed consent, *viz.*, that unless the potential donor has recorded his objection while alive, his consent to his organs being used for transplantation may be presumed. Such a system operates in Poland, where the family's consent to use of organs from a relative is not required. The same applies in Czechoslovakia, where consent is also presumed for autopsies and for the taking of organs for other purposes, such as extracting growth hormone from the pituitary gland.

The principle of presumed consent is also enshrined in an ordinance of the Hungarian Minister of Health, issued in 1988. That ordinance gives an indication of how organ transplantation is approached in Hungary, and also gives considerable detail of the definition of death that is used. Relevant sections are included below:

Section 5:
(2) Organs may only be removed if a medical commission made up of three members has ascertained that death has occurred. A physician designated by the director of the establishment shall act as chairman. The other two members of the commission shall be appointed in writing by the chairman of the commission from among the medical specialists of the establishment. One of the members shall be a specialist with wide experience in intensive care of cases of severe cerebral lesions. Physicians participating in the removal and transplantation may not be members of the commission. The document appointing the members shall be placed on the clinical file of the patient.

Section 6:
(1) It shall be considered that death has occurred, even in cases where blood circulation shows a progressive diminution or is artificially maintained, provided that all the conditions laid down in Annex 2 are met...

Section 8:
(2) An organ may not be removed for therapeutic purposes if the deceased person, during his lifetime, expressed an objection thereto in accordance with the law in force (Ordinance Number 14 of October 12, 1987 by the Minister of Health). The medical commission making the finding of death shall ascertain that the personal papers of and the clinical file on the deceased do not disclose such an objection. In the absence of an objection, the organ may be removed.

Section 10:
(1) An organ removed from the body of a living person or from a cadaver may be transplanted for therapeutic purposes into the body of a living person only in care establishments meeting the necessary requirements.

Annex 2 of the ordinance reads as follows:

Cerebral death, meaning the total and final cessation of all functions of the brain, shall entail the death of the person.

Cerebral death may be determined on the bases of defined criteria and symptoms. The cessation of cerebral functions, and its irreversibility, shall be determined by means of observation, clinical examinations, and, if appropriate, supplementary instrumental examinations.

The following are the conditions by which cerebral death may be confirmed:

(1) Cerebral death may be confirmed if the condition of the person in question has been caused by a very serious primary or secondary cerebral lesion which, in accordance with current scientific knowledge, may be considered incurable, definitive, and irreversible.
(2) Cerebral death may not be confirmed if, at the time of examining the person, symptoms are displayed on the basis of which the cessation of cerebral functions may be attributed to
- the effect of medicaments or chemical products (narcotics, hypnotics, tranquilizers, etc);
- hypothermia; or
- metabolic or endocrine complications; or
- neuromuscular blockage cannot be excluded with certainty.
(3) The following are the conditions on which irreversible cessation of cerebral functions may be determined:
- deep coma;
- absence of spontaneous respiration;
- medium or complete dilation of the pupils, with absence of reaction to light;
- absence of ocular cephalic reflex;
- absence of reaction to pain in the parts innervated by the trigeminal nerves; and
- absence of reflexes of the pharynx and trachea.
(4) Cerebral death may not be ruled out solely because it is possible to produce autonomic reflexes of the spinal cord.
(5) Examinations should be repeated in all cases when the conditions and symptoms cannot be interpreted with certainty.
(6) Electroencephalography need not be carried out to confirm cerebral death if the conditions referred to in items 1-4 can be determined with certainty. However, if the primary diagnosis is uncertain, an examination shall be carried out ... In addition, all examinations which the medical commission considers necessary and enable cerebral death to be determined in accordance with current scientific knowledge shall be carried out. The result of these examinations shall be evaluated in the same way as the result of the electroencephalographic examination.
(7) When it has been determined that cerebral death has occurred, the time of making and documenting the final diagnosis shall be considered the time of death ([11], pp. 588-590).

X. AIDS

Many of the laws and regulations that have been enacted in eastern Europe in the 1989-1991 period concern AIDS. In some countries, such as Bulgaria and the USSR, the regulations tend to be quite coercive in order to try to detect all cases of AIDS and to prevent further spread. Regulations in Romania, on the other hand, appear straightforward in their emphasis on the importance of confidentiality ([14], pp. 441-442). However, it is difficult to know whether a country such as Romania can develop administrative structures or a framework of continuing medical education rapidly enough for the regulations to be of any practical benefit. Order 1201 of the Romanian Minister of Health reads as follows:

1. It shall be the duty of all health units to provide medical care to HIV infected persons (asymptomatic or symptomatic), irrespective of the place of residence of the persons concerned.
2. HIV-infected persons shall be admitted to health units equipped with beds for adults and children, in keeping with the specific character of the opportunistic conditions involved.
3. Hospital units in which AIDS is diagnosed shall be required to fill in the coded card for the notification of new cases of the disease. The criteria applicable to the diagnosis of AIDS, and the conditions for disseminating information, shall be determined by the Commission of AIDS Specialists attached to the Ministry of Health.
4. From the date of this Order, all centers involved in the collection and preservation of blood must subject donated blood to tests for the detection of infection by HIV and hepatitis B, in order to prevent the risk of transmitting these diseases through transfusions, or through treatments using blood products.
5. Organ transplantation and sperm donation shall not be authorized unless the donors have previously undergone screening tests.
6. Laboratories in which the serological diagnosis of HIV infection is performed shall be required to communicate their results periodically. The criteria applicable to serological diagnosis, and the conditions for disseminating information, shall be determined by the Commission of AIDS Specialists attached to the Ministry of Health.
7. A Reference Laboratory shall be designated, within the framework of the Stefan S. Nicolau Virology Institute, for the purposes of confirming laboratory diagnoses of HIV and hepatitis B infection. It shall be the duty of this Reference Laboratory to coordinate activities carried out in the network of transfusion centers, and by blood collection and preservation centers.
8. A Reference Laboratory shall be designated, within the framework of the I. Cantacuzino Institute, for the purposes of confirming laboratory diagnoses of HIV and hepatitis B infection. It shall be the duty of this Reference Laboratory to coordinate activities carried out by all other laboratories working in this field.
9. All health units shall be required to take measures to prevent the transmission of HIV infection, and to protect staff.
10. The necessary measures shall be taken in all health units to ensure strict observance of professional confidentiality, and confidentiality in respect of each diagnosis, which shall be communicated only to the patient, or, if the patient is a child, his parents or legal representatives.
11. The Commission of AIDS Specialists attached to the Ministry of Health shall draw up, within 15 days of the signing of this Order, instructions and methodological standards, which shall be submitted for the approval of the Ministry of Health, and communicated to all health units for the purposes of implementation.
12. Any violation of this Order shall be punished in accordance with the health legislation in force [14].

A 1990 law in the USSR is a mixture of the coercive and supportive elements [18]. On the one hand, it compels citizens to undergo testing for human immunodeficiency virus (HIV) if ordered to do so by a health care agency on adequate grounds. The law also makes the deliberate exposure of another person to HIV infection, by one who knows he is HIV positive, a criminal offense. On the other hand, the law prohibits dismissal from work, refusal of work, refusal of admission to medical and educational establishments, refusal to admit children to pre-school establishments, or the refusal of other rights solely on the grounds that an individual is a carrier of HIV or has AIDS. It also provides that citizens of the USSR who become infected as a result of medical manipulations shall be entitled to a state pension.

Later in 1990, regulations were made under the above law to define who requires testing for HIV antibodies ([19] pp. 21-25). The following extract indicates how detailed the regulations are, and also shows concern for the maintenance of confidentiality:

2. The following shall be subject to testing:

2.1 Donors of blood, blood plasma, and other biological fluids and tissues, on the occasion of each donation;

2.2 Soviet citizens returning from service-related, business, and private journeys abroad of more than three months' duration;

2.3 aliens and stateless persons who have come to the USSR for study, work, or other purposes, in the 10 days following their arrival ... with the exception of aliens and stateless persons coming from countries whose certificates of testing for antibodies to the AIDS virus are recognized by the USSR;

2.4 Soviet citizens travelling abroad to countries whose requirements necessitate a certificate of testing for infection by the human immunodeficiency virus (AIDS);

2.5 Soviet citizens and aliens who have had sexual contacts with AIDS patients or carriers of the AIDS virus and who are identified during an epidemiological investigation shall be examined once every six months for one year;

[2.6 and 2.7 include patients with symptoms and signs suggestive of AIDS, which are listed in great detail.]

2.8 children born of HIV-infected mothers, at birth and 6 and 12 months after birth;

2.9 patients systematically receiving transfusions of blood and blood products (for hemophilia, Werlhof's disease, Willebrand's disease, anaemia of various origins, etc.), once a year;

2.10 pregnant women, upon being recorded as pregnant and at 30 weeks of pregnancy; where data from an AIDS test or medical records are not available, upon admission to the maternity unit;

2.11 pregnant women undergoing examination for artificial termination of pregnancy;

2.12 children admitted to intensive care, cancer, chest, and hematology units;

2.13 people suffering from sexually transmitted diseases, on applying for medical care and subsequently as indicated;

2.14 medical personnel working with the AIDS virus or providing diagnoses, treatment, and direct services for AIDS patients, once a year;

2.15 people in risk groups drug abusers, drug addicts, homosexuals and bisexuals, and people engaged in prostitution, twice a year;

2.16 people in risk groups placed in solitary confinement or in corrective labor establishments and preventive therapeutic/labor units, on admission and before release;

2.17 people of no fixed abode engaged in vagrancy ...

Section 12:

Medical workers and other persons who, in the course of their duties, become aware of the performance of medical testing for the detection of infection by the human immunodeficiency virus (AIDS) and the results thereof, shall be required to keep such information confidential [19].

XI. CONCLUSION

At the beginning of the period covered by this yearbook, there were nine states in the eastern European region. At the end of the period, there were eleven. By mid 1993, the end of the two years covered by the second regional yearbook in the series, there are likely to be at least seventeen states in the region. All are involved in the process of replacing communist regimes with democratic structures, a process ensuring that most state legislatures will be overwhelmed with political and economic problems for years to come. Bioethics is likely to remain well down on the list of priorities, even in those countries that are not beset by additional problems of division.

The study of bioethics presupposes that individuals can form their own conclusions as to what is ethically most appropriate in a given set of circumstances. That ability is not widely present in the former communist countries. Four or more decades of communist rule have left most people dependent on centralized decision-making, whether with regard to the simple things of life or to moral issues. In the summer of 1991, for instance, in Czechoslovakia, lunch was the same in every institution, even being served with the same cutlery on the same crockery. If no one has yet thought that there might be a way to serving lunch that differs from the requirements of a central edict given years before, how likely is it that doctors and patients are ready to discuss their obligations to each other as individuals? It seems inevitable that it will take years for most citizens to relearn the importance of an individual's moral values in a democracy. To legislate extensively on bioethical issues before that time might be to act in almost as authoritarian a fashion as the regimes that have been replaced. Thus, it is likely to be some years before a significant amount of legislative and regulatory activity on bioethics occurs in eastern Europe.

Bulletin of Medical Ethics
London, UNITED KINGDOM

BIBLIOGRAPHY[2]

1. Anon: 1991, "Abortion Debate Continues in Poland", *Bulletin of Medical Ethics* 66, 23-4.
2. Anon: 1991, "Abortion in Lithuania", *Bulletin of Medical Ethics* 66, 24.
3. Anon: 1990, "Poland Moves Against Abortion"', *Bulletin of Medical Ethics* 62, 3-4.
4. Baillie, K. and Szawarski, Z.: 1991, "Developing Abortion Rules in the USSR", *Bulletin of Medical Ethics* 69, 20-21.
5. Bulgaria: 1990, "Decree Number 2 of February 1, 1990 on the Conditions and Procedures for the Artificial Termination of Pregnancy", *International Digest of Health Legislation* 41, 624-625.
6. Bulgaria: 1990, "Instruction Number 1 of 1985 on the Procedures for Prenuptial Medical Examinations", *International Digest of Health Legislation* 41, 265-266.
7. Bulgaria: 1990, "Order Number 12 of May 30, 1987 on Artificial Insemination in Women", *International Digest of Health Legislation* 41, 72-75.
8. Hungary: 1990, "Ordinance Number 7 of March 22, 1989 of the Minister of Social Affairs and Health ... on Artificial Insemination", *International Digest of Health Legislation* 41, 75-76.
9. Hungary: 1989, "Ordinance Number 76 of November 3, 1988 of the Council of Ministers on the Termination of Pregnancy", *International Digest of Health Legislation* 40, 595.
10. Hungary: 1990, "Ordinance Number 30 of November 15, 1989 of the Minister of Social Affairs and Health on the Practice of Medicine...", *International Digest of Health Legislation* 41, 592-593.
11. Hungary: 1989, "Ordinance Number 3 of February 17, 1988 of the Minister of Social Affairs and Health ... for the Implementation of the Provisions Concerning the Removal and Transplantation of Organs and Issues of Law Number 11 of 1972 on Health", *International Digest of Health Legislation* 40, 588-590.
12. Kovacs, J.: 1991, "Terminating Treatment in Hungary", *Bulletin of Medical Ethics* 72, 13-19.
13. Poland: 1991, "Ordinance of April 30, 1990 of the Minister of Health and Social Welfare Concerning the Professional Qualifications Required by Physicians Performing Operations for Termination of Pregnancy and the Procedure for Issuance of Medical Rulings Permitting Such Operations", *International Digest of Health Legislation* 42, 463-466.
14. Romania: 1991, "Order Number 1201 of October 16, 1990 on

[2]Extracts from laws and regulations printed in the International Digest of Health Legislation are reprinted with the permission of the World Health Organization Office of Publications.

Epidemiological Surveillance, Prevention of Infection, and the Provision of Medical Care to Persons Infected by the Human Immunodeficiency Virus", *International Digest of Health Legislation* 42, 441-442.
15. Sandor, J.: 1991, "A Missing Sentence", *Bulletin of Medical Ethics* 66, 19.
16. Szawarski, Z. and Tulczynski, A.: 1988, "Treatment of Defective Newborns: A Survey of Pediatricians in Poland", *Journal of Medical Ethics* 14, 11-17.
17. USSR: 1989, "Decree Number 669 of May 13, 1987 of the USSR Ministry of Health on the Extension of the Experimental Application of the Method of Artificial Insemination by Donor on Medical Indications", *International Digest of Health Legislation* 40, 597-598.
18. USSR: 1990, "Law of the Union of Soviet Socialist Republics of 23 April 1990 on the Prevention of AIDS", *International Digest of Health Legislation* 41, 431-432.
19. USSR: 1991, "Regulations of October 4, 1990 on Medical Testing for Detection of the Human Immunodeficiency Virus (AIDS)", *International Digest of Health Legislation* 42, 21-25.

FRANCESC ABEL

BIOETHICS IN SPAIN: 1989-1991

I. INTRODUCTION

The Health Care General Law affirms the rights of patients and provides access to the Spanish health care system, which is a social welfare system. The provision of health care is generally good in the hospitals, but of mixed quality in primary centers of health care. Recent legislation on assisted reproduction (artificial insemination and in vitro fertilization) and research with gametes or embryos has imposed few constraints on a new reproductive technologies and practices. Abortion has been depenalized, but there are as yet no facilities at which abortions are performed. Although the law on transplantation of organs and tissues is good, administrative obstacles to organ donation remain. The recent Medication Law demands action on the part of Hospital Ethics Committees for clinical research and the issuing of a Royal Decree to set the endorsement norms for the same, is expected.

II. NEW REPRODUCTIVE TECHNOLOGIES

In 1988, Spain approved two laws which were considered at the time, the most comprehensive legislation on these issues. These laws were based on a report submitted by Marcelo Palacios, M.D., and socialist Member of Parliament for Asturias (1982-1989), President of the Special Commission for the study of the in vitro fertilization (IVF) and artificial insemination (AI) of the Lower House of Parliament (*Congreso de Los Diputados*), and author of the report fully approved by the Lower House of Parliament one year later (April 10, 1986) [8]. As member of the Spanish Delegation to the Council of Europe, Dr. Palacios participated in drafting Recommendation Number 1046 (1986) of the Parliamentary Assembly of the Council of Europe, on the use of human embryos and fetuses for diagnostic, therapeutic, scientific, industrial and commercial purposes. The "Palacios Report" was the basis for Recommendation Number 1100 on the same issues [9].

A. *Law on Assisted Reproduction*

This law covers issues related to AI, IVF, and other techniques, as well as by

B. Andrew Lustig (Sr. Ed.), Bioethics Yearbook: Volume 2, 251-264.

the donation of gametes and embryos for research and experimentation. The following is a brief summary of the law which highlights its most important provisions.

The law makes IVF and AI available to any woman, married or unmarried, who has reached the age of majority (18 years), is in good physical and mental health, and has sought and accepted such procedures freely and with full information in advance.

The fertilization of human ovaries for any purposes other than human procreation is prohibited. Research is allowed on non-viable surplus embryos. Experimentation on live pre-embryos obtained "in vitro", whether viable or non-viable, shall be prohibited until it is scientifically proven that animal models are inadequate for the purpose.

Donation of gametes and embryos is authorized with no more than six children allowed to be born from one donor. Donation shall be anonymous, with particulars of the donor's identity kept in strictest secrecy. The disclosure of the donor's identity (under judicial order) does not constitute legal determination of paternity. Semen and untransferred embryos from IVF may be kept under cryopreserved in authorized banks for a maximum period of five years.

The law decrees that a National Commission on Assisted Reproduction should provide guidance on these procedures and formulate standards for centers where assisted reproductive procedures are performed. The law also makes null and void any surrogacy contract. A husband may consent in a public document to the use of his gametes to fertilize his wife within six months after his death. Such procreation will have the same legal standing as legitimate paternity.

The law considers a number of procedures as very serious offenses, including the following: fertilizing human ova for purposes other than human procreation (the use of the hamster test is authorized); maintaining in vitro fertilized ovules, beyond the fourteenth day after fertilization; mixing semen from different donors to inseminate a woman or to perform IVF; creating identical human beings by cloning; and genetic manipulations to select sex for non-therapeutic purposes, or for therapeutic purposes that are unauthorized. None of these offenses, however, falls within the domain of the penal code which has caused concern among penal law experts. This lack of criminal sanctions is perhaps less surprising if one takes into account the principle of minimal intervention central to recent efforts to reform of the penal code.

B. *Law on the Donation of Human Embryos, Fetuses, or their Cell Tissues and Organs*

This law amends a previous law of October 27, 1979, on the removal and transplant of organs with provisions and concerns that law did not address. Law 42/88, as a complement to law 35/88 [10], deals with the donation and use of human embryos and fetuses and their use in transplant. Both laws are pragmatic and tend to authorize what the scientific community considers justifiable. However, the criteria of fetal viability or non-viability, and the required informed consent set limits to what is authorized or unauthorized. Articles 2 and 3 of Law 42/88 read as follows:

Article 2: The donation and use of human embryos and fetuses, or of other biological structures for the purposes foreseen in this Law, will be allowed, subject to the fulfillment of the following prerequisites:
a) That the donors be parents. b) That the donors give their prior free, explicit, and conscientious consent in writing. If donors are not emancipated or are incapacitated minors, it will be necessary to secure permission from their legal representatives. c) That the donors and, when applicable, their legal representatives be adequately informed of the consequences and the aims and ends to be served by the donation. d) That the donation and its later use should not be for profit-making or commercial purposes. e) That the embryos or fetuses subject to donation be clinically non-viable or dead. f) If the parents die, and there is no evidence of their expressed opposition. In the case of minors, the permission of the parents or guardians of the deceased will always be necessary. g) In case of death due to an accident, the donation will require the sanction of a judge acquainted with the cause.

Article 3:
1. The utilization of human embryos or fetuses, or of their biological structures, will be carried out by qualified biomedical teams, and in centers or services authorized and controlled by public authority.
2. The termination of pregnancy shall never have as its end the future donation and use of the embryos or fetuses or of their biological structures.
3. The medical team that carries out the termination of pregnancy will not participate in the utilization of the embryos or fetuses in the terms and ends foreseen by this Law.

The laws which we have reported were severely criticized by the Catholic Church, which already had voiced its opposition to them as bills. The reasons for the Church's opposion were that these laws denied the right to life of every human being from the moment of fertilization, that they interfered with the rights of the family and of marriage as an institution, and that they infringed upon the rights of the children to be conceived, brought to the world, and educated by their parents. According to the note from the Episcopal Commission for the Doctrine of the Faith, these minimal moral demands are replaced by a pragmatism inspired by an amoral conception of science and its technical applications, which will bring disastrous consequences for the future of man and of society [11].

III. ABORTION

In Spain, induced abortion is a punishable offense, except in those cases set forth by the Organic Law 9/1985 of June 5, 1985, in accordance with the sentence of the Constitutional Law Court Number 53/85 of April 11, 1985 [12]. The Court pronounced that certain aspects of the previous law which partially remove penalties for abortion did not conform to the Spanish Constitution, "not because of the assertion that abortion is not punishable, but because, with its regulations, certain demands" derived from Article 15 of the Spanish Constitution "do not find fulfillment". The three situations in which the practice of abortion is not a criminal offense are the following:

1. When it is essential in order to prevent a serious threat to the life or to the physical or mental health of the pregnant woman. No time limit is specified and a medical opinion is required from a doctor other than the physician who will perform the abortion.
2. When the pregnancy is a consequence of a criminal act of rape. In this case the abortion must be performed within the first twelve weeks of pregnancy.
3. When it is presumed that the fetus will be born with severe physical or mental defects. Two medical opinions are required, and the abortion must be performed within the first twenty-two weeks of gestation.

In addition the Royal Decree on health care centers for the voluntary termination of pregnancy of November 21, 1986, 2409/1986 simplified criteria allowing abortion for accredited private centers [13].

From 1986 to 1991, the vague concept of "depenalization", means that in those cases foreseen by Law 9/1985, neither the medical doctor who performs the abortion nor the woman who requests it will be punished. The Law does not imply provision by abortion services nor does it mean that "legal" abortion will be financed out of public funds. Although the media proclaim that individuals "have the right to abort", many hospitals make no provision for it, or if they do, they offer it in a restricted and insufficient manner. As a result, private centers have arisen which have accumulated large profits. At the same time, the poor have to socioeconomic disparities as a new ground to decriminalize abortion.

Forty-nine health care centers in Spain reported abortion procedures at their facilities in 1989. In 1987, the Federation of Family Planning in Spain estimated that 63,937 abortions had been performed that year. In 1989, the number of women undergoing a second abortion was about 17 percent. The statistics about the age and civil status of women who abort revealed a larger number of adolescents and unmarried women. Illegal abortion, in the majority of cases, occurs because of administrative irregularities [14]. Finally, medical abortion by means of anti-progesterone Mifepristone (RU-486), is not performed in Spain. The Ministry of Health has requested a rigorous clinical evaluation of the drug before endorsing it.

IV. DRUGS AND CLINICAL TRIALS

On December 20, 1990, a comprehensive law was passed to address all aspects of drug therapy including, drug research and development, drug safety and evaluation, clinical trials and the protection of human subjects, and regulation of the pharmaceutical industry. The law consists of 117 articles divided into ten sections. Two sections are especially noteworthy. Section II, on therapy, discusses the kinds of medication, their evaluation and authorization, the conditions for the issue of pharmaceutical specialties, health care, the special medications, and quality control of pharmaceuticals. Section III, discusses clinical trials, analyzes the appropriate conditions for research, the requirements of informed consent, questions of insurance, and the responsibilities of research promoters. Directives are given for the creation of Ethics Committees for Clinical Research under Article 64, which reads as follows:

Article 64. Ethics Committee for Clinical Research:

1. Not a single clinical trial shall be carried out without the prior report from an Ethics Committee for Clinical Research, independent of the promoters and researchers, and duly acknowledged by the proper health care authority who will report on it to the Ministry of Health and Consumer's Goods.
2. The Committee shall consider all methodological, ethical and legal aspects, and the foreseen risks and benefits of the essay.
3. The Ethics Committees shall be constituted by an inter-disciplinary team of physicians, hospital chemists, clinical pharmacologists, nursing staff, and persons outside the health care profession, of which at least one shall be a law expert.

A Royal Decree is expected shortly that will detail regulations to be followed by all hospital Ethics Committees, including guidelines on clinical trials.

V. CONFIDENTIALITY

Three aspects of confidentiality deserve special reflection: (1) the status of computerized information and data; (2) the status of data on persons who employ techniques of assisted reproduction; and (3) confidentiality for HIV-positive patients.

A. *Computerized Information*

The most common opinion among law specialists is that the law in general, and the penal law in particular, are not equipped to confront the challenge of the possible breaches of confidentiality with computerized data. With regard to health care, the "Six Safety First Principles of Health Information Systems" adopted by the Board of the European Union of General Practitioners (*UEMO*) are noteworthy. The UEMO believes that future developments in Health Information Systems:

- should be oriented to primary care;
- should be patient-centered so that the patient can benefit from medical developments and professional quality assurance monitoring. The patient, aided by general practitioner; should be able to have access to his record and control of that record both within and outside the health care system;
- should be capable of being used to support epidemiological studies and

health care planning services; and
- should be capable of being used to support the evaluation of diagnostic and therapeutic interventions.

B. *Confidentiality of Gamete Donation*

Article 5.5 of Law 35/1988 on assisted reproduction techniques is the subject of significant discussion. It reads as follows:

> The donation (of gametes) shall be anonymous, the particulars of the identity of the donor being kept in strictest secrecy and in coded form in the corresponding bank and in the National Register of Donors.
>
> The resultant children shall have the right, either personally or through their legal representatives, to obtain general information concerning the donors, although not ... their identity. Recipients of gametes shall likewise have this right.
>
> Only in exceptional cases, [either] in extraordinary circumstances that entail a verified danger to the life of the child, or under the law of criminal procedure, may the identity of the donor be disclosed; it shall be a condition that such disclosure is indispensable to avert a danger or to attain the legal objective referred to. In such cases ... [d]isclosure shall be limited in character and shall under no circumstances make public the identity of the donor.

Based on the Spanish constitution, different specialists in civil law believe that there are no arguments to support preserving the anonymity of the gamete donor if and when a child resolves to inquire into and discover his genetic origin. In such cases, and despite the provisions of the above mentioned article, these specialists believe that the child when mature, could be informed of the identity of the donor. This information would necessarily include a relationship of paternity [24].

C. *Confidentiality of HIV Data*

The confidentiality of data on HIV-positive patients is affirmed by health professionals and protected by professional codes within the boundaries of the right to intimacy and the duty of professional secrecy. The Health General Law also protects sub-data, and the already mentioned Resolution of the Catalonia Parliament (February 7, 1990) affirms confidentiality specifically with respect to HIV information.

VI. ACCESS TO HEALTH AND PATIENTS' RIGHTS

A. *From the Health Care General Law (14/86) to the Abril Martorell Report*

Although the Spanish health care system, from 1942 to 1980, lacked any systematic strategy for the delivery of medical care, the health of the general population remained high, despite major social inequalities.

A process of fundamental health care reorganization began in 1981 when the first democratic government created the Health Care and Consumer's Goods Ministry [1], which consolidated specialties that, in the past, had been assigned to more than ten different Ministries. The Heath Care General Law, passed in 1986, tried to correct deficiencies in the system by decentralizing administrative structures, by assuring health coverage for all people based on the National Health Insurance system (Social Welfare System) and by protecting the rights of the patients [2].

Decentralization has been partially achieved by blending a centralized structure under the Ministry of Health, the Internal Commonwealth Council of the National Health System, and the Health Service of the diverse Autonomous Communities. At the present time, the central administrative body (*INSALUD*) disburses 44.3 percent of the total health care budget, with the rest disbursed by the Autonomous Regional Health Care Services [3]. At the beginning of 1990, public funding for health care expenses constituted 75 percent of the total budget, with an average expenditure of 80 percent in countries belonging to to the Organization for Economic Cooperation and Development (OCDE).

Health care absorbed 6.3 percent of the gross domestic product (GDP), with 3.4 percent of the labor force engaged in that sector. While the GDP from 1975 to 1987 grew at an average rate of 1.4 percent per capita, the actual health care contribution per capita remained unchanged during the entire period, despite the preferences displayed by the citizens for health care services [4].

A strict budget policy, although it failed to introduce improvements and changes in the organization and supervision of health care policies, did achieve some success in limiting the unbounded growth of health care expenses. But the shortcomings of that policy were immediately felt: an increase in outpatient waiting lists in hospital clinics and poor primary health care. In short, as a result of cutting costs, the quality of health care worsened.

In 1990, National Health Insurance covered 96.14 percent of the Spanish populace. One is covered by social insurance either by paying income tax on wages or by being member of a special class *e.g.*, pensioner, unemployed, handicapped). Most people have a double system of both public and private insurance. They use public health care services when they require hospitalization and use private clinics for primary health care.

The Inter-Regional Council of the National Health System, was directed to review the efficiency and quality of the existing model. On February 13, 1990, the government nominated an Expert Commission for the Analysis and Evaluation of the National System of Health. Its Chairman, Mr. Fernando Abril Martorell, submitted his report one year later. This report ("The Abril Report") was not received favorably by the government or by political groups of the opposition, because it violates key tenets of the economic philosophy of the welfare state. The Commission recommended to charge users, including pensioners for some services, with later submission of bills. This process, normally known as "ticket moderation" would acquaint users with the actual costs of the services. As a primary strategy to improve efficiency, the Commission also suggested the separation of public funding from the provision of services (public and private) as a stimulus to free market competition among suppliers [5]. This initiative aroused complaints from the unions and from political opposition parties. They perceived it as a first step towards the privatizing of health services rather than as a way to improve public health care delivery.

The Health General Law represents a major advance in one important respect. In Article 9, the law articulates in great detail, the rights of patients to health care. Each person has the following rights in respect of different public health care allocations:

1. Regard for his person, human dignity and privacy, without discrimination because of ... any type.
2. Information regarding the health care facilities open to him and the necessary pre-requirements for their use.
3. Confidentiality of all information related to his treatment and to his stay in public and private health care institutions which collaborate with the public system.
4. Knowledge of whether procedures for early prognosis, diagnosis, and therapy ... applied to him may be used in relation to a teaching or research project which, under no circumstance, will pose additional danger to his health. In any case, the patient's prior consent will be strictly required in writing together with the doctor and the appropriate Health Center's Management.
5. ... complete, ... information regarding his care including diagnosis, early prognosis, and treatment alternatives.
6. A free choice between the options which the responsible doctor submits to him, and the user's written permission for any intervention, except in the following cases: (a) when the intervention entails a risk to public health or to the interests of the community; (b) when there is a legal mandate; (c) when one is incompetent, the right will rest in family members or close relatives; and (d) when an emergency situation does not allow for delays.
7. That a doctor be assigned to him, whose name will be made known to him, who will be his principal spokesman vis-a-vis the team that assists him. In that doctor's absence another member of the team shall assume the responsibility.
8. That a certificate with his case history be handed over to him, if there is a legal or a customary disposition which so establishes.
9. To refuse treatment, except for cases pointed out in (Provision 6)... .
10. To take part through the institutions of the community, in health care activities, in terms established in this Law and in the directives which developed it.
11. That the whole of his treatment be recorded in writing. At the end of the patient's stay in a health care institution, the patient, a relative, or a closely esteemed person will receive the Discharge Sheet.
12. To use, within the established period of time, the means for complaints or suggestions. In either case, one will receive a written answer within the periods which may be customary.
13. To choose his doctor and the rest of the accredited health care team, in accordance with the conditions as foreseen by this Law, the by-laws issued for its development, and those which monitor health care labor.

14. To obtain medical products instrumental to the promotion, preservation, and restoration of health, under conditions... established by the Administration of the State.

A general criticism of the Law, which consists of 113 articles and additional regulations, is that it does not clearly distinguish what might be considered a list of good intentions from what is juridically binding. Nonetheless, the Spanish Constitution and the Health Care Law, despite of its shortcomings, constitute the most important points of reference for the protection of the patients' rights.

VII. ON DEATH, DYING, AND HUNGER STRIKE

In Spain, a forceful movement to legalize active euthanasia and assisted suicide has emerged. Simultaneously, in medical circles, there is an increasing emphasis on palliative care, which adopts the fundamental ideas of the Hospice Movement. The Magistrate and Senator Cesáreo Rodríguez Aguilera worded a draft law to address passive euthanasia and the patient's right to refuse treatment [25]. Some believed that such legislation was unnecessary, while others argued that the law should include active euthanasia. All specialists agree that there is need to forego the adjectives "active" and "passive" and to employ the term "euthanasia"to mean only the incitement or the intentional acceleration of the patient's death.

There is no doubt that the present Spanish Penal Code is too outdated to answer the problems that our society faces. Although the Code does not punish suicide as a crime, it sentences an abettor of suicide to major imprisonment (from 6 to 12 years). If the abettor himself causes the death, the punishment will be imprisonment (from 12 to 20 years), *i.e.*, in the latter case, it is considered homicide.

Both the Association for the Right to a Dignified Death (supporting active euthanasia) and the Episcopal Spanish Commission (against active euthanasia) have written their own "Advanced Directives" [26]. Intense debates on legalization of euthanasia are likely to be waged during the coming year.

Another problem that has provoked heated debate among judges, and prosecutors during 1990 has been the legitimacy of prolonged hunger strikes by imprisoned terrorists. Some members of a terrorist group (*GRAPO*), while in jail, began an indefinite hunger strike to secure their confinement together in the same prison, rather than being transferred to different penitentiary centers all over the country. When their lives began to deteriorate significantly, enforced nourishment was ordered. The judicial sentence that ordered the enforced nourishment aroused dissenting opinions [27]. The debate has been interesting and the range of viewpoints extremely varied.

VIII. BRAIN DEATH AND ORGAN TRANSPLANTION

The present Spanish legislation is constituted by two laws on the procurement and transplantation of organs and two resolutions from the Secretariat of State for Health Care.

The Spanish Law Number 30 of 1979 was, for its time, considered to be the most progressive in Europe, although bureaucratic difficulties hindered efforts to obtain organs. This law, inspired by Resolution (78) 29 of the Council of Europe on the "Harmonization of Laws of Member States Relating to the Removal, Grafting and Transplantation of Human Substances", was adopted by the Council of Ministers on May 11, 1978 [17]. It espoused a brain death standard, and defined it as follows:

> Brain death ... is determined when cerebral and brain stem functions are irreversibly absent. Potentially reversible causes must be excluded. Drug intoxication, treatable metabolic disorders, hypothermia, shock and peripheral nerve, cranial nerve or muscle dysfunction due to disease or neuro-muscular blocking drugs must be excluded. Re-evaluation after six hours to ensure that the non-functioning state of the brain is persistent is mandatory for purposes of organ transplantation.

In further developments, the Ethics Committee of Saint John of God Hospital in Barcelona, published the first criteria for the determination of brain death in infants and children in November, 1988 [18]. On March 25, 1991, during the Seventh Winter Neurologic Seminar, a neurological statement on cerebral death and chronic vegetative state was developed. Its authors concluded that Spanish legislation on brain death is too closely linked to transplantation concerns, and insist on separate legal regulations for both cases [19].

This petition, in my opinion, transposes a fear "imported from the United States" and nurtured by some Spanish scholars of legal medicine. In fact, there is no ban in Spain, insofar as any medical practitioner may use his expertise and knowledge to declare an individual as dead.

IX. AIDS

From the legal perspective, the 1984 Bill of Rights and Duties of the Patients of the *Insalud Hospitals* [20] and Health Care General Law 14/1986 declare that a patient has the right to be provided with integral health care for all health problems, to be given complete information, and to have all confidentiality respected. A set of Decrees and Resolutions establish requirements for blood donation, for the production and importation of blood products, for blood and serum tests, and for procuring and storage of semen.

On June 26, 1987, The Medico-Collegial Organization (*OMC*), in accordance with Article Number 4 of the Royal Decree Number 1018 (1980) [21], reminded doctors that they must abide by the following rules of conduct.
First, under no circumstance shall a health professional refuse to attend a patient. Second, an AIDS patient has the same right to medical attention as any other victim of an infectious disorder. Third, he health professional is duty-bound to adopt all measures of protection and hygiene to avoid contagion and

transmission, as with all other infectious diseases. Fourth, professional secrecy is of the utmost importance as the basis of the doctor-patient relationship. Finally the team doctor of the hospital center will, in all cases, enforce these duties for the personnel under his authority.

On February 7, 1990, the Parliament of Catalonia passed a resolution urging the government of the "Generalitat" to issue rules on HIV-testing. The Parliament emphasized that such tests should not be carried out in systematic or indiscriminate manner, should not be performed as a precondition to employment, and should require the person's written authorization. In addition, the intimacy and confidentiality of the test results should be assured, with suitable information, and medico-psychological support to those affected, and affected individuals should have access to medical and psychological support. Finally, tests should not be required as a precondition to insurance coverage [22].

X. THE GENOME PROJECT

From October 24 - 26, 1988, Dr. Santiago Grisolía and the Foundation for Advanced Studies in Valencia organized a "Workshop on International Cooperation for the Human Genome Project". Invitation was sent to the prominent scientists interested in the problem of the Human Genome Mapping and Sequencing. The so-called "Statement from Valencia" resulted from the meeting [28]. We will conclude this report by quoting the Valencia Declaration on ethics and the Human Genome Project in its entirety:

1. We, the participants in the Valencia Workshop, affirm that a civilized society entails respect for human diversity, including genetic variations. We acknowledge our responsibility to help ensure that genetic information is used to enhance the dignity of the individual, that all persons in need have access to genetic services, and that genetics programs abide by the ethical principles of respect for persons, beneficence, and justice.
2. We believe that knowledge gained from mapping and sequencing the human genome will have great benefit for human health and wellbeing. We endorse international collaboration for genome research and urge the widest possible participation of countries throughout the world, within the resources and interests of each country.
3. We urge coordination among nations and across disciplines in the conduct of research and the sharing of information and materials relating to the genomes of human beings and other organisms.
4. Concerns about the use and misuse of new genetic knowledge have provoked debate in many quarters. In addition to discussions in professional circles, further public debate on the ethical, social, and legal implications of clinical, commercial, and other uses of genetic information is urgently needed.
5. We support efforts to educate the public, through all means including the press and the schools, about genome mapping and sequencing, genetic diseases, and genetic services.
6. In light of the great increase in prognostic and therapeutic information that will arise from the genome project, we urge greater support for training of genetic counselors and genetic education of other health professionals.
7. As a general principle, genetic information about an individual should be ascertained or disclosed only with authorization from the individual or his or her legal representative. Any

exceptions to this principle require strong ethical and legal justification.

8. We agree that somatic cell gene therapy may be used for the treatment of specific human diseases. Germ.line gene therapy faces technical obstacles and does not command ethical consensus. We endorse further discussion of the technical, medical, and social issues on this topic.

Institut Borja de Bioètica,
Sant Cugat del Vallès
Barcelona, SPAIN

NOTES

1. Real Decreto 2823/81 (November 27) "Creación del Ministerio de Sanidad y Consumo": 1981, *Boletín Oficial del Estado* 288, (December 2).
2. Ley No. 14/1986 (April 25) "Ley General de Sanidad": 1988, *Boletín Oficial del Estado* (April 29), 15207-15225.
3. Statement of the Minister of Health before the "Comisión de Política Social y Empleo del Congreso de los Diputados" (Lower House of Parliament), May 1991.
4. "Seminario sobre Análisis Comparado de Sistemas y Políticas de Salud". Ministerio de Sanidad y Consumo, Marzo 1991.
5. "Recomendaciones del Informe Abril Martorell": 1991, *Organización Médica Colegial* (Septiembre 1991) 14, 8-14.
6. Ley No. 35/1988 (November, 22) "Sobre Técnicas de Reproducción Asistida": 1988, *Boletín Oficial del Estado* 282 (November 24), 284 (November 26).
7. Ley No. 42/1988 (December 28) "Sobre Donación y Utilización de Embriones y Fetos Humanos o de Sus Células, Tejidos u Organos": 1988, *Boletín Oficial del Estado* 314, (December 31).
8. Marcelo Palacios: 1987, "Informe de la Comisión de Estudio de la Fecundación "In Vitro" y la Inseminación Artificial Humanas" *Congreso de los Diputados. Gabinete de Publicaciones.*
9. Marcelo Palacios (Rapporteur): 1988, "Rapport sur la Recherche Scientifique Relative a l'Embryon et au Foetus Humains" Document 5943 (September 13), Council of Europe, Parliamentary Assembly.
10. As an example, we quote Article 12 (2) of Law 35/1988 which envisions topics proper to Law 42/1988:
    Article 12 (2). No operation upon the embryo in utero or on the fetus, within the uterus or outside it, for diagnostic purposes on the live pre-embryo or live fetus, is legitimate unless its object is the well-being of the infant about to be born and the fostering of its development, or unless it is legally defensible.
11. "Nota de la Comisión Episcopal para la Doctrina de la Fe Acerca de las Proposiciones de Ley sobre 'Técnicas de Reproducción Asistida' y 'Utilización de Embriones y de Fetos Humanos o de Células, Tejidos u Organos'" *Ecclesia* 48, 503-6.

12. Ley No. 9/1985 (June 5) "Ley Orgánica Sobre el Aborto": 1985, *Boletín Oficial del Estado* 166 (July 12).
"Sentencia del Tribunal Constitucional Sobre la Ley del Aborto": 1985, *Boletín Oficial del Estado* 119 (May 18).
13. Real Decreto 2409/1986 (November 21) "Sobre Centros Sanitarios para la Interrupción Voluntaria del Embarazo": 1986, *Boletín Oficial del Estado* 281 (November 24).
14. Elvira Méndez: 1991, "Evolución de la Práctica del Aborto Legal en España", *Quadern Caps* 15 (Spring 1991), Barcelona.
15. Ley No. 25/1990 (December, 20) "Ley del Medicamento": 1990, *Boletín Oficial del Estado* 306, (December 22).
16. Ley No. 30/1979 (October 27) "Sobre Extracción y Trasplante de Organos": 1979, *Boletín Oficial del Estado* 266 (November 6).
Real Decreto 426/1980 (February, 22) "Desarrollo de la Ley Sobre Extracción y Trasplante de Organos": 1980, *Boletín Oficial del Estado* (March, 13).
Resolución 27/1980 (June, 20) *Boletín Oficial del Estado* (July, 1).
Resolución /1981 (April, 15).
Orden /1981 (April, 15) "Por la que se Regula la Obtención de Globos Oculares de Fallecidos, el Funcionamiento de Bancos de Ojos y la Realización de Trasplantes de Córneas".
17. Resolution (78)29 (May 11, 1978) "On Harmonization of Laws of Member States Relating to Removal, Grafting and Transplantation of Human Substances", in *Recommendations adopted by the Committee of Ministers and the Parliamentary Assembly of the Council of Europe on Bioethical Questions*,: 1989, Directorate of Legal Affairs, Strasbourg.
18. Comité de Etica del Hospital San Juan de Dios: 1989, "Criterios de Muerte Cerebral en el Niño" Documento 3, Serie Documentos (November 1988), *Labor Hospitalaria* 21(212):148-151.
19. Private communication.
20. Carta de los Derechos y Deberes del Paciente para los Hospitales del Instituto Nacional de la Salud, October 1, 1984 en *Plan de humanización de la asistencia hospitalaria*: 1984, Ministerio de Sanidad y Consumo, Instituto Nacional de la Salud, Madrid.
21. Real Decreto 1018/1980, "Conducta de los Médicos y Enfermos de Sida".
22. "Resolució 103/III del Parlament de Catalunya, sobre la Garantia de la Intimitat Personal i de la Confidencialitat en les Proves Diagnòstiques de Sida" February 7, 1990 (DSPC-C,109).
23. UEMO Board meeting, October 27, 1990.
24. Montés V.L. *et al*: 1991, *Derecho de Familia*. *Tirant lo Blanch*, Valencia.
25. Rodríguez-Aguilera C.: 1988, *Revista Jurídica de Cataluña* 88(4), 261-76.
26. Asociación Derecho a Morir Dignamente: 1990, "Testamento Vital", Barcelona; "Conferencia Episcopal Española: 1990, "Plan de Acción Sobre

la Eutanasia y la Asistencia a Bien Morir *Labor Hospitalaria* 22(216), 112-5.

27. *Tribunal Constitucional*
Sentencia 137/1990 de 19 de Julio. *Boletín Oficial del Estado* 181, suplemento del 30 de julio de 1990.
Sentencia 11/1991 de 17 de enero. *Boletín Oficial del Estado* 38, suplemento del 13 de febrero de 1991.
Sentencia 120/1990 de 27 de junio. *Boletín Oficial del Estado* 181, suplemento del 30 de julio de 1990.
28. Valencia Declaration on ethics and the Human Genome Project (November 14, 1990. Valencia, Spain), en *Proyecto Genoma Humano*: Etica: 1991, Bilbao: Fundación BBV.

Mª PILAR NUÑEZ AND FRANCESC ABEL[1]

BIOETHICS IN PORTUGAL: 1989-1991

I. INTRODUCTION

Since 1975 the Portuguese Health Care System has patterned itself on the British model. While the Portuguese system has rectified many of the shortcomings it has encountered, problems remain in several areas. Equity in access to health care resources has not been fully achieved; excessive numbers of health personnel remain concentrated in larger urban areas; and deficiencies persist in primary health care. On June 9, 1990 a National Ethics Committee for Life Sciences was established under the direction of the Council of Ministers. This Committee, which has already favorably reviewed a draft law on organ and tissue extraction, may accelerate a series of legislative projects on bioethics not yet considered by the Republic's Assembly. Especially noteworthy among them is a draft law on technologies for assisted reproduction, prepared by a special commission which completed its work on July 28, 1987.

II. NATIONAL COUNCIL OF ETHICS FOR THE LIFE SCIENCES

In Law 14/90 on June 9, 1990, the Republic's Assembly decreed the creation of a National Council of Ethics for the Life Sciences in accordance with Articles 164 and 169 of the Constitution. The National Council is an independent organ which collaborates with the Presidency of the Council of Ministers [3].

The Council is charged to analyze all moral problems arising out of the scientific progress in the fields of biology, medicine and health care in general; to offer opinions on those problems if so requested; and to submit to the Prime Minister a yearly report on the new technologies and their ethical and social implications, drafting recommendations as it sees fit.

The President of the National Council is nominated by the Prime Minister and the twenty incumber of the Council to a five-year term of office. In addition a permanent Executive Commission permanent character will be created, presided over by the same President and nine Council members. The President, the Assembly, Government and public and private centers using

[1] We wish to thank Carlos Ariza, M.D. for his able assistance.

B. Andrew Lustig (Sr. Ed.), Bioethics Yearbook: Volume 2, 265-271.

biological techniques which may pose ethical issues all have authority to request the opinion of the National Council. The Council will prepare and sensitize public opinion on ethical issues; create a documentation center, and request advice form relevant experts.

III. NEW REPRODUCTIVE TECHNOLOGIES AND EMBRYO RESEARCH

On July 28, 1987, the Commission nominated by the Ministry of Justice to draft a law regulating reproductive technologies completed its work. The draft law was approved by the majority of the members of the Commission, with two dissenting votes with regard to the legitimacy of heterologous artificial insemination. One of those dissenting votes also favored the right of a person conceived through artificial insemination to know the donor's identity. This draft law has not yet been submitted to the Assembly of the Republic. It was published in book form by the Faculty of Law of the University of Coimbra in 1990 [5].

At present official Portuguese legislation on assisted reproduction remains the regulations adopted in 1986. Those regulations included the following provisions. First, the spouse who consented to artificial insemination cannot later negate his paternity on the basis that the child was conceived in this manner (Article 1893/3 of the Civil Code, according to the Law Decree 496/77 of November 25, 1977). Second, the physicians who have been requested to perform an artificial insemination or sterilization may refuse based on grounds of conscience (Law 3/84 of March 24, 1984 on sexual education and family planning). Third, the collection, manipulation and preservation of sperm, or any other procedures required by the techniques of artificial procreation, may be performed only under the direct supervision of a physician in a private or public agency, specifically authorized by the Minister of Health (Law Decree 319/86 of September 25, 1986).

The draft law of 1990 on assisted reproduction addresses the following issues, which had not been considered previously.

- The techniques of assisted reproduction cannot be considered as an alternative method of procreation, but only as a recourse when other methods of treating infertility have not been successful. It is most important that [obtain to] insure the newly born a full human development.
- Only couples who are legally married or couples that live in a stable union [may have access to] of these techniques. The draft law excludes unmarried women, a restriction inspired by the right of the infant to a two-parent family.
- Surrogate motherhood is forbidden, because maternity, as a "primordial relationship", cannot be the subject matter of an a agreement or of a negotiation.

- Homologous artificial insemination (AIH) is allowed, though not after the death of the spouse.
- In heterologous artificial insemination (AID),
  - The beneficiaries must be women between 25 and 40 years of age men of nor more than 50 years of age without no physical impairments.
  - Beneficiaries must give free and informed consent, verbally or in writing.
  - The donor must be between 20 and 50 years of age, good health, and be free of congenital or infectious disease.
  - He must also give his consent in writing; if he is married, his wife's consent is also necessary.
  - Provision of sperm should not involve commercial compensation, may in no way be remunerated, although expenses incurred in transportation are allowed.
  - Secrecy is guaranteed for the beneficiaries and for the donor, and can be broken only if offspring claims the right to know the circumstances of conception especially for serious medical reasons.
  - No more than five successful gestations are allowed from the sperm of the same donor.
  - The legal father is the person who lives with the mother and who granted his consent to artificial insemination.
- With regard to in vitro fertilization:
  - Homologous IVF is allowed, provided that the stored gametes be used in the life, times of both spouses.
  - Heterologous IVF can be used with donor ova or donated sperm, but at least 50 percent of parental genetic endowment is necessary.
  - The donor of an ovum must be older than 18 years of age, and no more than thirty-five years old.
- With regard to embryo transfer, freezing, storage, and research:
  - The number embryos resulting from IVF must be strictly limited, in accordance with the present state of science, to assure the success of procreation. If not all embryos are implanted the surplus may be kept for a period of two years for the use of the proper beneficiaries. After this time, the center shall assign these embryos according to the wishes of the beneficiaries, for donation or for research purposes.
  - Donors must give their consent in writing.
  - Implantation of embryos that have been used for research purposes is not authorized.
  - Once implanted in the uterus, the embryo can be subject only to those interventions which would be lawful had it been conceived "in vivo".
  - The use of embryos for research is allowed only when the following three conditions are met:
    a) The research project has been favorably reviewed by the Bioethics Council.

b) The research has as its end preventive or therapeutic objectives, the diagnosis of grave diseases, or the improvement of techniques of assisted procreation that cannot be achieved by other means.
c) The research shall only be performed up to fourteen days after fertilization. In this calculation, time of preservation by freezing or any other means is not to be taken into account.

IV. ACCESS TO HEALTH AND PATIENT'S RIGHTS

A. *From the 1975 National Revolution to Health Care General Law of 1990*

With the 1975 revolution, everything in Portugal was subject to transformation; and with this transformation, positive changes in the health care structure were initiated. Law 56/79 of September 15, 1979 established the National Health Care Service in order to ensure the right to health care for all citizens independent of their financial and social resources. Law 28/48 of August 14, 1984 sets forth the bases for social insurance in accordance with the principles of universal participation, unified system, equal access to health services, decentralization, efficacy, solidarity, participation and judicial guarantee [1].

The Portuguese National Health Service covers all Portuguese, citizens not only those already registered in Social Insurance has transferred to the State almost all health care institutions from the private sector and from charitable organizations (*Misericordias*). Moreover, all health care professionals have in effect been granted the status of government servants.

The great concentration of health professionals in the large towns of: Lisbon, Coimbra, and Porto, with the consequent lack of resources in other areas, makes equity in the access to health services an ideal yet to be achieved. In addition, the free use of services is also the existence of the so-called "moderator ticket" (equity tax) to obtain ambulatory medical care, consultation, non-hospital emergency-care, home visits, tests, X-ray, and physiotherapies. Pregnant women, mothers with infants less than one-year old, invalids, and the elderly are exempted from this tax.

On August 24, 1990, the Republic's Assembly promulgated law 48/90 on the foundations of health [2]. Article 14 lists the rights of health care consumers. Their rights are as follows:

- to choose, within the National Health Service (*SNS*), the necessary care and appropriate professionals proportionate to existing resources and in accordance with the rules of the organization;
- to accept or to refuse the care that is offered, unless otherwise ordered by the Law;
- to be treated with adequate means; humanely and promptly, with technical correctness, respect, and privacy;
- to be accorded strict confidentiality over personal data

- to be informed of one's situation, of the possible alternative means of treatment, and of the probable prognosis;
- to obtain, if one desires, religious assistance;
- to register complaints about inadequate treatment and, if necessary, to receive compensation for injuries received.
- to create entities that will collaborate with the health system, be they associations for the promotion and protection of health or support groups for health care institutions.

In Article 17, the law encourages research within the framework proposed by the European Community, and emphasizes the principle of respect for human life under all circumstances. The policy norms for clinical tests state that tests must be carried out under medical supervision and according to rules to be specified in a subsequent document.

V. ORGAN TRANSPLANTATION

Law 1/70 of February 20, 1970 which authorized the procurement of "human biological products" (such as blood and milk, under conditions set down by the donor) did not include organ or tissue transplants from living persons.

Decree Law 553/76 of July 13, 1976 set forth the terms under which organs and tissue from a cadaver could be used for transplantation and other therapeutic ends. However, the legal framework for tissue and organ extraction from living persons remained unclear.

The problems raised by extracting tissue from living donors were addressed in a draft law on human tissue and organ extraction for transplant, therapeutic diagnosis, or scientific research [4].

The law includes the following provisions:

- Any person who has attained majority. [and is competent] is a donor within the terms provided by the law.
- A minor can be a tissue donor for one of his brothers or sisters only when this is strictly necessary to ensure physical survival.
- The consent must be the expression of the free, unequivocal and clear will of the individual and, can be revoked at any time before the extraction.
- The consent of a minor older than 14 years of age is valid only if [the minor clearly discerns] meaning of the consent at the moment he grants it.
- With a donor under 14 years of age, consent can only be presumed in the extraction of regenerable tissues that do not affect the physical integrity or health of the donor, with the consent of his legal representative, and without the opposition of the minor.
- The donation of organs or tissues must in all cases be free from compensation, and must always be performed by a physician in an institution have been authorized for the purpose.

- With regard to the organ extraction from cadavers, consent is always presumed unless the patient's opposition to extraction is stated in his health chart.
- Specific information about the source or recipient of organs or tissues shall not be given to the beneficiary or to the family of the donor.
- The Ministry of Health will prepare a form on which the blood type of the individual will be noted along with his authorization or non-authorization of organ or tissue extraction after his death.
- The Ministry of Health shall promote a campaign to educate the public about transplantation and to stress the importance of social solidarity in this matter.

The National Council of Ethics for the Life Sciences expressed its opinion on the Draft Law on Transplant on June 26, 1991. The Council praised the use of transplants, but also respect the integrity of the individual and the sensibilities of the family. The Council also emphasized affirmed the need to intensify a sense of moral solidarity on transplantation issues, and pointed out the lacunae in the present juridical system on these matters. In its assessment, the Council stressed the need to respect individual dignity, to obtain voluntary and informed consent for transplantation, and to assure that transplantation remains a non-commercial activity. In pursuit of these values, the Council offers the following general guidelines:

- Transplants shall be medical interventions of a scientific nature and with therapeutic ends.
- The recipient must be duly informed of all aspects of the transplant procedure prior to giving express consent.
- Transplants shall only be performed by duly qualified doctors in transplants institutions certified the health care authorities.
- The individuals wishes regarding donation will be honored.

Institut Borja de Bioètica.
San Cugat de Vallès
Barcelona, SPAIN

BIBLIOGRAPHY

1. Law 56/79: 1979 (September 15), Serviço Nacional de Saúde; *Diário da República*: 1979 (September 15), I Série - Number 214; Law 28/84: 1984, (August 14); Da Segurança Social, Diário da República: 1984 (August 14), I Série - Number 188.
2. Law 48/90, August 24, 1990: Lei de Bases da Saúde. *Diário da República*, August 24, 1990, I Série - Number 195.
3. Law 14/90, June 9, 1990: Conselho Nacional de Ética para as Ciências da

Vida. *Diário da República*, June 9, 1990, I Série -Number 133.

4. Arlindo de Carvalho, Ministro da Saúde: Consultation to the President of the National Council of Ethics for the Life Sciences on proposed the draft project of a law on "a colheita de tecidos e órgãos de origem humana para transplantação, diagnóstico ou terapêutica a fins de investigação científica". Ministério da Saúde, March 7, 1991.
5. Centro de Direito Biomédico. Facultade de Direito. Universidade de Coimbra: *Comissão para o enquadramento legislativo das novas tecnologias*. Utilizacão de Técnicas de Procriação Assistida (Projectos), 1990.

FRANCESC ABEL[1]

BIOETHICS IN ITALY: 1989-1991

I. INTRODUCTION

As a consequence of their prominence and widespread impact, the promulgation of the *Nuovo Codigo Italiano di Deontologia Medica* (July 15, 1989) and the institution of the *Comitato Nazionale per la Bioetica* (March 28, 1990), which has already issued four documents, possess an exceptional interest. The legal void with regard to the application of the new reproductive techniques persists. The Parliament, however, has issued various norms concerning transplantation, transfusion practices, the AIDS crisis, and problems of toxic dependence.

II. NEW REPRODUCTIVE TECHNOLOGIES

Despite the submission of many bills to Parliament, Italy still lacks a specific public policy on new reproductive technologies. The legal vacuum has not prevented universities and private hospitals from satisfying demands for homologous artificial insemination in Italy and for heterologous insemination in Sicily, all with government aid. In Catholic health institutions, techniques that are not clearly forbidden by the Instruction *Donum Vitae* of the Congregation for the Doctrine of the Faith (*e.g.*, gamete intrafallopian transfer) are often available.

Up to the present time, there has also been no legislation concerning research on embryos. The last attempt to regulate this research was the proposed Law 3485, "*Norme e Tutela del l'Embrione Umano*". That bill, which appealed to Recommendations 3486 (1982) and 1046 (1986) of the Council of Europe (Assembly of the Parliament) and to the fundamental principles of the Constitution, sought to protect the embryo from the moment of its conception and to penalize all research involving embryos except for those therapeutic interventions which, in due time, the Ministry of Public Health may allow. The bill also sought to penalize fusion of human and animal gametes, as well as cloning or predetermination of any human characteristic without therapeutic

---

[1] I wish to thank Dr. Manuel Cuyás, affiliated with the Universitá Gregoriana, Rome, and the Institut Borja de Bioética, Barcelona.

B. Andrew Lustig (Sr. Ed.), Bioethics Yearbook: Volume 2, 273-278.

purposes. The *Centro di Bioetica della Università del Sacro Cuore* (Policlínico Gemelli, Rome) issued a document on June 22, 1989 entitled: *Identità e Statuto del l'Embrione Umano*, which also argued for protection of the embryo from the moment of conception.

The *Comitato Nazionale per la Bioetica* (CNB), established on March 28, 1990, was charged, among other issues, to study problems related to fertility and to human procreation. The Committee has published one document on this subject, *Problemi della Reccolta e Trattamento del Liquido Seminale Umano per Finalità Diagnostiche* (May 5, 1991). In the document, the CNB appraises diverse methods for obtaining semen and the different processes used to test its fertilizing capabilities (seminologic diagnosis). In their discussion, members of the CNB, differed in their appraisals of the legitimacy of research on embryos, as well as in their evaluations of inter-specific fecundation as a test to ascertain the fertilizing power of spermatozoa. In a footnote, the document details the opinions that both groups held. It is very possible that either point of view will have sufficient strength in Parliament to prevent the approval of a proposal offered by the other group. Indeed, this may explain why no legislation has been forthcoming.

III. ABORTION

With regard to abortion, Law 194 of May 22, 1978, is still in force. It transcends the contradictory referendum of 1981 that permitted abortion free and on request in public hospitals when the mother's health is at stake, the fetus is pathologically affected, or conception results from rape, but with these provisos: prior counselling, a medical certificate, and at least a seven-day interval between the decision and the abortion itself.

In 1989, the journal "*Società e Salute*" issued a statement by the World Federation of Doctors Who Respect Human Life, that rebuked the procedures of those who, according to the Federation, distort data and manipulate the meaning of the terms in order to characterize intrauterine devices as contraceptives rather than abortifacients: "[An IUD] acts before hand to prevent the implantation of the embryo, and does not prevent conception".

IV. CONSENT TO TREATMENT AND EXPERIMENTATION

On May 28, 1991, the *Comitato Nazionale per la Bioetica* published the *Documento sulla Sicurezza delle Biotechnologie*. This document focuses on the ethical issues raised by research in biotechnology and its potential applications. The document urges the legislature to regulate biotechnology in accord with European and international recommendations, with special concern for nature and the ecological balance. It emphasized the need to improve and to standardize the criteria for assessing the risks of releasing genetically modified organisms into the environment. The document underscores the legislature to inform the public opinion of the risks and advantages associated with specific developments, and recommends that a national authority on biotechnology be established. This authority would coordinate the various disciplines involved in biotechnology, certify the institutions engaged in this work, gather necessary data, educate the public, and update technical-scientific experts on new developments.

The document draws these conclusions after first providing an overview of current knowledge and of debates. In addition, it has a useful bibliography and an appendix with information on present or recommended national and international procedural norms.

V. ACCESS TO HEALTH CARE

Since the Health Care Reform of 1978, Italy has offered a National Health Care Service to all without distinction as to individual and social circumstances. The responsibility for health care is assumed by the "local health care unit" at the centralized level of the municipal corporation. Beginning in 1978, the standard health care services, prevention and rehabilitation, were unified with public health services such as food quality control, water sanitation, environmental quality, housing, and work safety. The Italian Health Care Reform, which followed the British model, was implemented before the Spanish reform, and perhaps its most innovative aspect is found in its effort to include mental illness within the definition of generic illness, *i.e.*, to consider the mental patient an infirm person and not a social danger [1]. However, despite that effort, the law has failed to achieve its desired ends, because of the lack of coordination in making such services available.

VI. DEFINITION OF DEATH

In its "*Definizione e Accertamento della Morte Nel'Uomo*" (February 15, 1991), the *Comitato Nazionale per la Bioetica* addressed questions surrounding the definition of death, especially the criteria employed to obtain an early diagnosis of death [3]. The Committee defined death as "the total and irreversible loss of the capability of the organism to preserve autonomously its own functional unity". While acknowledging the value of other indicators the Committee considered the criteria of "brain death" to be acceptable and exclusive, if strictly measured. The Committee required the physician to retest all factors to confirm the certitude of death. With instruments and appropriate verifications, the time of observation required can be reduced to 12 hours, but particular criteria are needed to certify death in neonates and children. The Committee concluded by urging the legislature to define clear standards on brain death in order to clarify the uncertainties of the present situation.

VII. DONATION OF BLOOD, ORGANS, AND TISSUES

Law Number 107, passed by Parliament and the Senate on May 4, 1990, deals with "transfusion activities related to the human blood, and all its ingredients for obtaining products derived from the human plasma". It sets rules and blueprints at the national and local levels, describing these respective spheres of action, and discusses procedures for obtaining and treating blood and its byproducts. The law also sanctions the extraction of medullar marrow for allotransplants, even from children, provided that prior permission has been secured from parents or legal guardians.

A decree from the Health Minister, dated December 27, 1990, adds "Norms for Blood Donation". The decree establishes new criteria for the gathering, preparation, transport, and preservation of blood and blood products. It also sets rules for distribution, labeling, and quality control.

Another decree from the Ministry of Public Health, "*Protocolli per l'Accertamento della Idoneità del Donatore di Sangue ed Emoderivati*" (January 15, 1991), regulates various aspects of blood donation including the following: selection of donors, laboratory tests, confidentiality of the donor and recipient, and the requirements of informed consent. There are two appendices to the decree: first, a donor's consent form, together with information on illnesses that disqualify one from being a donor; and second, an appeal to a sense of civic responsibility in response to the public's need for blood.

In June 1991, the Health Minister forwarded to health centers and to appropriate doctors, guidelines developed by the *Commissione Nazionale per il Servizio Transfusionale* for the promotion of homologous blood transfusion. The guidelines are not meant to enforce compulsory auto-transfusion, but to embody practical and persuasive measures to render such transfusion habitual. By "auto-transfusion", the Minister understands "any blood transfusion and/or hemo-components previously extracted from the very same patient".

Encouraged by regional decrees to promote blood storage in order to decrease the infective and immunological dangers of transfusion, some health centers (*e.g.*, the Ospedale San Raffaele of Milan) now regard any patient who has to undergo a selective surgical intervention as eligible for a transfusion, from the moment of admission. When a doctor excludes someone from eligibility, he must record the cause in the patient's history.

On July 4 1990, the Parliament and the Senate approved Law Number 198, "*Disposizioni sul Prelievo di Parti di Cadavere a Scopo Di Trapianto Terapeutico*". This law amended several articles of Law Number 644, which had been enforced since December 2, 1975. These amendments facilitated procedures of organ procurement, and restricted transplants to those institutions with suitable technical facilities and qualified personnel.

On April 21, 1991, at the end of a Congress on "Ethics and Organ Transplants", the *Associazione Medici Cattolici Italiani* published a document entitled "*Troppo Pochi i Donatori di Organi*" [3]. The document called for a four-fold increase in organ donations to meet national needs. It criticized the lack of education about transplantation, as well as erroneous objections to donation based on the "sacredness" of the human body. The document urged a better organization of health services and a mobilization of efforts to increase the supply of organs.

VIII. OTHER ISSUES

On July 15, 1989, the *Federazione Nazionale degli Ordini dei Medici-Chirurgi e degli Odontoiatri* published the *Nuovo Codice Italiano di Deontologia Medica* [4], that updates the Code of 1978. It dicusses new problems that have arisen during the intervening decade. It also reaffirms the scientific and moral autonomy of the medical profession, the right of patients to obtain information, and the need to respect the dignity of the patient, even when only palliative care is offered.

On May 22, 1989, by means of an announcement to the press, the *Società Italiana per la Bioetica e i Comitati Etici* was established and its charter was made public. The primary objective of this society is to promote and encourage a balanced discussion of ethical issues posed by bioethics research, health care, and the allocation of resources. The Society admits as members, individual with recognized expertise in these areas.

Regional Law Number 12 of April 10, 1989, addresses the religious aspects of patient care in the residential structures of the local health units (*Regione Emilia-Romagna*). The Law affirms the autonomy of these units over the services they provide and emphasizes each citizen's right to exercise religious freedom.

Circular Letter Number 14 (February 13, 1988) from the Ministry of Health updated the measures to control HIV-infection in children and adults. Eight appendices defined categories of AIDS infections.

Law Number 135 (June 5, 1990), *Priogramma di Interventi Urgenti per la Preventione e la Lotta Contra l'AIDS*, itemizes measures to fight the spread of AIDS together with suitable methods of care, in accordance with the directive norms framed by the *Commissione Nazionale per la Lotta Contra l'AIDS*. Those norms addressed aspects of education, prevention, research, and control of the AIDS epidemic. They emphasized ways to restructure hospital services, to recruit personnel with skills in personal care and laboratory work, to improve facilities that treat venereal diseases, and to broaden the range of authority of the *Instituto Superiore di Sanità*.

In Article 5, the law cares for afflicted individuals and protection of their anonymity. No one must submit himself/herself to testing. No one should subject to discrimination because of test results. According to Article 6, employers cannot subject applicants or employees to AIDS testing. Article 7 imposes a duty to the State to promulgate safety norms to prevent possible contamination in the healthcare environment. Article 8 details how the Inter-ministerial Committee against AIDS is to be constituted. Article 9 discusses the coordination of regional and provincial centers to fight against the AIDS epidemic.

Law Number 162 (June 26, 1990) updates and modifies Law Number 685/1975 on drug abuse and broadens its concern to toxic dependence. Article 1, with the approval of the "Presidenza del Consiglio dei Ministri", and the Anti-drug National Coordination Committee, create and describe its spheres of influence, and sets limits to activities vis á vis other government agencies.

The document *Terapia Genica* ("Gene Therapy"), published on February 15, 1991 by the *Comitato Nazionale per la Bioetica*, provoked significant discussions. The first part underscores the fundamental difference between somatic cell gene therapy and germ line gene therapy. It notes that germ line therapy still faces major technical problems and has profound to say nothing of the legal, ethical, and social implications. The document recommends that somatic cell gene therapy follow general norms of experimentation with human subjects. In addition, it calls for the establishment of an authority to monitor the illnesses compatible with this therapy, to define operative criteria for such therapy, and to collect comprehensive data on research.

Part Two provides an overview as well as a discussion of the present state of technical progress in gene therapy, and its ethical implications. Part Three details the specifics of gene therapy for adenosine-deaminase (ADA). This therapy has already been performed in the "Ospedale San Saffaele" of Milan, in close connection with the similar experiences first initiated at the NIH Clinical Center of the U.S.A. The document also includes a general bibliography, a list of original research articles, a brief dictionary of technical terms, and useful bibliographic references.

Institut Borja de Bioètica.
Sant Cugat del Vallès
Barcelona, SPAIN

BIBLIOGRAPHY

1. Instituzione del Servicio Sanitario Nazionale. Gazzeta Ufficiale della Republica Italiana Number 360 (December 28, 1978).
2. Comitato Nazionale per la Bioetica. Presidenza del Consiglio dei Ministri. Via dei Villini 15, 0161 Rome.
3. Medicina e Morale: 1991, 41 (3): 558-560.
4. Medicina e Morale: 1989, 39 (5): 983-1002.

REIDAR K. LIE AND JENS ERIK PAULSEN

BIOETHICS IN SCANDINAVIA: 1989-1991

I. INTRODUCTION

In this report, we will present an overview of important events in medical ethics in the three Scandinavian countries, Denmark, Norway and Sweden. We have, unfortunately, not been able to report on the activities in Finland and Iceland, but will do so in the next edition of this yearbook. Our report is divided into four main sections. In the first section, we give an overview of the ethics committee systems in the three countries. Although some of these committees were established before 1989, they have issued various recommendations during the period covered by this yearbook. Since the recommendations will be discussed below, we have felt a brief overview of the history and function of these committees to be essential. In the second section, we will introduce the Scandinavian debate on the issues of abortion, prenatal diagnosis and fetal research. Then follows a presentation of the Danish and Swedish developments regarding criteria of death and transplantation. The fourth section deals with a number of issues which involve adult patients: treatment of patients in a persistent vegetative state, genetic testing, and HIV/AIDS.

In all of the sections, we will give extensive reviews of the various government reports. We have chosen to deal with each country separately. We have also chosen to give a fairly detailed summary of the official documents, since they are in languages which are not read by most people. There will then necessarily be some overlap in the kinds of arguments used in the various reports. We will also refer to some of the discussion in the professional journals.

In a report such as this one, it is of course impossible to review all the material that has been published. We have concentrated on the official documents and the discussion which has followed their publication. We have therefore not reviewed many excellent articles on medical ethics published in Scandinavian language journals. The material covered has been published in 1989, 1990, and the first half of 1991, with a few exceptions such as the Danish report on the criteria of death which was published at the end of 1988. The discussion, however, took place in the period covered by this report. We have also chosen not to report on a very important document published in June 1991 on protection of the rights of people with mental disabilities [47]. This document also contains an important and extensive overview of Norwegian

B. Andrew Lustig (Sr. Ed.), Bioethics Yearbook: Volume 2, 279-307.

thinking on informed consent. It will be discussed in the coming year in Norway and will be covered in our next report.

In all countries, there have been certain institutional developments regarding medical ethics. Medical ethics has been introduced in the teaching of medical students at the Universities of Copenhagen and Oslo. A Center for Medical Ethics has been established in Oslo, and a national professorship in medical ethics has been established in Sweden (Professor Göran Hermerén).

II. COMMITTEES

Each of the three Scandinavian countries has a system of research ethics committees that reviews all biomedical research involving human subjects. We will not here report on any details concerning how these committees function. In addition to these review committees, there are national committees in all three countries. These may have a connection to the review committees, or they may be independent of them. All national committees function as advisory bodies to their respective governments with regard to policy issues in medical ethics. During 1989-1991, they have issued several policy reports, which will be discussed below. First, however, we will give a brief overview of the organization of these committees.

A. *Denmark*

Det Etiske Råd (The Danish Council of Ethics) was established by law in 1987. The Council consists of 17 members. Eight are chosen by the Minister of Health, and nine by the Danish Parliament (*Folketing*). The members are elected for three years and can be re-elected only once. The members must have publicly documented professional credentials on ethical, cultural, and social matters and may not be members of the *Folketing* or the municipal or county councils.

The Council's two main assignments are to promote public debate and to submit proposals for new legislation in the areas of medical research and development. The Council is also advised to start with the judgment that human life begins at the moment of conception. Thus, from the moment of conception, choices must be submitted to ethical and legal considerations, and may be limited. The Council's task, however, is not to give an explicit definition of human life and its beginning.

The Council is asked to make suggestions on the following matters: (1) laws to protect fertilized human eggs, living embryos and fetuses, as well as genetic experimentation on human gametes; (2) the admissibility of undertaking genetic treatment of human gametes used for fertilization, human fertilized eggs, embryos, and fetuses; (3) the admissibility of using new technology for detecting congenital defects or diseases in human gametes, human fertilized eggs, embryos, and fetuses; and (4) rules for freezing human gametes, intended for future fertilization, and fertilized human eggs.

Beyond these specified tasks, the Council can also give advice to the research ethics committees about general ethical problems raised by experimentation on human beings, to authorities of the health organizations on questions within their purview, and to national authorities about questions of registration, dissemination and application of genetic technologies. Within its field, the Council can also consider subjects on its own initiative. The Council will inform the public about the progress of its work and initiate public debate over ethical problems which may occur.

Denmark has seven regional research ethics committees which review all biomedical research. In addition, there is a Central Research Ethics Committee, which functions as a body of appeal for the seven regional committees. The seven regional committees and the Central Committee have semiofficial status, but no specific legislation covers the field at present. In November 1988, however, the Minister of Health established a committee to consider the need for legislation in certain areas of biomedical research involving human subjects. This committee finished its work in 1989 by proposing a statutory two-tier system of ethics committees very similar to the existing system. Thus far, its proposal has not led to any changes in legislation. The two independent central bodies, the Danish Council of Ethics and the Central Research Ethics Committee, are asked to cooperate, although they remain independent, autonomous councils.

B. *Norway*

Norway has a system of five regional research ethics committees. They review all biomedical research involving human subjects within their region and are fully autonomous. Unlike the Danish system, there is no possibility of appeal. In June 1989, the Norwegian Parliament (*Stortinget*) endorsed the recommendation of a 1988 White Paper from the Ministry of Education and Research to establish National Research Ethics Committees to study the following three subject areas: (1) medicine in a broad sense (health and the life sciences); (2) normative academic disciplines, *i.e.*, the social sciences and the humanities, including law and theology; and (3) natural sciences and technology, including those fields of biotechnology which are not subsumed under medicine.

Great importance is placed on securing representation in the national committees from the fields of ethics and law, and all of them have lay members as well. An important function of the committees is to provide information to politicians, civil servants and the general public. They will also issue policy recommendations concerning research ethics.

The National Committee of Medical Research Ethics has 12 members: three physicians, three members trained in ethics, two lay members, and four from relevant disciplines such as biotechnology, the social sciences, personal data registration, and law.

According to the mandate laid down by the Ministry of Education and Research on May 16, 1990, the main assignments of this National Committee are the following:

1) to keep itself continuously informed on current and potential questions of research ethics in the field of medicine;
2) to be the coordinating and advisory body for the five regional ethics committees;
3) to inform researchers, the administration, and the public on current and potential questions of research ethics in the field of medicine;
4) to submit reports on matters of principle relating to medical research ethics, and to comment on specific matters of special significance relating to research ethics;
5) to report on its activities during an open meeting at least once a year, and, in whatever way it finds suitable, to promote informed societal discussions of ethical questions relating to medical science and knowledge; and
6) to keep other national and international research ethics committees informed of its activities, and in cooperation with them, to seek to establish a platform of principles of research ethics which extends beyond the boundaries of the respective research subjects.

C. *Sweden*

Sweden has a system of six regional ethics committees. Each committee reviews all biomedical research involving human subjects in its region. The Swedish Council on Medical Ethics was established by Parliament (*Riksdag*) and given the status of National Council in March 1984. The National Council on Medical Ethics is made up of seven politicians and eleven expert members representing medical science, philosophy and the arts, law, the Catholic and Protestant churches, and one member from an organization for the disabled. The Council's principal assignment is to maintain a continuous interchange of information and opinion concerning research and medical treatment of critical consequence to human integrity or capable of influencing respect for human dignity. The Council is to act as an advisory body to the government and the *Riksdag* on questions of medical ethics. Its proceedings, which were made public, are aimed at encouraging debate, with particular emphasis on human equality and the right to physical and psychological integrity. The Council should also act as an intermediary between the scientific community, politicians and the general public. There is, however, no formalized cooperation between the National Council and the system of regional ethics review committees.

The Swedish government issued two reports concerning the research ethics committees in 1989 ([59];[60]). The reports recommend that the composition of the committees should be changed so that fully one-fourth of the committee members are lay persons. A lawyer should also serve on the committee. The committees should continue to be advisory bodies, but in reality, it should be virtually impossible to conduct research without approval from a research ethics committee. The report also recommends that courses in research ethics should be made compulsory in the education of researchers.

III. INFORMED CONSENT AND RESEARCH ETHICS

There have been a number of developments with regard to informed consent in the three countries. A report was published in Denmark claiming that research subjects were poorly informed about the proposed research [53]. The report was based on participant observation of 15 patients who took part in six different research protocols. In Norway, the regional ethics committees decided, after much debate, that informed consent to research should be in writing. The revised Declaration of Helsinki recommends that written consent be given, but in Norway and Denmark it has been argued that a written consent would make the research subjects more reluctant to withdraw from the project should they so desire. This argument has not been found convincing by the ethics committees. They now require that patients be given written information, which, among other matters makes it clear that the subjects can withdraw at any time. The committees also require written consent [5].

During the last part of 1991, a discussion has begun in Norway concerning ethical guidelines for epidemiological research. We will not report on that discussion here, but only mention that a working group issued recommendations for Sweden in 1989. One of the recommendations is that informed consent need not be sought for the use of already existing data about patients. Researchers should, however, inform the public through the news media about proposed research using such data [2].

IV. ABORTION, PRENATAL DIAGNOSIS, AND FETAL RESEARCH

There are a number of interrelated issues which have dominated the medical ethics debate in the Scandinavian countries in the period 1989-1991: the question of abortion, especially of late abortion after diagnosis of a fetal abnormality, the legitimacy of fetal research, and new reproductive technologies. In addition, there has been a debate about the ethics of organ donation and transplantation, which in Denmark has been coupled with a discussion of the definition and criteria of death.

There are many similarities in the ways that the three Scandinavian countries have dealt with these issues, but there are also important differences. In the following exposition, we will treat each country separately, concentrating on the various government reports which have been issued. We will also refer to some of the discussion in the professional journals.

A. *Abortion and Prenatal Diagnosis*

All three Scandinavian countries have abortion laws, passed during the mid 1970s, which grant a pregnant woman the right to freely choose an abortion until the end of the twelfth week of pregnancy. There have not been any serious attempts to repeal these laws. There has, however, been concern expressed over late abortions. With new diagnostic tools such as amniocentesis and ultrasound, it has become possible to diagnose diseases and abnormalities in the fetus. Such a diagnosis may lead to a decision for an abortion. During the last few years, there has been a fierce debate in the Scandinavian countries over the justification for abortions based on such diagnoses.

In all three countries, working groups have issued recommendations to their respective governments concerning the regulation of such practices. There are marked similarities between these reports. All stress that the abortion decision should be made by the woman herself in consultation with her physician. The decision should be based on an individual evaluation of the hardships expected by having to raise a child with severe disabilities. The use of invasive diagnostic tests such as amniocentesis should never be mandatory, but should be voluntarily offered in cases of high risk pregnancies. There has been a marked resistance to drawing up lists of indications for the use of diagnostic tests or lists of conditions which would justify abortions. It has been felt that such lists would imply that society has decided that some human lives are not worth living.

There has been a debate in all three countries about whether a time can be determined during pregnancy after which no abortions should be permitted. The reason for this debate has been the ability to keep ever younger newborns alive. It has been felt that it is morally problematic to abort a fetus which could live outside the body of the mother. Although the Norwegian abortion law includes a clause which forbids an abortion in cases where the fetus is "viable", only Sweden has introduced a specific time in pregnancy after which no abortions can take place (22 weeks).

1. *Norway*

In the current Norwegian abortion law a woman can freely choose to have an abortion until the twelfth week of pregnancy. After the twelfth week of pregnancy, abortion can be performed if there is great danger that the fetus may suffer from serious genetic disease or may develop a disease during pregnancy. There are also other reasons involved to permit abortions after 12 weeks, *viz.*, if the pregnancy, birth or care of child can represent hardships for the woman. After the eighteenth week, abortion can only be performed if the reasons are especially serious (*tungtveiende*). Decisions to terminate a pregnancy after the twelfth week have to be approved by a committee of at least two physicians.

One troubling issue in Norway has involved the appropriate basis for abortion decisions after the twelfth week, and particularly after the eighteenth week. There has been particular debate about the justification for terminating pregnancy if the fetus suffers from a genetic condition or a chromosomal abnormality, after that diagnosis has been made by procedures such as amniocentesis or ultrasound.

There has been no legal regulation of prenatal diagnosis in Norway until now. The use of diagnostic tests have been limited to pregnancies where there is an increased risk of certain genetic diseases. This includes pregnancies in women over the age of 35 or pregnancies where parents have already had a child with a chromosomal abnormality. In a discussion in the Norwegian Parliament in 1982, it was argued that such a list of indications, and an automatic offer of diagnostic tests based on them, would imply that society accepted the notion that some human lives were not worth living. Members of Parliament therefore urged that the use of these tests and subsequent abortion should be individualized decisions in cases where the care of a disabled child would be a severe burden on the family and/or the woman. Members also stressed that the final decision about whether one should perform an abortion should rest with the woman herself or with her family.

Despite this debate in the Norwegian Parliament, a list of indications for prenatal diagnosis has been used in Norway, in part, because of limited capacity to perform such diagnostic tests and a consequent need to give priorities to women with increased risks. Nevertheless, there has continued to be a certain uneasiness among members of Parliament about the practice of aborting fetuses with certain genetic diseases. In a question to the Minister of Social Affairs on November 15, 1989, a member of Parliament wanted to know the percentage of abortions performed after the twelfth week of pregnancy due to a genetic condition of the fetus. Another representative asked the same Minister, on February 7, 1990, whether abortion based on a diagnosis of Down syndrome in a fetus was in accordance with the law. In her answer, the Minister pointed out that it was the woman and the family who made these decisions after they had evaluated their abilities to care for a disabled child, and that such a practice therefore was in accordance with the law [23].

The use of ultrasound as a diagnostic tool during pregnancy has accentuated the ethical debate. In a consensus conference in 1986 in Norway, it was recommended that all pregnant women should be offered one ultrasound investigation during pregnancy, on a voluntary basis. Through an ultrasound investigation, one may suspect a chromosomal abnormality in the fetus and consequently desire amniocentesis. This has led to an increase in the number of diagnosed abnormalities and to an increase in the number of abortions performed for this reason. Concern has also been expressed that little documentation exists concerning the psychological effects on the pregnant woman after receiving information from ultrasound investigation, and that the use of this diagnostic tool may lead to the abortion of healthy fetuses, based on misdiagnosis [52].

As a result of this uneasiness over abortions performed after the eighteenth week of pregnancy, two working groups have been charged with the task of providing guidelines. The first group issued its report to the Norwegian Directorate of Health in July 1990 and offered the following recommendations:

1) As before, abortion after the eighteenth week should only be performed if the reasons for the abortion are especially serious. After the twenty-first week, permission for an abortion cannot be granted unless the committee clearly finds that the fetus has a condition which is incompatible with life.
2) A special committee should review all applications for an abortion which is desired because of serious disease or abnormality in the fetus.
3) A "second opinion" should be sought in cases where an abnormality is diagnosed by ultrasound.
4) The committee reviewing the applications shall establish that there has been a full review of the seriousness of the disease or abnormality, including possible prenatal and postnatal treatments, the prognosis for survival, and the possibility of disabilities [23].

The second working group was appointed to draw up guidelines covering the aspects of research and development in biotechnology and gene technology related to human beings. A substantial portion of its report, issued in November 1990, is devoted to issues raised by prenatal diagnosis [46]. The Committee points out that within the next few years, a technique using a combination of tests of the mother's blood will be able to reveal 60 percent of fetuses with Down syndrome in the sixteenth week of pregnancy. When this technique is combined with ultrasound investigation, an even greater percentage of such fetuses will be diagnosed. A majority of the committee want to prohibit the use of such tests for women not at risk. A majority of the committee would, however, also permit abortions on the basis of a diagnosis of Down syndrome. They argue that this does not involve a denigration of the human worth of the fetus so diagnosed because the reason for the abortion is the intolerable burden such a child may represent for the family as a whole. A minority of the committee favor an amendment to the current abortion law providing that non-fatal developmental abnormalities should not in themselves be indications for abortion. Abortion should only be permitted if there is a risk that the child itself will have to live with a great deal of pain and suffering.

2. *Sweden*

In 1989, the Swedish government issued a report concerning abortion and prenatal diagnosis [57]. With regard to abortion, this report upholds the 1975 Swedish abortion law, which allows a woman to have an abortion without interference in any way before the twelfth week of pregnancy. Between the twelfth and the eighteenth week, a so-called "pastoral" investigation will be made. After the eighteenth week, permission of the National Board of Health and Welfare is required. The report also proposes that an obligatory offer of counselling be made after an abortion, as well as pastoral counselling before the intervention.

The main ethical standpoint of the report is that the fetus and the woman are two different individuals, although the fetus depends on the woman. A potential conflict of interest between these two individuals exists from the very beginning of the pregnancy, and the interests of the fetus grow increasingly important as the pregnancy proceeds.

The report has led to an extensive debate in the professional journals, because of its insistence that the fetus is a separate individual with at least some moral standing. Torbjörn Tännsjö praises the thoroughness of the report, but criticizes the justification for its conclusions [61]. He finds that the philosophy of the report is based upon natural rights and the sanctity of life, and that, on this basis, the fetus is considered to be a moral subject. Thus, it would seem to possess rights equal to those of the pregnant woman, and the conclusion should be that every abortion is an intolerable infringement of the fetus' rights. The report, however, does not draw this conclusion, although Tännsjö notes that its discussion imposes guilt upon a woman seeking an abortion.

In a reply, the Catholic theologian Erwin Bischofberger argues that the fetus indeed does have an inherent right to existence, but interestingly enough does not draw the conclusion that this entails that one should not perform abortions [6]. In certain cases, the rights of the fetus should be weighed against the rights of the pregnant woman and the family.

As in Norway, the new diagnostic tools used during pregnancy have raised new ethical, psychological, and legal dilemmas in Sweden, which are discussed in the Swedish report. The dilemmas arises because prenatal diagnosis, carried out routinely at the maternal clinics, sometimes reveals diseases or malformations which cannot be treated. Some see few problems with these techniques, and emphasize that they prevent human suffering (*e.g.*, [8]). Others, while agreeing that these techniques should be used to diagnose severe defects, point out that these techniques are already used to abort fetuses with less severe conditions, and can lead to devaluation of the life of disabled human beings [20].

The Swedish report published in 1989 suggests that, prior to prenatal diagnosis, the woman should be informed of available diagnostic options, what they can achieve, and their limitations and risks. The pregnant woman, together with the treating physician, should then determine which examinations she wants carried out. It is crucial that correct and detailed information be given. There are limits, however, to the procedures which can be offered. There must be some reason for carrying out the diagnostic test, compatible with science and experience. Unsubstantiated worries do not justify an examination, as long as the worries do not affect the woman's general state of health.

The Committee recommends that these examinations should not be used as general screening tools for two reasons: first, because such use could be regarded as a "hunt" for deviations, and thus degrade the value of human life; and second, because there is limited value to prenatal diagnosis in discovering fetal damage. Thus, the permission for screening examinations should be made on a case-by-case basis.

Prenatal diagnosis should only be permitted to check the progress of the pregnancy or to determine the possible presence of any serious injuries or diseases. It should not be used in order to determine the sex of the fetus. The woman herself is the one who should decide whether to have an examination, and the current indications for examinations should consequently be abolished.

Women who do not belong to any risk group, and who are below the age of 35, should be given detailed information concerning prenatal diagnosis only on request. Women above the age of 35, or who belong to a risk group, should always be offered information.

When examined, the woman is entitled to receive all possible information concerning the health of the fetus. Information about abnormalities should be given in the presence of a responsible gynecologist and representative of the Women's Clinic, and if possible, a geneticist and a representative of the disabled should also be available. These recommendations of the *Sveriges Offentlige Utredninger* have not yet resulted in any alterations in Swedish law.

In 1990 the health authorities issued new guidelines governing cremation and burial of aborted fetuses [54]. It recommended that fetuses aborted after the twelfth week should be routinely cremated or buried. It should also be possible to do this for fetuses aborted before the twelfth week. Earlier, only viable fetuses could be buried (fetuses after the twenty-eighth week, or those born alive). These guidelines have been criticized by some who have pointed out that such a practice is a symbolic criminalization and stigmatization of those women who seek abortions ([38]; [39]). These arguments have been countered by the health authorities who do not find any inconsistency between the current abortion law and the new recommendations [35].

3. *Denmark*

To stimulate the debate over prenatal diagnosis, prenatal screening, and genetic counselling, the *Etiske Råd* issued the booklet "*Prænatal Diagnostik, Prænatal Screening, Genetisk Rådgivning*". It surveys the different types of diagnostic tools available and also covers the practical organization of testing and its legislative aspects. The booklet discusses diagnostic tests and their possible connections to subsequent abortion decisions in light of two opposing principles.

The first principle maintains that the human embryo is a human life from the moment of conception. Given this status it has the same right to protection as other human beings. This principle does not, however, exclude the possibility of an abortion in circumstances where the embryo is viewed as similar to a terminally ill patient. In such cases, the child, if carried to term, will only live for a short period, purposelessly, and under extreme suffering. This exception, however, excludes abortion based on the interests, economic or otherwise, of the parents or society. This principle might be invoked, therefore, when describing the defects and diseases which justify abortion and the precise criteria by which these are to be judged.

The second principle is based on the wish to avoid suffering. Again, this would include the desire to avoid purposeless suffering for the newborn, but it would also consider the situation of the family into which the child is born. For instance, if the diagnosis indicates that the child will suffer a severe handicap with which the family is unable to cope, abortion might be viewed as an alternative.

In a later publication, the Council of Ethics offers its recommendations [12]. The goal of prenatal diagnosis is to identify serious diseases in order to prevent such diseases in the expected child. It is therefore unacceptable to perform prenatal diagnosis of sex and normal traits. Eugenic or cost-benefit considerations should also not be the main reasons for introducing prenatal diagnosis. In cases where the child can be expected to be born with extreme suffering or disabilities, abortion is an acceptable alternative. Other disabilities are of lesser character and do not ethically justify an abortion. Nonetheless, information should not be withheld, nor should the right to abortion be taken away. The woman's right to decide for herself remains the most important consideration.

In an article in the *Danish Medical Journal*, the Minister of Health affirms the principles which should govern the use of these tests [34]. The tests should be voluntary, should not be used for screening purposes, and the decision to use them and decisions to perform abortions should be made by the mother/parents, within the limits of the law.

B. *New Reproductive Technologies and Fetal Research*

All three countries have accepted the various new reproductive technologies, and in vitro fertilization in particular. These techniques are governed by legislation in the three countries. In Norway and Sweden, for example, it is forbidden by law to perform in vitro fertilization with donated sperm or donated eggs. In Norway, the same law forbids research on fertilized eggs and fetuses. In Sweden, a new law was proposed in November 1990, and passed in March 1991, which would allow research on pre-embryos up to 14 days after conception. Fertilized eggs which have been used for research cannot be reintroduced into the woman's body, and can only be used for research if the woman has given her permission. Denmark passed a preliminary injunction against fetal research in 1987, until the *Etiske Råd* (Danish Council of Ethics) could issue its report and the matter could be debated in the Danish Parliament.

1. *Denmark*

In the Fall of 1989, the report from the Council of Ethics was issued [10]. The majority of the Council would allow some research on fertilized eggs and fetuses. Before such research can be performed, however, it must be approved by a committee and meet the following conditions. The research must not pose any unnecessary harm for the fetus, and there must be justified hope that one can achieve results which will be significant for the future prevention and treatment of diseases. The research also must not harm the health of the pregnant woman. Informed consent to such research must be given by the pregnant woman and her husband or partner. A minority of the Council felt that any research on fertilized eggs or fetuses should continue to be forbidden.

As a result of this disagreement, the Minister of Health decided in February, 1990 to continue the preliminary injunction against fetal research for one year. In April, 1991, a bill was introduced in the Danish Parliament concerning research on human subjects, which also included the provision allowing research on fertilized eggs during the first 14 days after conception.

2. *Norway*

As mentioned above, any fetal research, including research on fertilized eggs, is currently forbidden by Norwegian law. However, the National Medical Research Ethics Committee in Norway issued policy recommendations on fetal research in 1990 [50]. Fetal research is defined as all research involving unborn fetuses from the moment of conception until birth. Although the recommendations would allow some research on fetuses, they are still quite restrictive. They would only allow research which is of possible benefit to the research subject (the fetus) itself. Specific recommendations include the following:

1) The same ethical principles governing research on born human beings should apply to research on fetuses.
2) Fetuses as research subjects should be protected in the same way as other groups of research subjects which cannot give their informed consent to research.
3) Research on fetuses which leads to the death of the research subject or irreversible damage in the fetus is unacceptable. It is also ethically unacceptable to produce fetuses in order to do research on them.
4) Non-therapeutic research which involves short- or long-term risk or discomfort for the fetus or the mother is ethically unacceptable.
5) Risk-free, nontherapeutic research on fetuses which cannot be done in another way (*e.g.*, systematic monitoring of the fetal heart beat) can be ethically acceptable.
6) Therapeutic research on fetuses can be ethically acceptable if this research directly or indirectly is of value to the fetus itself, knowledge is available concerning the possible short- or long-term risks of the research, and there

is no risk of irreversible damage to the fetus or the mother.

7) Therapeutic research on fetuses where there is no extant data concerning the possible risks is ethically unacceptable. This includes gene therapy on germ cells, and gene therapy on somatic cells from fetuses.

The Ethics Committee of the Norwegian Ministry of Health and Social Affairs also recommended that the current ban on research on fertilized eggs should continue in Norway [47]. The Committee acknowledged that information and methods based on research on fertilized eggs may lead to significant improvements in the methodology of in vitro fertilization and that such treatment is currently accepted in Norway. The Committee recognizes a possible inconsistency here, but concludes that the ban should not be lifted. According to the Committee, "Norwegian society will simply have to live with this inconsistency". The Committee does, however, allow changes in the procedures used during the extraction and fertilization of eggs in order to improve them, so long as the eggs are returned to the woman's uterus. It upholds the Norwegian ban against using donor sperm and donor eggs for in vitro fertilization. Eggs which are not implanted must be destroyed or can be frozen for up to three years for later implantation. The Committee also would not allow selective reduction of fetuses in order that the remaining fetuses may continue to develop.

Research on spontaneously aborted fetuses should be permitted after normal review by regional research ethics committees. The Committee recommends, however, that research on fetuses before induced abortions should be prohibited. Cells and tissue from aborted fetuses, however, can be used for diagnostic purposes in approved medical laboratories (as is the practice today) and can also be used for research. The majority of the Committee did not draw firm conclusions about transplanting organs from aborted fetuses, because such procedures are still in the research stage. A minority of the Committee would prohibit all such use of organs from aborted fetuses. With regard to the use of organs from born anencephalic children, the Committee recommended that this question be reviewed at a later date in conjunction with a possible revision of the criteria of death.

A working group within the Church of Norway has reached the same conclusions opposing fetal research, although a minority of this group would accept such research if it could result in alleviation of human suffering ([24];[30]).

The disagreement within the Danish Council of Ethics, the delay in the revision of the Danish law, and the restrictive policies recommended in Norway reflect deep disagreement among professionals and the public in all three Scandinavian countries concerning these issues. During the past several years, there has been a debate in both the professional journals and the popular media on the ethical issues raised by in vitro fertilization and by research on fertilized eggs. Many have pointed out certain inconsistencies in official attitudes about these issues. Abortion without any restrictions is allowed before the twelfth week of pregnancy, but no research on fertilized eggs is allowed. The technique of in vitro fertilization is accepted, and even given high priority in Norway, yet no research is allowed which may improve the methodology of this treatment ([31];[32]). Some commentators, however, have noted other problems with in vitro fertilization: its low success rate, uncertainty about possible negative effects, and the unrealistically high expectations of infertile couples which may be fostered ([9];[21]). In Denmark, the proposed changes in the law concerning fetal research have led some to complain that the whole issue of assisted fertilization has not been properly analyzed ([43];[44]). There have also been discussions about whether it makes sense to speak of the right to have children, or about the rights of the unborn [64]. Finally, one commentator has rejected an ethics based on rights as irrelevant, and instead has proposed that an "ethics of care" is more appropriate [17].

V. BRAIN DEATH, ORGAN DONATION AND TRANSPLANTATION

A. *Brain Death and Transplantation, Denmark*

In 1988, Denmark was the only country in Western Europe which had not yet adopted brain death criteria. Kidney transplants were performed in Denmark, but organs were removed only after the heart had stopped beating. When brain death was diagnosed, it was a general practice to withdraw therapy such as the use of respirators. The lack of brain death criteria, however, made heart and liver transplantations impossible to perform.

The law in effect in 1988, which had been passed in 1976, allowed death to be declared immediately after cessation of heart function in a person with irreversible loss of whole brain function. The law in effect before 1976 had demanded a period of observation after the heart had stopped before organs could be removed.

In a government report submitted in 1985, it was recommended that brain death be introduced as a parallel criterion of death along with the cessation of heart function. The report initiated a fierce debate over whether Denmark should adopt brain death criteria.

In 1988, The Danish Council of Ethics issued its own report on the topic ("*Dødskriteriet*"). (For a presentation of the report in English see [52]; for a criticism, see [16].) The report discusses the ethical aspects of the concept of brain death. The common experiences of death are viewed in relation to the biomedical concept of death. The report argues that death should be seen as a process, rather than as an event, in accord with both common usage and scientific knowledge. This process is not completed until the following three functions can no longer be traced or restored: (1) heart function/blood-circulation, (2) breathing, and (3) brain function. The total loss of brain function may initiate the process of death. The process, however, is not complete until all the above functions have ceased.

The report then considers various medical definitions and their legal meaning. Even if death is regarded as a process, it may be necessary to introduce certain *criteria of death* that are associated with a *moment of death*. Both terms are necessary, because the fact that a person is dead, and fixing the time of this event, is of crucial significance for the person himself, his next of kin, and the interests of other persons. The report considers what sort of criteria one should introduce, based on the interests of these three groups.

One might argue that, in considering the dying or dead person's interests, criteria based on some variants of brain death should be introduced. On this interpretation, if there is no possibility of experience, one has crossed the border between life and death. The Council argued, however, that one's attitude toward death also involves interpersonal aspects. For one's relatives, it is not the case that "death" coincides with the absence of brain function. Rather, for them it is important that all stages of the process of death have passed. According to the Council, the only reason for wanting to introduce brain death criteria is to serve the interests of third parties, specifically those of the potential recipients.

If one regards death as occurring at a specific moment, one will face the ethical dilemma of having to choose between the interests of the dying person and his or her relatives, on the one hand, and of potential recipients of organs, on the other. If, however, death is regarded as a process, one might argue that total brain death indicates that this process has irreversibly started, and cessation of breathing and heart function indicates that the process has ended. Lack of brain function indicates that a person is *within* the process of death. When within the process of death, all efforts to sustain life artificially should cease, not in order to cause death, but to effect the end of the process. The moment of death is that particular instant when the process is finished, the moment when heart and respiratory function have irrevocably stopped.

Despite this, the Council argued that the process may, under special circumstances, be prolonged artificially in order to remove organs for transplantation, if the person (or the next of kin) has registered as a donor. This intervention causes the end of the process of death, but it is not what causes the death of the person.

The Council concluded that under no circumstances should a potential organ recipient be viewed as having a right to receive organs from others. Society, therefore, is not free to use the organs of dead or dying people as it pleases.The Council also rejected the earlier proposal that brain death criteria should be used in parallel with cessation of heart function. It argued that it is important to have one concept of death in order to protect the interests of the dying person and his or her relatives. This concept should not, however, prevent dying persons from donating their organs if they so desire, or prevent the removal of respirator therapy. The Council, therefore, proposed that after two physicians have diagnosed irreversible cessation of brain function, all measures that prolong the dying process should be withdrawn. This action can be postponed for a maximum of 48 hours.

The Council further proposed that a directory containing information about donors should be created and administered by the Ministry of Health. Anyone who has reached the age of 18 is allowed to register. The permission can, however, be withdrawn by the donor at any stage. All information to the general public concerning the directory should be provided by the Minister of Health in cooperation with "*Det etiske Råd*". Any payments in conjunction with registration or donation should be criminalized.

The report by the Danish Council of Ethics has been criticized. Some critics have focused on the report's insistence that death is a process, rather than an event ([42];[69]). Others have pointed out that the Council does not wish to introduce brain death criteria but does wish to permit transplantation, goals which may be in tension [19].

Denmark adopted a new law on the criteria of death in 1990 (Law Number 402, June 13, 1990). According to this law, death can be established either by the cessation of breathing and heartbeat or the irreversible loss of all brain function. The law, therefore, accords with the earlier government proposal and opposes the recommendations of the Council of Ethics. The same law also regulates transplantation. Organs may be removed from deceased persons who have given permission, or if there is no evidence that the deceased person would object to organ donation, when relatives have given their permission. Organs cannot be removed if the relatives cannot be consulted, thus establishing a presumption against organ donation. This particular issue has been debated in Sweden (see below).

B. *Transplantation, Sweden*

The Swedish Law dealing with transplantation dates back to 1975. Since then, the criterion of death has been changed (January 1, 1988). Together with new technology, this change has led to a wide range of possibilities in transplantation medicine, but has also caused various problems. The 1989 report of the *Sveriges Offentlige Utredninger* (SOU) suggests changes that should be made in the law of 1975 and emphasizes guidelines for obtaining consent [60].

1. *The Dead Body as a Donor*

The Transplantation Law of 1975 states as a guiding principle that the deceased should have given written consent for an organ to be removed. If such a document does not exist, an organ can be removed only if the deceased had expressed an opinion consistent with such an act. If there is disagreement about the view of the deceased, his or her next of kin can make the decision. If no next of kin is available, or if relatives disagree, consent to organ removal is presumed. After the introduction of the new definition of death on January 1, 1988, the transplantation law was changed; it is now the case that under the latter circumstances, refusal to donate organs should be presumed.

According to the SOU report published in 1989, the first principle that should govern our relationship to a dead body is to show piety towards it. This is not a matter of the integrity of the dead body, but of the psychological integrity of the living. The Committee's aim has not been to increase the number of organs provided, but to determine conditions under which organs may be procured. The paramount rule is the right of the individual to decide about one's own body; no one else should have the power to overrule a person's decision about whether or not to be a donor. A written document of intention, therefore, is to be preferred; but since very few people document their opinion in this manner, written consent cannot be an absolute requirement. Indeed, such a requirement might in many cases be at odds with the will of the deceased. The donor's own wishes should, if possible, be determined even in the absence of written consent. If his or her opinion cannot be ascertained, the next of kin should have the opportunity to refuse transplantation in order to protect themselves against adverse psychological reactions. However, since such decisions normally are made at times of grief, they must be voluntary, not obligatory. Thus, in cases of uncertainty, transplantation can occur if close relatives do not refuse it. Relatives should, of course, be informed of the planned transplantation, and offered an absolute right of veto. A reasonable time for decision-making must also be provided. If the next of kin are impossible to reach, or if the deceased had none, the Committee suggests that no organs should be removed.

2. *The Living Person as a Donor*

The procurement of organs and tissues which do not regenerate should be limited. If a transplant involves risk of serious damage to life or health, it should not be done. The current shortage of kidneys, however, indicates that an absolute prohibition against organ donation by adults would be unreasonable.

With regard to tissue such as bone marrow, skin, or blood, permission should be granted as long as the donor is an adult. If the person, because of age (under 18) or mental incapacity, lacks the ability of giving consent, no action should be taken. Under special circumstances, however, when it is crucial that tissue come from a brother or sister who is under age or mentally incompetent, a transplantation can proceed with the consent of the parents and the assent of the child. The necessity of the transplantation should in such cases be controlled by *Socialstyrelsens Rättsliga Råd.*

Within the context of present legislation, small surgical interventions such as the taking of blood, skin, and corneas, do not require consent. The report proposes that this presumption come to an end, and that written consent should be required from a living donor for such things as blood and skin. With regard to deceased donors, the general rules of consent should be followed.

3. *Information and Registration*

Information is the key word if the right of self-determination is to be maintained. The Committee urged that information should be widely spread, and especially to target groups such as students and military conscripts. However, the Committee concluded that the cost of a directory which contained all available information would be disproportionate to its benefit. (The Committee report claims that only about 30 percent of the population would be willing to register). The Committee therefore recommends that an official system using donor cards should be introduced.

The Committee also suggested that after death has occurred, the relatives of the deceased should routinely be asked about the attitude of the deceased towards transplantation, regardless of whether a transplantation is pending. This would not, however, be a part of any legislation.

With regard to buying and selling organs, the Committee stated that an organ is not a product. Thus, no profits should be made in the procurement or distribution of organs and tissues. The report suggested that the 1980 law concerning secrecy should be altered so that it explicitly mentions transplantation as a relevant area covered by the law.

The report proposed that in the absence of explicit consent, one should presume refusal to organ donation. This has been the topic of some debate in Sweden. The Swedish Medical Association issued guidelines in 1989 which differ from those in the Committee's report [58]. These guidelines also affirm that respect for the deceased wishes is central, and that no organ donation should take place if refusal is explicit. However, if there is no explicit refusal from the deceased or from the next of kin, one should presume consent to organ donation. The Swedish Medical Association offers the following reasons for this point of view:

1) Population surveys have shown that a majority of people are in favor of organ donation.
2) In light of the need for organs, it seems worse not to use an organ from a person who would have wished to donate than to use an organ from a person who would have refused.
3) It is often difficult for relatives to consider questions of organ donation in the period after the death of a loved one. Sometimes it may be hard to reach relatives, or they may have different opinions.

This reasoning has been criticized by some (*e.g.*, [27]) and supported by others (*e.g.*, [49]).

The population survey to which the Swedish Medical Association refers is quite interesting [18]. Sixty-eight percent of those surveyed expressed no objections to organ donation after death, and 45 percent would give permission for organs to be used from a deceased relative in cases where the opinion of the deceased concerning organ donation is unknown. Thirty-one percent were not able to answer this question. Eleven percent were against organ donation.

Sixty-eight percent felt that it would be more difficult to consent to donate a heart than to donate a kidney or liver.

There has been a similar debate in Sweden concerning autopsy, which also involves issues of consent and of weighing the interests of the deceased against the interests of others [45].

C. *Brain Death and Transplantation, Norway*

As is evident from the above discussion, there has been a lively debate in Denmark and Sweden about brain death and organ transplantation. Similar discussion has not occurred in Norway, because brain death criteria have been widely accepted there. The Norwegian law concerning brain death and transplantation dates from 1973. The number and results of transplants at the Norwegian National Hospital in Oslo are comparable to those of the main international transplantation centers. Organs can be removed from deceased persons in hospitals if they or their next of kin have not explicitly refused, or if one has no reason to believe that it would be against the religious views or the values of the deceased person. While there has not been any discussion in Norway about the ethical issues raised by organ donation and transplantation, questions have been raised about whether transplantation is an effective use of scarce resources [51].

VI. ISSUES INVOLVING DIAGNOSIS AND TREATMENT OF ADULT PATIENTS

A. *Human Genome, Denmark*

In 1990, the Danish Council of Ethics issued a working paper on the mapping of the human genome [15]. Large projects has already begun in the United States and Japan. In the summer of 1990, the European Community (EC), of which Denmark is a member, initiated a program called "Human Genome Analysis". The working paper of the Danish Council describes international progress on genome mapping and the organization of present projects, and offers a perspective on areas to which mapping knowledge has been applied. The paper also evaluates legal aspects of the use of genetic tests in the workplace for insurance purposes.

A predecessor program of the EC project "Human Genome Analysis" was entitled "Predictive Medicine". That project suggested that common diseases such as diabetes, cancer, and heart disease might result from interactions among a number of genes, which would make these diseases difficult to predict merely by considering heredity. Such diseases might occur when genetically sensitive individuals are exposed to specific environmental influences. The project concluded that environmental components would prove difficult to cope with. "Predictive medicine" therefore is understood as the best means to protect the individuals from their genetic predispositions and to prevent the passing of genetic sensitivity to the next generation.

A screening of the population might make early intervention with certain illnesses possible, but genetic information might also be misused by employers and insurance companies. Furthermore, the "Preventive Medicine" program stated that, with improved fetal diagnostics, it might be possible to satisfy parents' wishes for certain qualities in their children. The program stressed the crucial need for society to balance the right of the individual to decide about personal information against the health benefits for the society as a whole.

The program was heavily criticized, especially by the Danes and the West-Germans. The latter described the project as neo-eugenics. "Det Etiske Råd" argued that the program should concentrate on diagnosis and treatment of serious hereditary diseases. The Ministry of Health directed an official critique of the project for its failure to consider sufficiently the legal, ethical, and social issues raised by genetic screening.

The program was revised, and the new project was called "Human Genome Analysis". The purpose of this project is to apply and develop biotechnology in the mapping of the human genome with regard to prevention and treatment of diseases. All gene therapy research on gametes and embryos, leading to hereditary changes, is abolished. The program also does not include development of gene therapy on somatic cells.

The "Human Genome Analysis" maintains that individual rights must be protected and respected as the genome is mapped. All genetic data should be considered an aspect of the individual's integrity and treated confidentially. The project also calls for a parallel investigation of the ethical, social, and legal issues raised by the Human Genome Project by a committee of lawyers, geneticists, philosophers, ethicists formed for that purpose.

B. *Gene Technology and Ethics in Norway and Sweden*

Related issues have been discussed by the committee appointed by the Minister of Social Affairs. In its report issued in 1990, the committee takes a strict view regarding the use of genetic data by anyone other than the person whose it is. No other individual or institution, including the authorities, present and future employers, educational institutions, medical institutions, pension funds, or life insurance companies may request, possess, receive or make use of such information, or attempt to learn whether such tests have been carried out.

The committee is, in general, critical of the use of screening methods to register the prevalence of genetic disease or individual predispositions towards disease. Tests which may detect risks of developing future disease should not be introduced unless there is a specific medical objective, adequate evidence of the efficiency of preventive measures, and safeguards of privacy in place. The committee focuses on screening for cystic fibrosis, and argues that a general testing of the population to establish carriers of the cystic fibrosis gene should not be introduced in Norway. The committee argues against such screening on several grounds: (a) because it would lead to stigmatization of persons with cystic fibrosis and persons who are carriers of the disease; (b) because it would require large expenditures and extensive resources; (c) because the number of falsely positive test results would create problems; and (d) because a testing program would lead to unnecessary fear and worry among many people [48]. A Swedish expert group is also skeptical about general screening for cystic fibrosis, but recommends that diagnostic methods based on gene technology should be introduced, that close relatives of sick individuals should be tested after gaining their informed consent, and that prenatal diagnosis should be offered to carriers [67].

The committee also considers the use of predictive testing for Huntington disease. In early 1990, the Ministry of Social Affairs had introduced testing on adults who are at risk at their request, and detailed guidelines have been published as an appendix to this report. The issue of testing for Huntington has also been discussed in Sweden [3]. In one article, authors conducted intensive interviews with ten people at risk for Huntington disease. Only six of the ten favored being tested. Those who were most interested in being tested either had had little contact with family members with the disease, or had passed the normal age of disease manifestation [43].

The Norwegian committee would allow somatic gene therapy but not germ cell therapy, because of the risk of man-mediated changes in the genetic makeup of the human species. In Sweden, a law was passed in March 1991 which, in addition to the consent of individuals to be tested, required permission from the health authorities (*Socialstyrelsen*) before genetic tests can be done.

C. *Ethics, HIV, and AIDS*

The three Scandinavian countries differ markedly in their legal responses to the AIDS epidemic. Sweden has adopted some of the most restrictive measures in the Western world, while Denmark has no laws which apply specifically to HIV/AIDS. Norway also has not yet adopted any legislation. We will first discuss the situation in Sweden and Denmark, and then the proposed law on infectious diseases in Norway.

HIV in Sweden is covered as a venereal disease by the Infectious Disease Act. An overview of the Swedish response to HIV/AIDS is given by B. Henriksson and H. Ytterberg [26]. According to the provisions in this law, infected persons must be registered (although not by name), contact tracing must be carried out, and infected persons can be quarantined in a hospital if a physician feels this is necessary to prevent disease transmission. Any individual who believes that he or she has been exposed to an infectious disease must report to a physician to be examined. The physician must then determine the identities of those who may have been exposed, and these individuals can be subject to compulsory testing.

When individuals test HIV-positive, they must seek regular medical examination and follow guidelines recommended by the physicians concerning the prevention of transmission. If it is suspected that a person does not follow these guidelines, the local Infectious Disease Officer must be informed, who will determine whether the person has failed to follow the rules.

The health authorities recommend psychiatrists to supervise HIV-positive individuals who are isolated according to this law, arguing that since these persons do not have any physical conditions, psychiatry is the proper supervising specialty. This proposal, however, has met with marked resistance among psychiatrists. The Chairperson of the Swedish Psychiatric Association has argued that psychiatrists cannot take part in such measures of social control because it is not in keeping with the Hawaii declaration, which states that the only legitimate purpose for treatment against a person's wishes is the presence of a psychiatric disorder [4]. According to the Chairperson, it is not enough that an individual may fail to know his or her own interests, or may have an unusual opinion concerning what is in his or her own interests. The Chairperson also rejects the view of the health authorities that the willingness of these individuals to engage in behavior that may endanger others is in itself a sign of the presence of a psychiatric disorder. The Chairperson sees the desire by the health authorities to use psychiatrists for supervising HIV-positive individuals as no different in principle from the misuse of psychiatry in the Soviet Union and the misuse of physicians in Nazi Germany. In contrast, some have defended the use of psychiatrists in this role, arguing that many HIV-positive patients show signs of personality disorders [37].

It has also been pointed out that the law is counterproductive, because people will be reluctant to be tested if they fear quarantine. Moreover, the law should not allow infringements of individual freedom merely on the suspicion that someone may endanger others (*e.g.*, [57]). The government has recently proposed that it is not necessary for a physician to be in charge of the involuntary isolation.

Denmark has taken a very different approach to the AIDS epidemic [1]. As in Sweden, issues concerning infectious diseases are governed by the Infectious Disease Act. This Act distinguishes between infections which are "generally dangerous" and those which are not. For example, "generally dangerous" infectious diseases include such diseases as the plague and smallpox, but diseases such as HIV/AIDS or tuberculosis are not included in the "generally dangerous" category. Restrictive measures may only be carried out with regard to infectious diseases in this category.

The most recent Danish Venereal Disease Act was adopted in 1973. After the spread of the AIDS epidemic, there was heated debate about whether HIV/AIDS should be covered by this stringent Act, which allowed for such measures as contact tracing and prison terms for people who exposed others to a venereal disease. However, the Act was repealed by the Danish Parliament in 1988 [41].

Several arguments were made by those who wanted the law repealed. They pointed out that there was no relationship between the number of sexually transmitted diseases and changes in legislation. They also argued that measures other than legal regulation would be effective in combatting infectious diseases, and that legal regulation might prompt a false sense of security. Finally, they pointed to experience from abroad which indicated that the use of force, isolation, and punishment in response to AIDS is counterproductive. Even if such measures are aimed at "totally irresponsible" persons, they tend to weaken the possibilities of effective public health measures.

The Danish health authorities have been very reluctant to suggest that the principle of confidentiality should be set aside in cases involving HIV/AIDS. In a case discussed in the *Danish Medical Journal*, in 1985, the health authorities (*Sundhedsstyrelsen*) recommended that a physician should not break confidentiality in one case where a man who was HIV-positive refused to tell his pregnant wife of his HIV status. According to the health authorities, although the recommendation might lead to detrimental consequences in this particular case, a policy of breaking confidentiality would result in other patients losing trust in the health care system, ultimately leading to an increase in the number of people infected. Recently, however, the Minister of Health recommended that physicians should tell the partners of HIV-positive patients when the patients refuse to do so. The Minister referred to the general exception to the legal requirement of confidentiality in cases when one needs to protect the interests of others [29].

Norway has not yet adopted any specific legal regulation concerning HIV/AIDS, but a proposal for an Infectious Disease Act was submitted to the Department of Social Affairs by the Directorate of Health in 1990 [47]. As with the Danish Act, the Norwegian proposal is also a general law, covering all infectious diseases.

Some of its restrictive measures will only be applicable to what the law defines as "generally dangerous" infectious diseases. In the Norwegian proposal, this class is much broader than in Denmark and includes the following: plague, tuberculosis, HIV/AIDS, meningococcal disease, viral hepatitis B, gonorrhea, genital herpes, genital chlamydia infection, and syphilis. For this class of diseases, the following measures can be used:

1) A physician may, against the wishes of a patient, inform another person about the infectiousness of the patient when there is a reason to believe that there is a danger that the disease may be transmitted to this person.
2) A physician has a duty to inform another person of an obvious and immediate danger that the person may become infected by the patient.
3) A health care institution can test a person for an infectious disease as a condition for diagnosis and treatment, if the presence of disease would necessitate extraordinary measures to protect other patients or personnel from infection. Examples of situations where one may demand a test

include surgical interventions and use of respirators and dialysis.

4) An infected person has a duty to inform the physician about the person who may be source of the infection, and about whom he or she may have infected. There are, however, no sanctions attached to this part of the proposed law.
5) Contact tracing is made obligatory for infectious diseases.
6) Infected persons may be isolated in hospitals if it is very likely that they may transmit the disease to others and if there are no other means of preventing the disease from being transmitted to others.
7) An infected person may be prohibited from certain kinds of work if he or she poses a significant threat of infecting others.
8) Testing of health care workers for infectious diseases is not mentioned in the proposed law.
9) No attempts are made to introduce legislation which would explicitly protect the rights of infected persons in terms of insurance or work.

The proposal has not yet been submitted to the Norwegian Parliament.

All pregnant women are offered screening tests for HIV in Norway. As of July 1990, 244,608 women have been tested, and 19 HIV-positive cases have been identified. All but one of the 19 women had already tested positive or belonged to a clear risk group. It is estimated that 100 women tested falsely positive [38].

D. *Blood Transfusion to Members of Jehovah's Witnesses*

It may seem surprising to some North American readers that the issue of how to deal with refusals of blood transfusions by Jehovah's Witnesses has been the topic of significant debate in the Scandinavian countries. The legal situation is somewhat unclear. On the one hand, it is clearly against the law to impose a treatment on another person against her or his will. On the other hand, physicians have a clear legal duty to do what they can to save a human life. There are no court decisions that give guidance about what to do in cases where these two principles may conflict. The legal situation is similar in all three Scandinavian countries. In Denmark, there is a proposal that would give legal status to advance directives, but this has not yet led to any changes in the law [34].

The Swedish and Danish health authorities have issued guidelines concerning how physicians should deal with such cases. The Swedish health authorities and the Ethics Committee of the Swedish Medical Association issued guidelines in 1989 which emphasize that the wishes of the patient should be the basis for treatment decisions [56]. Competent patients who refuse blood transfusions should not, therefore, receive any blood products. However, if the slightest uncertainty exists about what the patient really wants, blood should be given in life-threatening situations.

The Danish Directorate of Health (*Sundhedsstyrelsen*) issued an announcement in 1991 on this matter [58]. According to the Directorate, if a competent adult refuses a blood transfusion under any circumstances, the physicians must decide whether they wish to treat the patient under these conditions. If the physicians accept these conditions, they cannot give the patient any blood products. If, however, significant new developments have occurred after the patient gave his or her directive, and if the patient no longer is competent, the physicians must decide whether these new conditions would have influenced the patient's decisions about blood transfusion. For Jehovah's Witness patients who are temporarily unconscious, who have not unambiguously stated their wishes about blood transfusions, and who are in life-threatening situations, transfusion must be given if necessary to save their lives, regardless of whether relatives state their opposition to blood transfusion on behalf of the patient.

These directives have been criticized by members of Jehovah's Witnesses for failing to protect the rights of temporarily incompetent patients [8]. Moreover, the announcement by the Danish health authorities does not resolve the legal issue about what to do with a temporarily incompetent patient who has previously expressed a desire not to have a blood transfusion.

E. *Treatment of Patients in a Persistent Vegetative State and Terminally Ill Patients*

A 1987 case generated debate in Sweden about the treatment of patients in a persistent vegetative state. The case involved a three-year old boy brought into the hospital after a drowning accident. The physicians decided that there were no possibilities of meaningful life and on that basis withdrew artificial feeding. The health authorities then criticized the three physicians involved.

One Swedish lawyer has concluded that, according to Swedish law, such withdrawal of feeding would be classified as murder ([68];[69]). He argued that from a legal standpoint, to allow a patient to starve to death is the same as to kill a patient by lethal injection. In contrast, others have pointed out that a physician will not be charged with murder if he or she follows standard medical practice, *i.e.*, if treatment is in accord with science and confirmed experience ([30];[66]). On this interpretation, if there is consensus in the medical community that treatment should be withdrawn for a particular condition, this withdrawal does not constitute murder.

Because Sweden's highest court has not reviewed any relevant cases, it is uncertain how the law would be interpreted. Interestingly enough, however, the issue in the Swedish debate has not been about what the patient would decide, as has been the case in the United States.

A survey of Swedish internists regarding do-not-resuscitate (DNR) orders revealed that the interpretation of these orders varies considerably, and that the orders were often associated with withdrawal or withholding of life-sustaining treatments other than cardiopulmonary resuscitation. The survey also found that physicians rarely discussed DNR orders with their patients or the relatives

[57].

Finally, E.J. Husabø discussed the changes that have occurred in the Netherlands with regard to active euthanasia. Although there have not been changes in Netherlands law, the courts have found arguments couched in terms of a person's freedom of choice and of compassion to be persuasive. Husabø argues that these arguments would not be persuasive in Norway, because Norwegian law explicitly forbids killing a terminally ill person out of compassion [27]. Norwegian law also explicitly forbids killing people who consent to active euthanasia. These particular laws give Norwegian courts much less room for interpretation than courts have in the Netherlands.

Center for Medical Ethics
University of Oslo
Gaustadalleen, Oslo
NORWAY

BIBLIOGRAPHY

1. Albaek, E.: (In Press), "AIDS: The Evolution of a Non-controversial Issue in Denmark", in R. Bayer and D. Kirp (eds.), *Passion, Politics and Policy: AIDS in Eleven Democratic Nations*, Rutgers University Press, New Brunswick, N.J.
2. Allebeck, P. and Westrin, C.G.: 1989, "Etik i Folkhälsoforskning – Förslag til Riktlinjer", *Läkartidningen* 86, 3050-3052.
3. Andersson, K.: 1990, "Etiska Spörsmål Kring Huntingtons Sjukdom. Prediktiva Test – Som Att Spela Rysk Roulette", *Läkartidningen* 87, 4290.
4. Åsberg, M.: 1990, "Psykiatrerna, HIV-smittade Och Tvång – Håll Isär Medicinsk Terapi Och Social Kontroll!', *Läkartidningen* 87, 1383-4, 1387-8.
5. Asplund, K. and Britton, M.: 1990, "Do-Not-Resuscitate Orders in Swedish Medical Wards" *Journal of Internal Medicine* 228, 139-145.
6. Bergsjø, P.: 1991, "Etikkomiteene og Det Informerte Samtykke", *Tidsskrift for Den Norske Lægeforening* 111, 558-559.
7. Bischofberger, E.: 1989, "Fostret är en person Med Inneboende Rätt til Existens", *Läkartidningen* 86, 4585 - 4586.
8. Bølling, S.: 1991, "Blodtransfusion til Jehovas Vidner", *Ugeskrift for læger* 153, 2578-2581.
9. Brody, S.: 1989, "Den Moderna Fosterdiagnostiken Förebygger Svårt Mänskligt Lidande", *Läkartidningen* 86, 2839.
10. Dahlquist, G.: 1989, "Barnets Intressen Väger Tungt Mot Det Barnlösa Parets Behov", *Läkartidningen* 86, 1058-1061.
11. Etisk Råd: 1989, *Beskyttelse af Menneskelige Kønsceller, Befrugtede Aeg, Fosteranlæg og Fostre. En Redegørelse*, København.
12. Etisk Råd: 1988, *Dødskriteriet, En Redegørelse*. København.

13. Etisk Råd: 1989, *Fosterdiagnostikk og Etikk: En Redegørelse*, København.
14. Etisk Råd: 1990, *Foster Diagnostik og Etik. En Redegørelse*, Købenavn.
15. Etisk Råd: 1990, *Kortlægningen av Menneskets Arvemasse*, København.
16. Etisk Råd: 1989, *Prænatal Diagnostik. Prænatal Screening. Gentisk Rådgivning. Et Debatoplæg*, København.
17. Evans, M.: 1990, "Death in Denmark", *Journal of Medical Ethics* 16 191-194.
18. Frost, L.: 1989, "Reproduktionsteknologi: Rettigheds, Procedural Eller Omsorgsetik?", *Tidsskrift for Rettsvitenskap* 2, 89-111.
19. Gäbel, H. and Lindskoug, K.: 1989, "Två Tredjedelar Redo Att Ge Organ. Besked Till de Anhöriga Från en rtedjedel", *Läkartidningen* 86, 3681-3692.
20. Gjerris, F.O.: 1989, "Hjernedød", *Ugeskrift for Læger* 151, 2454-2457.
21. Grunewald, K.: 1989, "Fosterdiagnostik Som Trend Och Hot", *Läkartidningen* 86, 3440-41.
22. Hagenfeldt, K.: 1989, "Modern Teknologi Ger Möjligheter Som Vi Inte Alltid Bör Utnyttja", *Läkartidningen* 86, 1055-1058.
23. Hellum, A. *et al.* (ed.): 1990, *Menneske, Natur og Fødselteknologi. Verdivalg og Rettslig Regulering*, Ad Notam Forlag, Oslo.
24. Helsedirektoratets Utredningsserie: 1990, *Abort Etter 18. Svangerskapsuke På Grunn av Fosterskade*, Helsedirektoratet, Oslo.
25. Henriksen, J-O. and Aarre, T.F.: 1989, "Mer Enn Gener – Men Mer Enn Et Menneske", *Tidsskrift for Den Norske Lægeforening* 109, 3621-3624.
26. Henriksson, B., and Ytterberg, H.: (In Press), "Swedish AIDS Policy: A Question of Contradictions", in R. Bayer and D. Kirp (eds.), *Passion, Politics and Policy: AIDS in Eleven Democratic Nations*, Rutgers University Press, New Brunswick, N.J.
27. Husabø, E.J.: 1990, "Legalisering av Aktiv Dødshjelp i Nederland", *Tidsskrift for Rettsvitenskap* 4, 672-707.
28. Ifvarsson, C.A. *et al.*: 1989, "Läkaresällskapets Delegation för Medicinsk Etik Rekommenderar Brott Mot Transplantationslagen!", *Läkartidningen* 86, 1681.
29. Jungersen, D.: 1991, "Ikke Belæg for Brud På Lægers Tavshedspligt i HIV-sager", *Information*, July 26.
30. Karlsson, Y.: 1991, "Medicinsk Kontrovers om Avbrytande av Behandling. Riskerar Läkaren Att Bli Åtalad för Dråp? Endast om Han Avviker Från Kårens Konsensus", *Läkartidningen* 88, 1417-1420.
31. Kirkerådet: 1989, *Mer enn Gener*, Oslo.
32. Kjessler, B.: 1989, "Embryonalforskning – En Etisk Styggelse Eller en Chans för Mänskligheten?" *Läkartidningen* 86, 1610-1613.
33. Kjessler, B.: 1989, "Skall Osäkerhet Inför Det Nya Och Okända Behöva Tvinga Oss Avstå Från Viktig Forskning?", *Läkartidningen* 86, 1063.
34. Knudsen, F. and Guldager, H.: 1991, "Blodtransfusion og Jehovas Vidner. Etiske og Medikolegale Aspekter", *Ugeskrift for Læger* 153, 632-636.
35. Larsen, E.: 1991, "Fosterdiagnostik i Danmark", *Ugeskrift for Læger* 153,

2247-2248.
36. Larsson, Y. and Thorén, M.: 1991, "Abortlagen Ifrågasätts Inte i Almänna Råden om Foster!", 88, 384.
37. Lassenius, B.: 1990, "Tvångsisolering av HIV-smittade. Psykiatrerna Måste ta Sitt Ansvar", *Läkartidningen* 87, 1556.
38. Lindemann, R. *et al.*: 1991, "Perinatal HIV-smitte. Epidemiologiske, Sosiale og Etiske Aspekter", *Tidsskrift for den norske lægeforening* 111, 434-436.
39. Löfgren, M. *et al.*: 1991, "Aborterade Foster – Etik Eller Etikett?" *Läkartidningen* 88, 1265.
40. Löfgren, M. *et al.*: 1991, "Abortlagen Ifrågasätts Genom Nya Rutiner för Omhändertagande av Foster Efter Abort", *Läkartidningen* 88, 12-13.
41. Lovforslag: 1988, "Forslag til Lov om Ophævelse af Lov om Bekæmpelse af Kønssygdomme", *Lovforslag. Folketingsåret 1987-88*, København, pp. 3151-3160.
42. Lübcke, P. and Norup, M.: 1989, "Døden Som Process?", *Ugeskrift for Læger* 151, 2816-1818.
43. Mattson, B. and Almqvist, E.W.: 1990, "Prediktivt Test för Huntingtons Sjukdom – Djupintervjuer Med Personer i Riskzonen", *Läkartidningen* 87, 3204-3206.
44. Møller, J. and Nielsen, B. H.: 1991, "Tiden er Inde til Eksperimenter Med Menneskelige Aeg", *Information* July 29.
45. Nielsen, B. H., and Møller, J.: 1991, "Lov om Humane Aeg", *Information* October 2.
46. Nilstun, T.: 1991, "Svensk Förening för Medicinsk Etik. Samhällsnyttan av Obduktion Måste Vägas Mot Respekten för Den Avlidnes Integritet", *Läkartidningen* 88, 1603-1604.
47. Norges Offentlige Utredninger: 1990, *Lov om Vern Mot Smittsomme Sykdommer*, NOU 1990:2, Oslo.
48. Norges Offentlige Utredninger: 1991, *Mennesker og Bioteknologi*, NOU 1991:6, Oslo.
49. Norges Offentlige Utredninger: 1991, *Rettssikkerhet for Mennesker Med Psykisk Utviklingshemming*, NOU 1991:20, Oslo.
50. Persson, I.: 1989, "Förutsätt Samtycke Till Organtagande i Oklara Fall. Den Modellen Ä bäst ur Allas Perspektiv", *Läkartidningen* 86, 1394-1395.
51. Rasmussen, K.: 1991, "Transplantasjoner, Prioriteringer og Etikk", *Tidsskrift for Den Norske Lægeforening* 111, 295-296.
52. Rådet for Medisinsk Forskning: 1990, *Forskning På Fostre*, Oslo.
53. Rix, B.A.: 1990, "Danish Ethics Council Rejects Brain Death as the Criterion of Death", *Journal of Medical Ethics* 16, 5-7.
54. Schei, B.: 1991, "Mellom Mor og Barn – Tidlig Fosterdiagnostikk Som Rutine Innen Svangerskapsomsorgen", *Tidsskrift for Den Norske Lægeforening* 111, 2118-2121.
55. Scocozza, L.: 1990, "Forsøgspatienter Informeres Dårligt", *Sygeplejersken*

3, 8-14.
56. Socialstyrelsen: 1989, "Nej Till Blodtransfusion Bör Respekteras om Patienten Ar Klar Över Följderna", *Läkartidningen* 86, 450-451.
57. Stiernstedt, G.: 1990, "Avskaffa Tvångsvårde Enligt Smittskyddslagen. Ett Hot Mot Rättssäkerheten", *Läkartidningen* 87, 2405-2406.
58. Sundhedsstyrelsen: 1991, "Meddelse om Patienter Der Afviser at Modtage Blodtransfusion", *Ugeskrift for Læger* 153, 2583-2584.
59. Svenska Läkaresällskapet: 1989, "Självbestämmandet Avgör om Organ Får Tas. Samtycke Bör Förutsättas i de Oklare Fallen", *Läkartidningen* 86, 980-981.
60. Sveriges Offentlige Utredninger: 1989, *Den Gravida Kvinnan Och Fosteret –Två Individar*, SOU 1989:51, Stockholm.
61. Sveriges Offentlige Utredninger: 1989, *Etisk Granskning av Medicinsk Forskning. De Forskningsetiska Komitteernas Versamhet*, SOU 1989:75, Stockholm.
62. Sveriges Offentlige Utredninger: 1989, *Forskningsetisk Prøvning. Organisasjon, Information Och Utbildning*, SOU 1989:74, Stockholm.
63. Sveriges Offentlige Utredninger: 1989, *Transplantation / Etiske, Edicinska Och Rättsliga Apsektar*, SOU 1989:98, Stockholm.
64. Tännsjö, T.: 1989, "Utredningen om Det Ofödda Barnet - Bra Förslag Men Illa Motiverade", *Läkartidningen* 86, 3437 - 3440.
65. Tranøy, K.E.: 1989, "Den Nye Fruktbarhetsteknologien – Nye og Gamle Rettigheter", *Tidsskrift for Rettsvitenskap* 2, 112-126.
66. Vängby, S.: 1990, "Förkortande av Liv Under Adekvat Vård Inte en Fråga som Lämpar Sig for Lagstifning", *Läkartidningen* 87, 641-642.
67. Wahlström, J. *et al.*: 1990, "Diagnostik av Cystisk Fibros Med Hjälp av Genteknik – Riktlinjer Från Expertmöte", *Läkartidningen* 87, 3429-3430.
68. Wennergren, B.: 1989, "Diskrepans Mellan Samhällets Och Medicinens Etik i Synen På Olika Former av Eutanasi?" *Läkartidningen* 86, 1708 - 1709.
69. Wennergren, B.: 1990, "Förkortande av Liv Under Adekvat Vård – Lagstifta för Att Undanröja Oklarheter!", *Läkartidningen* 87, 640-641.
70. Wiingaard, P.: 1990, "Det Forvredne Dødsbegrep", *Ugeskrift for Læger* 152, 247.

ISHWAR C. VERMA

BIOETHICAL DEVELOPMENTS IN INDIA: 1989-1991

I. INTRODUCTION

India is a vast country with a huge population, having 16 percent of the world's population packed in 2.4 percent of the global area. In 1991 the population of India was recorded as 843.9 million ([3], pp. 1-5). The other demographic data ([2], pp. 36-71;[10], pp. 71-89) relevant to bioethical considerations are: birth rate 30.5 per 1000, death rate 10.2 per 1000, life expectancy at birth 59 years, infant mortality rate 91 per 1000, under five mortality rate 142 per 1000, and literacy 52.1 percent. Such rapid growth in the population puts a tremendous pressure on the resources needed to meet basic needs, such as food, housing, education and health. Rapid growth is associated with a high mortality at all age groups, a high cost of living, and low income levels (the per capita annual income being only 340 U.S. dollars). It is not surprising, therefore, that in bioethical decisions, the sanctity of life is deemed less important than the quality of life. This sort of judgment poses many ethical dilemmas for the people, which in practice are resolved in the context of cultural traditions interacting with socioeconomic considerations.

II. REPRODUCTIVE TECHNOLOGIES – PRENATAL DIAGNOSIS OF SEX AND FEMALE FETICIDE

The most important ethical problem which has surfaced in India in the last five years is prenatal diagnosis of sex for social reasons, and abortion of the fetus if it is determined to be female. This ethical problem has resulted from the development of reproductive technologies in the past decade, which allow determination of fetal sex from the chorionic villus samples at 9-12 weeks of pregnancy, or from amniotic cells withdrawn at 14-18 weeks of pregnancy; and ultrasonographic equipment, which establishes the sex of the fetus at 16-18 weeks of pregnancy. Moreover, socio-cultural and traditional considerations make the presence of male offspring in the family highly desirable, in order to ensure family lineage, to keep property, wealth, and business within the immediate family, to have a son to help in the fields, to augment the earnings of the family, or to look after the parents in old age since the state provides no social security.

The opinions of doctors and geneticists, as well as of laypersons, are divided

B. Andrew Lustig (Sr. Ed.), Bioethics Yearbook: Volume 2, 309-341.

on the issue. In a survey of 27 geneticists in India, 37 percent said they would grant the request of a couple for prenatal diagnosis for sex selection unrelated to X-linked disease, 15 percent would refer the case to another geneticist, while 48 percent would refuse [37]. In another survey of 1,212 laypersons in five cities in India (Delhi, Bombay, Calcutta, Hyderabad and Chandigarh), 70 percent responded that they would not resort to prenatal diagnosis of sex to ensure having a boy, because they believed this would constitute sex discrimination, but 14.7 percent agreed to do so. However, 56.6 percent considered prenatal diagnosis of sex to be justified when parents have only daughters, are poor, or belong to a caste where dowry is essential [39]. Thus, there is no consensus that prenatal diagnosis of sex for social reasons is unethical, since most people recognize the social benefits of having a male child. The advocates of sex determination tests argue that India is wedded to the idea of family planning, which presupposes planned parenthood. According to proponents, it therefore becomes the duty of each parent to restrict the number of children. Now that this new technology is available, people should have the right to use it to determine the sex of a child in order to plan the "ideal family". Amniocentesis is new knowledge and knowledge cannot be banned ([28], pp. 28-29).

Some proponents also recommend that the use of prenatal diagnosis for sex selection should only be made by women. Unfortunately, most women in India have little choice in any matter. Before marriage, their parents decide about their lives, and after marriage, their identities and patterns of life are determined by their husbands and husbands' families ([28], pp. 28-29). In fact, with sex determination tests freely available, husbands and in-laws often pressure women to undergo the test. Refusal to do so may invite hostility, ill treatment, divorce, and even death.

Pointing to the overwhelming discriminatory treatment toward female offspring in the Indian society, many protagonists of sex determination tests are in favor of female feticide ([28], pp. 28-29). They argue that, because significant social change within less than two generations is unlikely, the persecution of unwanted girls should not be allowed to continue. Such thinking presupposes that social evils like dowries and women's lower status are immutable, and that methods like female feticide, even if dangerous to women, must therefore be allowed.

Women's activists argue against female feticide in the following terms:

> Do these people want to tell us that the world would be a better place after the elimination of women? The basic question is who demands the dowry? Then why punish women? It is very important for people to know that sex determination tests do not guarantee a male child. It is a trial each time. How many times one would want a woman to go through a test-abortion cycle? Is it right to abort fetus after fetus until the parents have a child of the preferred sex? Advocates of these tests should try to think of the pain, torture, and guilt that a woman goes through while undertaking sex determination tests and abortions, not to mention the risk to her life itself. By asking women to undergo sex determination tests to abort female fetuses, the society is criminalizing woman against woman ([28], pp. 28-29).

Amidst debate on this issue, the advocates of female feticide have been characterizing the woman as an economic drain, a liability, and a burden on parents and are rationalizing the use of sex determination tests for these reasons! Left unaddressed by their arguments is the likely impact of female feticide on the sensitive minds of young girls. After hearing about the worthlessness of being a woman, how will the young girls be taught the equality of sexes and the importance of women's role in society?

Most people appreciate that sex determination and female feticide are ethically wrong, but economic and social realities override their moral objections. For these reasons, the practice of prenatal diagnosis of sex exploded during the 1980s. In a survey in Bombay, 84 percent of the obstetricians in private practice admitted to performing prenatal diagnosis of sex for social reasons ([17], p. 99), while during 1978-1982, several thousand cases of post-amniocentesis female feticide occurred. The situation became almost intolerable. As a result, during 1989-1990, the Ministry of Health and Family Welfare of the Government of India convened meetings of medical specialists, health administrators, voluntary organizations for women's welfare, and legal experts to consider a proposal for legislation to curb this practice. In the meantime, the state of Maharashtra, whose capital is Bombay, passed a bill to regulate prenatal diagnostic tests to prevent their misuse for prenatal sex determination leading to female feticide (Number VIII, 1988). However, the committee called for the central government to pass legislation applicable to the entire country to obviate the need for every state to pass a bill of its own. On September 6, 1991, Mr. M.L. Fotedar, the Minister of Health and Family Welfare, tabled a bill entitled "The Prenatal Diagnostic Techniques (Regulation and Prevention of Misuse) Bill" ([19], pp. 1-19). It proposes to prohibit prenatal diagnostic techniques for determining the sex of the fetus leading to female feticide. Such abuse of techniques discriminates against the female sex and affects the dignity and status of women. The bill provides for: (i) prohibition of the misuse of prenatal diagnostic techniques for determination of sex of fetus, leading to female feticide; (ii) prohibition of advertising of prenatal diagnostic techniques to or determine sex; (iii) permission and regulation of the use of prenatal diagnostic techniques to detect specific genetic abnormalities or disorders; (iv) permitting the use of such techniques only under certain conditions by registered institutions; and (v) punishment for violation of the provisions of the proposed legislation.

The proposed bill provides for the regulation of genetic counselling centers, genetic laboratories, and clinics. It requires them to be registered with the appropriate authority, to have trained manpower and specified equipment, and to maintain records of each woman or man undergoing any tests (Chapters II and III). It lays down in broad terms the conditions under which prenatal diagnostic tests are permitted (for chromosomal abnormalities, genetic metabolic diseases, hemoglobinopathies, sex-linked genetic diseases, and congenital anomalies). It forbids relatives or husbands of the pregnant woman to encourage the use of prenatal diagnostic techniques for sex determination. It envisages the establishment of a central supervisory board to advise the government on policy matters, to review the implementation of the bill, to create public awareness, and to lay down a code of conduct to be observed by health professionals. Each state will have an "appropriate authority" to grant registration, to suspend or cancel registration of a genetic center, to enforce standards prescribed for genetic counselling, and to investigate complaints of breaches of the bill.

The bill lays down stiff penalties. For example, advertisement of any kind regarding facilities for prenatal diagnosis of sex (which has occurred openly in the newspapers so far) or any contravention of other provisions of the bill shall be punishable by imprisonment up to three years, and with a fine of up to Rs. 10,000 (400 U.S. dollars). Any subsequent contravention is punishable with imprisonment which may extend to five years, and with a fine which may extend to Rs. 50,000 (2,000 U.S. dollars). The name of the registered medical practitioner who has been convicted by the court shall be reported to the respective state medical council for necessary action, including the removal of his/her name from the medical register for a period of two years for the first offense, and permanently for any subsequent offense.

A woman who seeks prenatal diagnosis of sex, unless she was compelled to do so, is punishable with imprisonment up to three years, with a fine up to Rs. 10,000. Interestingly, the court shall presume, unless the contrary is proven, that the pregnant woman has been compelled by her husband or the relatives to undergo the prenatal diagnostic technique. Such persons shall be liable for abetting the offense and punishable by the same range of penalties. The bill, is due to be considered by the Lok Sabha (the lower house of Parliament) in 1992. It is well-meaning and demonstrates the interest of the government in the welfare of women, but how effective it will be in practice remains to be seen. Those in favor of sex determination tests argue that banning the tests would only cause clinics to go underground and that the government should not pass laws which cannot be implemented. However, most people feel that banning these tests would have the effect of curbing the rampant commercialization and misuse of medical technology. Moreover, ugly advertisements would no longer be present to make women feel inferior and useless. There are laws banning dowry, bigamy, child labor, and although they are being violated, they are not repealed. By the same token, the aim of the proposed bill is to have a society which is just and fair.

III. DEFINITION OF DEATH

The definition of death currently in use in India is unsatisfactory. Section 46 of the Indian Penal Code states that "Death denotes the death of a human being unless the contrary appears from the context". Section 21(b) of the Registration of Births and Deaths Act, 1969, defines death as the "permanent disappearance of all evidences of life at any time after live birth has taken place". Under these definitions, medical practitioners in India declare death when there is an absence of spontaneous heart beat and cessation of breathing, *i.e.*, cardiorespiratory failure. These traditional criteria have been in use for well over a century, although they have not been legally defined in India. Considerable advances have been made in medicine in recent years, especially in the fields of resuscitation and in the artificial maintenance of cardiac and respiratory functions. As a result, many lives have been saved.

However, as is often the case, the introduction of new methods into clinical medicine has also produced new problems. These include the attempted resuscitation, for long periods of time, of individuals who are probably dead. The crux of the problem is the determination of the moment of death. If death has occurred, there is no reason whatever to "ventilate a corpse". On the other hand, if the afflicted individual is not dead, it is only right that all the necessary medical and nursing attention should be provided ([7], pp. 1-29). Defining death as cardio-respiratory failure is clearly unsatisfactory, and it is necessary, for medical, legal, social and religious purposes (as indicated in Table 1) to determine clearly the precise moment when "death" has occurred.

At present it is universally accepted that all human organs are controlled directly or indirectly by the brain. The brain is therefore called the "master organ" or controller of all vital human functions. On the one hand, under normal conditions, spontaneous activity of heart, lungs and brain must be present for life to exist. Machines can maintain, of course, heart and lung activity, but not brain function. On the other hand, serious accidents or medical catastrophes sometimes have resulted in the initial, but irremediable and irreversible destruction of the brain, yet heart and lung activity have been maintained with the help of modern equipment. Such episodes have led to the recognition of the concept of primary brain death.

It is well-known that the integrity of the brain stem is essential for alertness; in a sense, the brain stem generates the "capacity for consciousness". The vital centers of human function which control the "breath of life" and "vasomotor activity" are also found in the lower part of the brain stem. All brain stem functions can be tested by a skilled person at the bedside of the patient without using sophisticated or expensive equipment. At the same time, human abilities to think, talk, communicate, and interact are all dependent on a functioning cerebral cortex, which therefore generates the content of consciousness.

International medical opinion has progressed from the concept of classical death to total brain death (*i.e.*, death of the whole brain), and eventually to brain stem death as the physiological core of brain death. Thus "the irreversible cessation of heart beat and respiration implies death of the patient as a whole. It does not necessarily imply the immediate death of every cell in the body" ([7], p. 7). The irreversible absence of brain stem functions, therefore, would be tantamount to the death of the individual, absent any evidence of hypothermia, drug or alcohol intoxication, or metabolic or endocrine disease.

The concept of brain stem death would also permit removal of the organs for transplantation, as all human tissues do not die simultaneously at the moment of death. However, in employing a brain death standard, there is need to incorporate appropriate safeguards in order to prevent abuse as a result of possible collusion between the doctor who performs the transplantation and the patient and/or relatives of the patient who may require organ transplantation. The need for such safeguards has been discussed at a number of scientific forums in India. The FIAMC Biomedical Ethics Centre held a consultation workshop on "Criteria of Death" in Bombay in 1985. In addition, a national seminar on "Determination of Death" concluded that "the concept of brain stem death needs to be introduced in India, like the 47 other countries in the world who have enacted legislation for this purpose" [7]. This would allow for the optimal utilization of life saving equipment and increase the availability of organs for transplantation.

Internationally, a broad consensus on the concept of brain stem death has been achieved, based in large measure on the reports of two august bodies: the Conference of Royal Colleges and Faculties of the United Kingdom (U.K.) in 1976, and the United States (U.S.) President's Commission for the Study of Ethical Problems in Medicine and Biomedical and Behavioral Research in 1981. In the U.K. Code, it is mandatory that certain preconditions and exclusions be satisfied before undertaking an examination to determine brain stem death in a deeply comatose patient on a respirator. Preconditions include the assured presence of known irremediable and irreversible brain damage. The exclusions require that reversible causes of coma such as hypothermia, drug and alcohol intoxication, and metabolic and endocrine disorders be conclusively excluded. The U.K. Code's explication of preconditions and exclusions has been very influential for two reasons: it employs simple criteria without the need for instrumental corroboration, and it prevents diagnostic mistakes when heeded carefully.

On the other hand, the U.S President's Commission defined death as follows:

> An individual who has sustained either (1) irreversible cessation of circulatory and respiratory functions, or (2) irreversible cessation of all functions of the entire brain, including the brain stem, is dead. A determination of death must be made in accordance with accepted medical standards.

Criticisms have been directed at this definition on three grounds. First, it postulates two types of death. Second, it requires that there be "irreversible cessation of all functions of the entire brain, including the brain stem". Few will deny that brain stem functions can be satisfactorily examined in a deeply comatose patient on a respirator, but none will dispute that the functions of the remainder of the brain, such as the basal ganglia, thalamus, hypothalamus and cerebellum, cannot all be tested in an unconscious patient. This being so, recourse must be had to sophisticated, expensive, and laborious instrumental methods of demonstrating non-viability of structures that cannot be clinically tested. A third criticism relates to accepted U.S medical standards for the determination of death which do not require the preconditions and exclusions specified in the U.K. Code before undertaking an examination for the determination of brain death. It is precisely the absence of these preconditions and exclusions that has contributed to some errors made in determining brain death using the U.S. criteria.

The national seminar in India recommended the recognition by Indian society of the concept of brain death, with the criteria of brain stem death as practiced in the U.K. However, at the present time, it recommended that these criteria should be restricted to adults and children over the age of three years ([31], pp. 735-742). The seminar did not require the confirmation of brain stem death by electroencephalogram (EEG). If an EEG is performed, however, it should conform to internationally accepted standards. The seminar participants happily noted that the concept of brain death and the criteria of brain stem death they discussed were in no way incompatible with Hindu, Islamic, Christian, Zoroastrian, or Judaic thought. The national seminar advised medical personnel and organizations as well as governmental and paragovernmental agencies to adopt the following resolutions as guidelines for the determination of death in India:

> An unconscious individual will be considered dead at the time when in the opinion of a registered medical practitioner, based on the ordinary standards of medical practice, there has been an irreversible cessation of spontaneous respiratory and circulatory functions; or, should the use of artificial means of support preclude the determination of irreversible cessation of spontaneous respiratory and circulatory functions, the unconscious individual will be considered dead at the time when in the opinion of two registered medical practitioners, based on ordinary standards of medical practice, there has been an irreversible cessation of brain stem functions. However, death must be pronounced as having occurred before artificial means of supporting respiratory and circulatory functions are terminated and before any organ is removed for the purpose of transplantation. It is imperative that no declaration of death be made by any individual connected directly or indirectly with any "transplant team", if there is any intention of donating any part or organ of the deceased individual for transplantation purposes. Likewise, any individual interested in obtaining any part of the body of the deceased for medical or scientific research, shall not be permitted to make a declaration of death [9].

IV. DONATION OF HUMAN ORGANS FOR PURPOSES OF TRANSPLANTATION

A. *Commercial Trafficking in Organs and Tissues*

Recent developments in the fields of medicine and surgery have made it possible to transplant human tissues and organs such as corneas, ear-drums, kidneys, hearts, lungs, livers, pancreas, and skin and bone-marrow. Corneal grafts and transplants of bone-marrow and kidneys have become quite common, and a number of institutions in India perform these procedures.

Numerous reports have indicated a large scale "trading" in human organs, which the magazine *India Today* termed "the organ bazaar" ([4], pp. 60-67). All parties concerned apparently benefit from this "transaction". The patient receives the transplant which prolongs life and improves function; the donor gets money which augments the family income; the broker receives his commission; and the surgeon is able to help the patient and makes money in the bargain. In a country where per capita incomes are very low, and almost 30 percent of population is below the absolute poverty line, the desire to make money by donating of a bilateral organ tissue is strong. In Vallivakkam, a suburb of Madras, with about 5,000 impoverished residents, signs of relative prosperity are evident in an estimated 100 families. These are families whose members have made money by selling their kidneys, so much as that the colony has been nicknamed "the kidney colony" ([5], p. 62). The trade in kidneys has expanded so rapidly that an estimated 2,000 or more kidneys are sold every year in India ([5], p. 61). And while this trade was initially restricted to hospitals in Bombay and Madras, it has now expanded to Calcutta and Bangalore, and even to smaller cities like Pune, Jaipur, and Madurai, with a total sales volume of Rs. 40 crores (16 million U.S. dollars). At the same time, what doctors have feared most is actually occurring: the trade is now other organs and tissues like eyes and skin. Present prices average Rs. 30,000 per live donor kidney, Rs. 80,000 per live cornea, and Rs. 1,000 per skin patch. ([4], p. 60). The live organ trade has developed into a vast, smoothly run, but shady business. In major cities, brokers frequent private hospitals and contact patients with renal failure to find suitable donors for them. With other countries banning the trade, India has become an international center for kidney transplants. In the past, Indian donors were actually flown to countries like the United Kingdom to sell their kidneys to foreign buyers. After these countries banned such operations, foreigners needing transplants now flock to India. Wealthy Arabs come regularly to Bombay, while people from Singapore and Thailand come to Madras. The doctors performing transplants ask whether they "should turn patients away and consign them to death". The "buy-or-let-die debate" has turned acrimonious. A large number of prominent physicians and surgeons have spoken openly against this "sacrifice of morals to expediency", this "exploitative business of transferring the health of the poor to the rich". They call for a strict law penalizing trafficking in human organs, and for urgent steps to launch a full-fledged viable cadaver donor program, the first step being to enact legislation brain stem death.

The question of trading in human organs has been discussed in many forums in India, including the Parliament. In view of the current deplorable scenario, it is not surprising that the Government is actively considering the enactment of comprehensive legislation to regulate the removal of organs and tissues from cadavers for transplantation in suitable subjects. It is expected that such legislation would discourage trading in human organs and increase the availability of organs through voluntary donation.

Currently there is no comprehensive legislation regulating the removal of human organs from deceased persons, although regional legislation has been passed. In the Union Territory of Delhi, two enactments called the "Eyes (Authority for Use for Therapeutic Purposes) Act" (1982) and the "Ear Drums and Ear Bones (Authority for Use for Therapeutic Purposes) Act", (1982) regulate the removal and transplantation of corneas and ear drums/ear bones in that region. In Maharashtra, the "Kidney Transplantation Act" (1983), the "Corneal Graft Act" (1986) and the "Maharashtra Live Donor Kidney Transplantation and Prohibition of Commerce in Kidneys Bill" (1990), regulate the transplantation of kidneys and cornea.

The Ministry of Health and Family Welfare, therefore, has appointed a committee to examine a proposal to enact legislation on the removal of human organs, tissues, and transplantation for therapeutic purposes. The committee, under the chairmanship of eminent jurist Dr. L.M. Singhvi made the following recommendations:

(1) The concept of "brain stem death" may be operationalised in the context of the comprehensive legislation on organ transplantation.

(2) The proposed legislation on organ transplantation should incorporate a provision making trading in human organs a punishable offense.

(3) The legislation must provide for removal of organs from live donors in special circumstances, subject to appropriate safeguards. These could include a provision restricting the removal of organs from live donors who are related to the person requiring transplantation. It is well established that the probability of successful transplantation is relatively higher if the organ is extracted from a related donor. However, an enabling provision could be incorporated permitting donation by non-related donors on account of (i) special attachment/affection to a person requiring transplantation, and (ii) for any other special reason. The Maharashtra Bill permits transplantation from an unrelated donor only if the donor has a close, enduring, and emotional relationship with the recipient and [acts for] altruistic considerations only. Further, there shall be collateral evidence of such a relationship. In both cases, there should be an affidavit of consent which should indicate the reason for donation by unrelated persons. A further additional provision should be incorporated for scrutiny by an empowered Committee so that the possibility of trading in human organs could be minimized or eliminated.

(4) With regard to donation by children, the concept of voluntary donation in advance is not recommended, since children are not mature enough to decide on such matters and are likely to be influenced by elder members/relatives in taking such a decision. This is particularly relevant in the case of destitute, physically handicapped and mentally retarded children. However, the legislation may permit removal of organs from children below the age of 18 years subject to the condition that irreversible brain death has occurred and a written consent of the parents

has been obtained.

(5) In the case of unclaimed cadavers and accident victims the Committee recommend[s] removal of organs subject to safeguards, namely: (i) it should be established from hospital records that there is no immediate relative or any responsible person to whom the "body" has to be handed over by the hospital authorities; (ii) if the name of any relative or responsible person is mentioned in the hospital records, efforts should be made to contact the relative of the responsible person for the purpose of handing over the body, (iii) if after making reasonable efforts by radio broadcast and publication in the newspapers no claim is made within a stipulated period then it [may] be presumed that the cadaver is unclaimed and removal of organs in such cases [may] be authorized.

(6) There should be a mechanism for investigating complaints alleging malpractice on the part of doctors while removing and transplanting human organs and tissues. This would help in fostering professional ethics in the medical community as well as provide an opportunity for conducting a fair trial in cases where unsubstantiated allegation are made against doctors.

(7) The legislation should have a provision for authorizing hospitals which have the capability in the form of skilled manpower, equipment and associated facilities for removal [and] transplantation of specific human organs; and periodical inspection by competent authorities of hospitals authorized for removal and transplantation of organs with a view to examine the quality of transplantation and follow up medical care to persons who have undergone transplantation and persons from whom organs and tissues have been removed [26].

In the Maharashtra Bill, a person is guilty of an offense if he makes or receives any payment for supplying a kidney, makes an offer to supply any kidney, makes or receives payment for a kidney, solicits a person willing to donate his kidney for payment, initiates or negotiates any arrangement which involves payment for supply or an offer to supply a kidney, or distributes or knowingly publishes any information or advertisement about sale of kidneys. Contravention of any of these provisions shall be punished with imprisonment of up to six months, and also a fine not exceeding Rs. 5,000. The surgeon who removes the kidney or who transplants the organ into another unrelated person is also guilty of an offense, unless the conditions stipulated for such a transplantation are fulfilled. Contravention of these provisions shall be punished with imprisonment of up to one year, and a fine not exceeding Rs. 10,000.

B. *Implantation of Brain Tissue*

Human brain transplantation was discussed at an international consultation workshop organized by FIAMC Biomedical Ethics Centre in Bombay in 1990 [32]. Participants included neurologists, neurosurgeons, basic scientists, researchers in neurophysiology neuropathology, and forensic science, as well as philosophers, theologians, hospital administrators, and lawyers and ethicists. The group judged the reported story of a "total brain transplant" to be a piece of journalistic fiction but concluded that experimentation in this direction should not be banned.

The FIAMC workshop also discussed the recent use of neural grafts. The implantation of neural tissue has been increasingly undertaken as an experimental procedure over the past few years. Initially, it was performed with autologous adrenal medullary tissue with mixed results. These procedures were deemed ethical because the criteria justifying human experimentation were usually satisfied. Later, in an attempt to achieve better results, the implantation of fetal neural tissue and cells began. These developments generated considerable anxiety and a call for the scientific community to be accountable to society. The workshop participants concluded that heterologous and homologous fetal neural tissue obtained after spontaneous abortions would meet criteria required for the ethical performance of such experimental procedures. However, participants were highly critical of the use of fetal materials obtained from elective and induced abortions. Participants also recommended that the scientific community should desist from therapeutic brain implantation using fetal materials at present, on two grounds. First, it remains a totally experimental procedure, about which little is known; and second, other less ethically troubling methods and materials be found for the treatment of patients with progressive neurological disorders. There was a clarion call for "patience rather than patients".

In October 1989, Hungarian scientists reported at the World Congress of Neurosurgery, held in New Delhi, India, that they had performed brain transplantation for Parkinson's disease using stellate ganglia tissue with good results. If these findings can be confirmed, the ethical constraints on the use of fetal tissue obtained from aborted human fetuses will not arise. A remarkable breakthrough in medical research involving the culture of human brain cells in the laboratory has recently been reported in the United States. This promising development, if successful, would eliminate many of the ethical problems associated with transplants from live donors. There should be no ethical objection if the primary source for the tissue culture be obtained from spontaneously aborted fetuses.

V. UNEQUAL DISTRIBUTION OF HEALTH SERVICES AND RESOURCES

In all countries, there is an unequal distribution of health services between the poor and the rich, rural and urban people, and men and women. In India, disparities also exist between tribal and nontribal people, and between "scheduled" castes and higher castes. The existence of such inequalities in India is a stark reality, and has led to considerable social, political and ethical turmoil.

A. *Poor vs the Rich*

Even in the United States poverty exists in the midst of plenty. The various indicators of health reveal glaring disparities between the rich and the poor there. It is, therefore, expected that such differences would exist in developing countries like India. Health services provided for fees by private practitioners and specialists is of very high quality comparable to countries in the West. However, because of costs such services are beyond the reach of the poor. The latter usually patronize government hospitals, which provide health services of reasonably good quality at little or no cost. As might be expected, these hospitals are overcrowded, involve long waiting periods, and often lack cleanliness and courtesy. Sick patients may on occasion be refused admission. Moreover, since and the best treatments may not be affordable, the treating doctor often must rely on the second or third best line of therapy. The physician faces constant ethical dilemmas in the choice of treatment, and the choice of the patient who should receive the available treatment. For example, an intensive care unit may have only one respirator, thus requiring the doctor to select from among the many patients who may require it. However, to the credit of the doctors, they try to chose patients according to medical need, and would certainly prefer to use limited resources for those who have treatable disorders and reasonable chances of full recovery without handicap.

Infants who suffer from disorders where normal mental function cannot be ensured, and which require very expensive treatment, are often given only restricted treatment, or treatment is withheld. Although this appears unethical, there seems to be no other choice. Moreover, it is unlikely that reporting such parents to the police or juvenile courts would provide any benefit to the concerned child.

B. *Rural vs Urban Areas*

It is paradoxical that, although 79 percent of the Indian population lives in rural areas, only 20 percent of the health budget is allocated for rural use. Table 2 summarizes the differences in health indicators between rural and urban people. That table shows that in all sectors – health, education, sanitation, or social welfare – the rural areas are undeserved. The health system has inherited three characteristics from 250 years of British rule: it is predominantly based in urban areas, in hospitals, and is curative in its objectives. Although there have been changes since Indian independence in 1947, the disparities persist, despite the vast network of rural health services which has been established. There are almost 20,537 primary health centers, and 130,390 sub-centers. If these rural centers were to function properly, there would be a virtual revolution in health care in India, and the country would dramatically improve its heath care indices.

C. *Bioethics and the Caste System*

Caste in India, although sanctioned by scriptural texts, is more a sociological than a religious phenomenon. The five castes represented functional divisions, with only shades of racial purity. The *brahmins* were the scholars and priests, the *kshatriyas* the warriors and rulers, the *vaishas* the commoners and tradesmen, and the *sudras* the workers, artisans, farmers, and herdsmen. There were also those beyond the pale of caste and society, the "untouchables" who performed the most unpleasant tasks. The British encouraged the caste system to create divisions in Indian society and to prevent unity among different groups. However, when India became a republic in 1950, differentiation based on the caste system and "untouchability" were abolished by the Indian constitution (Articles 15 and 17 respectively). Nonetheless, because of the backwardness of the lower castes, special reservations (22.5 percent) in jobs and educational institutions were made for the lower castes, the "scheduled castes" and the "scheduled tribes" (so-called because they were listed in a government order or schedule). The rest of the population reconciled themselves to the reality of Sveh Scheduler. However, in 1990 the issue became highly politicized, when the Janta Dal ruling party for partisan reasons announced that it would accept the recommendation of the Mandal Commission to reserve another 27 percent of jobs and seats in educational institutions for "other backward castes" in addition to scheduled castes and tribes ([1], pp. 98-103). The Mandal Commission's methodology to identify the other backward castes was faulty and inadequate, for its survey covered only 1 percent of the Indian population, and was based on the 1971 census ([22], pp. 100-103). The Janta Dal announcement provoked outrage; in several instances, anti-reservation demonstrators who were college students faced with the prospect of unemployment despite having better grades than those belonging to lower castes, killed themselves by burning or taking poison. The Janta Dal party, which had accepted the Mandal Commission report, was subsequently voted out of power. However, the succeeding political party, rather than reducing the reservation, increased it by another 12.5 percent for "backward" groups to be determined according to economic criteria. The reservation issue is currently before the Supreme Court, and a great deal hinges upon its judgment. Those from the backward classes contend that expanded reservation is the only way to end centuries-old disparities. However, when the constitution had been drafted, it allowed reservations only for a forty-year period. As the issue became increasingly politicized, the government extended the reservation policy for another ten years in 1990 in order to retain the votes of those in the reserved category. Overall, this issue divided the Indian people more than any other. The danger looms large that enforcement of the Mandal report would legitimize and perpetuate the caste system, which is against the proclaimed secular character enshrined in the preamble to the Indian Constitution.

VI. THE STATUS OF WOMEN AND ITS ETHICAL IMPLICATIONS

Women in India have traditionally held a low status in society. The Indian Constitution, however, strongly asserts the rights of women. Article 15 declares "The state shall not discriminate against any citizen on grounds of religion, race, caste, sex, [or] place of birth". Article 16 states that "No citizen shall, on grounds only of religion, race, caste, sex ... be ineligible for, or discriminated against, in respect of any employment or office under the State". Article 51A (e) of the Constitution emphasizes that "it is a fundamental duty of men to renounce practices derogatory to the dignity of women". The Dowry Prohibition Act (1961) states: "If any person gives, takes, or abets the giving or taking of dowry, he shall be punished with imprisonment for a term which will not be less than five years, and with fine not less than Rs. 15,000". Even demand for a dowry is a punishable offense. Every offense under this act is nonbailable and non-compoundable. The burden of proof that one not commit an offense is on the person prosecuted. In response to reports of a large number of "dowry deaths", where married women were harassed and led to their death, an amendment was enacted in 1986. The amendment states:

> where the death of a woman is caused by any bodily injury within seven years of marriage, and it is shown that she was subjected to cruelty or harassment by her husband or any relative, such a death shall be called a dowry death, punishable with imprisonment of not less than seven years, extendable to life.

Despite such stringent laws, during 1991 (January – November) 57,141 crimes against women, including rapes and dowry murders, were committed throughout India. Of these 4599 were dowry deaths. The department of Women and Child Development, therefore, is reviewing four proposed amendments during 1992 parliamentary session which would make laws on crimes against women even more stringent [30]. Laws under consideration to be amended are the Dowry Prohibition Act (1961), the Commission of Sati (Prevention) Act (1987), the Indecent Representation of Women (Prohibition) Act 1986, and the Immoral Traffic (Prevention) Act (1958).

Women are at a disadvantage for all demographic and health indicators (Table III). *Sex ratio*, which reflects the status of women and mortality among them, reached its lowest point in this century in the 1991 census. At 929 females per 1000 males, it was down by five from the 1981 census. The ratio reflects the preference for male children and the neglect of female children. One reason for the low ratio could be the undercounting of females, but that is unlikely to be the only reason. Higher sex ratios have been recorded in five states – Haryana, Punjab, West Bengal, Himachal Pradesh, and Kerala. Only Kerala has a sex ratio (1,032) 1,000 more favorable to women than men, and it also has the best health indicators ([2], p. 44;[15], p. 1).

Literacy, which is a basic requirement for socio-economic development and has a crucial role to play in the improvement of health status of women and children, is considerably lower among females than males (39.4 percent vs 63.9 percent in persons aged 7 years and older in the 1991 census). Although this represented an increase of 32.5 percent from 1981, as compared with an increase of 13.3 percent among males, it is still poor consolation, and highlights the need to increase educational programs. It is heartening to note that during the latter half of 1991, Kerala reached a 100 percent literacy level, as did West Bengal a few months later.

The discrimination against the female child begins even before she is born. The female fetus is often aborted when its sex is determined by prenatal diagnosis. In Bombay, a study revealed that in 1984, 40,000 female fetuses were aborted. Another survey showed that out of 8000 abortions, only one fetus was male [22].

Although infanticide was banned in 1870 by the British, in recent years, the practice has been reported in villages of Tamil Nadu and Rajasthan. The magazine *India Today* highlighted this issue in a report entitled "Born to Die" about the Kallar community in Usilampatti Taluk of Tamil Nadu ([35] pp. 26-33). In this martial subcaste of 200,000 members, female infanticide has increasingly been accepted as a means to escape the burden of dowry. An estimated 6,000 female babies have been poisoned to death during the last decade. Few such deaths are recorded. As a matter of practice, the first child is not killed, even if it is a daughter, but killing begins with the second female child. Many Kallar families understand they are committing a crime, but they are convinced that, because of their difficult circumstances, they are taking the only course open to them. Most of the Kallars are landless laborers. The rest eke out an existence as marginal farmers. This scenario may be unfolding in only one community, but it is a grim reminder of man's in humanity to woman in India, even in the twentieth century.

In general, young girls have poorer nutritional status than boys, as shown by several microlevel studies and hospital data ([9], p. 15). A recent study in Punjab showed that 24 percent of females were malnourished in the privileged group, but 74 percent in the underprivileged group, as compared with 14 percent and 67 percent respectively in the males ([21], pp. 301-303). In many Indian cultures, bonding at birth is denied to the female child, perhaps in the hope that her eventual leaving, through either death or marriage, will be less painful than the loss of a son ([27], pp. 295-299).

Higher mortality among girls is reflected not only in their lower sex ratio in India, but unlike most other countries, also in the higher male-female sex ratio in all age groups. From birth to four years of age, the ratio is 1,044. From five to nine years of age, it is 1,061. From ten to fourteen years of age, it is 1,067. In the group from birth to five years of age, female mortality is 20 times higher than in males. The higher mortality in females continues until forty years of age, even though the female is considered biologically stronger. Females also have higher rates of morbidity, lower attendance and retention at school, lower wages for equal work, and outright exploitation by family, communities and entire segments of society. Indeed one of the greatest ethical issues at the present time is this pervasive neglect, apathy, and undervaluation of girls and women in India.

With the widespread recognition of the importance of women to India's future development there have been major efforts to create projects and programs which will counter the disparities so obviously afflicting females. Some have suggested that legislation should be introduced to aid women's causes, especially bills to ensure the protection of women's right ([12], pp. 15-17). Once it is clearly established that woman is entitled to all the rights and responsibilities of a human being, practices such as female feticide, infanticide, and deprivation of nutrition and education to women will gradually diminish, if not completely disappear.

In order to make legislation on such matters more useful and effective legislative methods must be changed. First, the law must be implemented as soon as the President assents to the bill, and rules and regulations must be framed at the same time. Second, the machinery to implement the law, and specified punishments for not implementing the law, must be provided in the statute. Third, a program to increase public awareness must be planned as a part of the statute, in order to enable the illiterate masses to understand the law. If this model is accepted for this legislation in general as well as for laws affecting women and children, the law may become a meaningful instrument of social change ([13], pp. 15-17).

There is need to remove glaring legal discrimination between men and women and between different classes of women. Such discrimination violates the Constitution of India, which guarantees fundamental rights to all citizens of the country. This discrimination is most evident in personal and family laws which affect women most deeply. A woman in India, for example, is not the natural guardian of her children.

Moreover, a woman in India is either a Hindu, Muslim, Christian, or Parsi, and legal discrimination varies depending on the religious family into which she is born. A Hindu woman is discriminated against by law. She is not entitled by birth to a share in the coparcenary property, nor can she ask for a partition of the property. This aspect of Hindu law is now the subject of debate and challenge in the Supreme Court of India. However, it is difficult for the Supreme Court to strike down sections which leave legal lacunae, and easier for Parliaments to amend the law. The state of Andhra Pradesh, for example, has enacted legislation to remove this form of discrimination.

A Muslim woman is discriminated in against matters of marriage, divorce, inheritance rights, and so on. Because of the present misinterpretation of religious precepts, it is said that a Muslim male can marry four wives while monogamy remains the rule for women. Properly interpreted, polygamy is not really permitted under Muslim law, because a Muslim is supposed to treat all his wives equally. In fact, this provision is sometimes abused when a Hindu, in order to avoid the penal consequences of a bigamous marriage, converts to

Islam. Such fraudulent conversions should be prohibited. A provision such as Section 4 of the Parsi Marriage and Divorce Act, which prohibits marriage in the lifetime of the spouse by a change of religion, should be enacted.

The Christian Laws of Marriage and Divorce are archaic. The Indian Divorce Act of 1869 and the Christian Marriage Act of 1872 employ the language of British colonialism. A Christian woman faces significant discrimination. For example, a man may ask for divorce on the ground of "adultery", whereas a Christian woman must prove "incestuous adultery" under Section 10 of the Indian Divorce Act. Or again, the Travancore Christian Succession Act denied to women born in a particular area of Kerala the right to inherit equally with their brothers the intestate property of the father. The Supreme Court of India has now declared that the Indian Succession Act, which grants equal rights to both sons and daughters, will apply to all Christian women.

A detailed review of legislation affecting women is overdue, but pending this review, the glaring discrimination against women must be removed. Article 372 of the Constitution provides for the continuance of laws in force until they are modified or repealed, while Article 13 states that such laws are void to the extent they are inconsistent with the provisions of Part III of the Constitution (on fundamental rights). Laws inconsistent with fundamental rights, therefore, should be formally deleted.

While such efforts are laudable and necessary, society must address the underlying attitudes and social conditions that have given rise to the plight of women in India today. Only then will India's national development reach its full pace and potential.

VII. BIOETHICAL ISSUES RELATED TO THE DISABLED

Precise data on the prevalence of various disabilities in India is lacking. The best estimates are those provided by a survey carried out by the National Sample Survey Organization of the Registrar-General, Government of India in 1981. The survey found that about 1.8 percent of India's population suffer from some disability. Extrapolating this to the population in 1991, there are about 15.2 million people with disabilities, excluding those with mental retardation. Those with locomotive disabilities are the largest group (6.9 million), followed by those with visual disabilities (4.4 million), hearing disabilities (3.8 million) and speech disabilities (2.2 million). The survey covered totally blind, totally crippled, and totally deaf persons, but did not include mild to moderate grades of these abnormalities, nor did it cover other disabilities including mental retardation. Based on four community surveys, the prevalence of mental retardation is estimated to be 26.9 per 1000 persons of all ages ([36], pp. 99-106).

We have recently concluded a study to ascertain the views of 1,212 common people on hereditary handicaps and related ethical issues [39]. The distribution of subjects was as follows: Delhi 275, Chandigarh 300, Hyderabad 300, Bombay 200, and Calcutta 137. There were 334 pregnant women, 535 normal persons with no handicapped person in the family, and 343 parents with a handicapped child. Subjects were questioned about handicaps in general. Seventy-seven percent of the respondents were female, and 23.2 percent were male. Seventy-five per cent responding were Hindu, 8.5 percent were Muslims, 5.5 percent were Christian, 8.8 percent were Sikh, 2.4 percent belonged to other religions.

A majority of the respondents (86.4 percent) agreed entirely that the welfare of a handicapped child should not be considered secondary to that of a healthy child, and a majority (81.8 percent) also agreed that they would spend money equally on a healthy or handicapped child. A majority (85.7 percent) stated that parents and siblings should take care of the handicapped, while 41.7 percent gave secondary responsibility to local/state government. A majority of respondents (53.9 percent) believed that rehabilitative facilities in their locality were very poor. A majority (82.6 percent) agreed that the handicapped should have contact with healthy children, in addition to their own group. A majority (75 percent) gave correct responses regarding factors that cause birth or developmental defects. Thirty-seven percent responded that they would bear a baby only if there was no risk of giving birth to an abnormal child, while 25.1 percent would do so only if the risk was less than 10 percent. Thirty-eight percent would prefer an abortion even with only 1 percent risk of abnormality in the fetus, while 17.9 percent would do so only if the risk was 100 percent.

Thirty-two percent of respondents agreed that a fetus with an abnormality has a right to survive. The most common reasons for this belief were that life is a gift given by God (58 percent), or that life should be preserved and respected (47.4 percent). However, 48.8 percent of subjects disagreed with this view. The most common reasons for the latter view were that it is not good for an abnormal child to come into this world or that "a handicapped child has low value as a human being (41.6 percent)". In cased where disorder in the fetus is suspected, most people (40.8 percent) agreed that abortion should be recommended, while 34.3 percent responded that abortion should be avoided in some cases. Even if facilities for rehabilitation were to improve, 34.8 percent of respondents would be willing to abort an abnormal fetus.

If a couple already had one child with Down syndrome, most women said they would opt for prenatal diagnosis in the next pregnancy, but would continue the pregnancy if prenatal diagnosis was not available. If the fetus was diagnosed to be abnormal through prenatal diagnosis, a majority (72 percent) said they would opt for abortion, while 11.2 percent said they would bear the child. Surprisingly, a majority (70.3 percent) would not resort to prenatal diagnosis of sex to have a boy, because they believed that this would constitute discrimination between the sexes (94 percent). However, 14.7 percent agreed to do so. Their reasons for agreeing to diagnosis for sex selection were to continue family lineage (67.7 percent first preference), to have a male child to support family in old age (51.6 percent second preference), or to reduce the economic burden of a dowry (25 percent third preference). Prenatal diagnosis of sex was considered to be justified by 56.6 percent when parents have only daughters, are poor, or belong to a caste where dowry is essential.

When asked what they would do if the person they chose to marry had a handicapped family member, 37.7 percent replied that they would consult a doctor, while an almost equal number (37.5 percent) said they would marry, as the prospective spouse was normal. Only 10.1 percent said they would not marry. A majority (62 percent) said they would be very worried if they were carriers of a deleterious gene. Most agreed to be tested for carrier status (85.5 percent). A majority also agreed that they would recommend that siblings be tested for carrier status. As for the timing of carrier testing, first preference was given to testing before marriage (32.2 percent), but an almost equal number (28.8 percent) opted for testing after marriage but before pregnancy. The most common third preference was for testing during pregnancy. In the event of a high risk of recurrence, 33.9 percent would accept adoption, 26.4 percent would use contraception, while 24.8 percent would pursue artificial insemination / ovum donation.

The conclusions which emerged from preliminary analysis of the data are that on most issues, people held similar ethical viewpoints. Most subjects expressed support for prenatal diagnosis and for abortion of an abnormal fetus. A majority did not favor prenatal diagnosis of sex for social reasons, although they considered this justified under special circumstances. Because of the heavy socio-economic burden of disease in India, most favored carrier testing.

Because facilities for the rehabilitation of persons with disabilities are totally inadequate, the care and rehabilitation of the disabled is a source of ethical dilemmas for both professionals and parents. The Government of India is keenly aware of the problems of the disabled in India, and has formulated a number of welfare measures in response. In 1981, the International Year of the Disabled, four national institutes were opened: one for the visually handicapped at Dehradun, one for the orthopedically handicapped at Calcutta, one for the hearing handicapped at Bombay, and one for the mentally handicapped at Secunderabad. From 1985-1990, outlays were increased to Rs. 100 crores (40 million U.S. dollars). During the years 1990-1991 and 1991-1992, Rs. 46 crores (11.5 million U.S. dollars) and Rs. 47 crores (11.75 million U.S. dollars) were allotted, respectively. The government provides assistance up to 90 percent of the cost to voluntary organizations of educational training and rehabilitation of the disabled. It provides free aid and appliances to the disabled persons whose income is less than Rs. 1200 per month, and a 50 percent subsidy to those whose income is between Rs. 1200-2500 per month. It funds an artificial limbs manufacturing corporation, which operates 35 limb-fitting centers, and 152 implementing agencies. Because the economic rehabilitation of the disabled is necessary for their integration into society, the government has promulgated special rules to reserve 3 percent of the vacancies in central government and central public sector undertakings for the disabled, 1 percent respectively for those with visual impairments, speech and hearing, disabilities and locomotor disabilities. Placement is provided through 23 special employment exchanges, and 55 cells in regular employment exchanges. In addition, 18 vocational rehabilitation centers providing training have been established in different parts of the country. Thus far, almost 5,000 disabled persons have been provided employment. (Moreover, scholarships for free education are provided beyond the ninth class in school as well as for college. The Ministry of Welfare also provides funds for science and technology projects in order to develop new aids and appliances for the disabled. To increase public awareness a program to educate parents of the mentally retarded was launched in 1990. A four-day long film festival on the theme "Welfare of the Disabled" was organized, while an exhibition depicting various aspects of disability was held.

Three new laws are being introduced. A Rehabilitation Council was established in 1986, and a bill (Number 155 of 1990) has been introduced in Parliament to grant the council the following statutory powers: to enable it to regulate the training of rehabilitation professionals and to enforce standards of training, to maintain a central rehabilitation register, to oversee professional conduct, to take note of offenses, and to provide immunity for actions taken in good faith. The bill envisages that no person other than the rehabilitation professional enrolled in the register shall hold office in government or in any institution maintained by a local or regional authority. Only registered professionals will be entitled to practice in any part of India and to charge fees for their services. It is expected that granting there statutory powers to the Rehabilitation Council will improve the standard of service for the disabled.

The second legislative proposal (Bill Number 1 of 1991), pending before Parliament, concerns the constitution of a Board for Welfare and Protection of the Rights of the Handicapped. The Board shall perform the following functions: (a) investigate and examine all matters relating to the rights and safeguards provided for handicapped under any laws issued by the government; (b) present reports to the central government, annually and at such other times as the Board may deem fit, upon the implementation of those rights and safeguards; (c) make recommendations for the effective implementation and protection of the rights of the handicapped by the Union or any State; (d) look into complaints and, on its own authority, to take notice of matters relating to deprivation of rights of the handicapped, non-implementation of laws enacted for the welfare and protection of the rights of the handicapped, and non-compliance with policy decisions, guidelines or instructions aimed at mitigating hardships, ensuring the welfare of and providing relief to the handicapped, and to take up such issues with appropriate authorities; (e) undertake promotional and educational research on ways to ensure the welfare and rights of the handicapped in all spheres and to identify factors responsible for impeding such advances; and (f) evaluate progress in the development of the welfare of the handicapped under the Union and any State.

VIII. MENTAL RETARDATION AND THE LAW

Most parents of the handicapped, especially of the mentally retarded, are worried about the future of their children after their own deaths. To address this felt need, the Ministry of Welfare has drafted a bill to constitute a "National Trust for the Welfare of Persons with Mental Retardation and Cerebral Palsy", which is currently before Parliament (1991). The trust will have the following functions : (a) to arrange and provide care and rehabilitation to persons with mental retardation and cerebral palsy; (b) to set up homes and service institutions; (c) to provide guidelines, aid, and assistance to the organizations providing care and rehabilitation services; (d) to provide guardianship and foster care; (e) to strengthen and support the welfare programs of families, foster families, parent associations and voluntary organization; and (f) to receive from the parents the properties bequeathed by them for the maintenance of their children with mental retardation or cerebral palsy [25]. The trust shall maintain a fund to which shall be credited monies received from the government, and all monies received through grants, gifts, donations, bequests, or transfers. It will use there funds to fulfill its objectives.

Currently there is no specific legislation for those with mental handicaps, and the provisions applicable to them are scattered among different bills. These bills have tended to be discriminatory, because the mentally retarded are included with other handicapped such as the blind or the deaf, on the one hand, and with the mentally ill and lunatics on the other. Until recently in India, the mentally retarded were considered under the Lunacy Act of 1912, which included mental retardation as a form of mental illness by defining "lunatic" as an "idiot, or a person of an unsound mind." As a result, all persons with mental retardation were categorized as mentally ill. Only in 1986 did the government pass the Mental Health Act (1986), which excluded the condition of mental retardation from the definition of mental illness. The Act, however, includes no enforcement mechanism, thus leaving a legal vacuum. Furthermore, there have been significant changes in the concept of mental retardation, which is now viewed more as a condition than as an illness. Persons with mental retardation are those who have reduced intellectual function, coupled with deficits in adaptive behavior. Professionals believe that in order to sustain a consistent policy for dealing with the mentally retarded, a single comprehensive law is mandatory, which should reflect the experiences of concerned persons and the realities of mental retardation.

A. Dhanda states that, in accordance with international instruments on rights of the mentally retarded, comprehensive legislation should guarantee to the mentally retarded as normal a life as possible. The limited capacities of the handicapped should be recognized, but legal protections should not impede the opportunities of the mentally retarded to exercise autonomy. The thrust of the legislation, therefore, should be towards autonomy rather than protection. The legislation should also have integration of the mentally retarded into society as an important focus, by specific provisions on anti-discrimination and vocational rehabilitation.

The bill should also recognize that the mentally handicapped have varying intellectual capabilities, learning abilities and social adaptability, so that the deprivation of rights is individuated, *i.e.*, only if a person is professionally assessed to be incapable of exercising the right in question should rights be limited. Under Section 5 of the Hindu Marriages Act, an idiot is not entitled to enter into a contract of valid marriage. However, experts believe that there should be no express prohibition in the law with regard to a mentally retarded person's right to marry. Annulment of marriage because one of the parties is mentally retarded also should not be permitted. However, when the retardation has been concealed, the marriage can be annulled on the ground that consent to the marriage had been fraudulently procured. Divorce because the respondent is so severely retarded that the petitioner cannot reasonably be expected to live with him/her should also be permitted.

Under current contract law, parties must give their free and full consent to bind themselves. It is presently accepted that there cannot be such free and full consent by "idiots". However, a contract by a mentally retarded person could be characterized similar to that of a minor, *i.e.*, void *ab initio*. A person contracting with a mentally retarded person only for necessities should, however, be entitled to compensation. A mentally retarded person should have the right to own and inherit property. As soon as the mentally retarded person becomes a property owner, a trust for the management of such property should be established. Such trusts could be supervised by the National Trust.

The Code of Civil Procedure (Order 32, Rule 15), which makes the provision applicable to minors, shall extend to persons adjudged to be of unsound mind and to persons who, though not so adjudged, are found by the court to be incapable of protecting their interests when suing or being sued. Experts suggest that when a mentally retarded person is accused of a crime, he should at once be professionally assessed to determine whether he comprehends the nature of the trial and can conduct his defense. If the person is found incapable of defending himself, the trial should not be postponed, but a public defender should be appointed to take up his defense. With regard to a legal defense of insanity, experts suggest that because efforts to expand the defense of insanity have been rejected by the Law Commission of India, Section 84 of Indian Penal Code should apply at present only to individuals with moderate to severe degrees of retardation. That section states that "nothing is an offense which is done by a person, who by reason of unsoundness of mind is incapable of knowing the nature of the act or that it is wrong or contrary to law".

Article 45 of the Constitution affirms that the state shall endeavor to provide, within a period of ten years from the commencement of the Constitution, free and compulsory education for all children up to fourteen years of age. However, for children who are mentally retarded, special educational facilities should be provided. With regard to livelihood, mental retardation should not disqualify an individual if he or she has the capacity to perform the job in question. Nor will a worker's retardation be a defense available to the employer in a workman's compensation claim. An employment registration bureau for the mentally retarded should be established to facilitate placement in jobs reserved for them. Presently, the government of India reserves 3 percent of jobs for the disabled, but the mentally retarded are excluded from this. Voluntary organizations have brought on the government to allow 1 percent of jobs be reserved for the mentally retarded. If the government agrees to this suggestion, it will be necessary to relax the essential qualifications.

It must be realized that legislation is vital and complementary to other welfare measures for the mentally retarded. Legislation at the hands of a welfare state can achieve more than individual efforts and sympathy, because the resources and authority of the state are vast. However, legislation has its own handicaps. Social evils cannot be entirely eradicated by legislation. The mentally retarded deserve the respect and understanding of society, and the legislation cannot substitute for this larger work.

IX. ETHICAL CONCERNS IN SUICIDE AND EUTHANASIA

Suicides are on the increase in India. In 1990, almost 60,000 people killed themselves, while in 1986 the number was 46,000 ([33], pp. 3-116;[21], pp. 134-141). Nearly half of those who committed suicide were between 18 and 35 years of age, and about 14 percent were below the age of 18. Every 2 minutes someone in India attempts suicide and every 10 minutes someone succeeds [21]! Behind these cold figures lies an endless variety of human misery – poverty, unemployment, miserable marriages, debt, incurable diseases, dowry harassment, alcoholism, depression, and broken love affairs. Psychiatrists believe that breakdown of the joint family system and the emergence of the nuclear family are linked to the distress and loneliness that may provoke suicidal feelings. The fundamental cause of most suicides in India is rapid social change, which takes a heavy toll on the emotional well-being of individuals.

A consultation workshop was organized in Bombay by the FIAMC Biomedical Ethics Centre ([33], pp. 3-116). The group defined suicide as "any act with the direct intention of killing oneself". Under Section 309 of the Indian Penal Code (IPC), the attempt to commit suicide is a criminal offense and shall be punished by imprisonment for up to one year, or by fine, or both. Aiding and abetting a suicide is also an offense punishable with imprisonment for up to 10 years and is liable to fine. Recently, there have been attempts to decriminalize suicide in India, as has already occurred in many western countries. Justices P.B. Sawant and Kolse-Patil in their judgment in writ petition Number 641 of 1986 ruled that Section 309 of the IPC contravened Articles 14, 19 and 21 of the Constitution of India ([24], pp. 87-91). These articles are concerned with personal liberties. Article 14 reads, "The state shall not deny to any person equality before the law or the equal protection of the laws within the territory of India". Article 19 protects the rights of freedom of speech, of peaceful assembly, of association, of moving freely throughout India, and of choice of profession trade. Article 21 states that "No person shall be deprived of his life or personal liberty except according to procedures established by law". One commentator quotes the judgment of the court as saying

> ... Logically it must follow that the right to live recognized by Article 21 will include also a right not to live or not to be forced to live.......One's life, one's body with all its limbs, are certainly one's property, and one is the sole master of it. One should have the freedom to dispose of it as and when he desires ([12], pp. 92-94).

The counsel for the Government of India, however, argued that neither Articles 19 or 21 created or recognized the right to life as such. All they do is to prevent the state from depriving an individual of his right to life, otherwise than by a just, fair and reasonable procedure established by law. What remains unclear is whether the existence of a positive aspect of a right under the Constitution automatically permit the existence of the negative aspect of the right. Does the right to life and personal liberty also include the right to take one's own life? Can the "desire to die" be equated with "right to die"?

It is worth noting that Section 309 of the IPC has not been struck down. Thus far the High Court in Bombay has only stayed prosecution under Section 309. Moreover, Justice Sawant has stated at public meetings that he wishes to have public opinion generated on this issue. Nonetheless, the judgment of the Bombay High Court does not countenance euthanasia, for the court said:

> Euthanasia or mercy killing on the other hand, means and implies the intervention of other human agency to end the life. Euthanasia or mercy killing is nothing but homicide, whatever the circumstances in which it is effected. Unless it is specifically excepted, it cannot but be an offense. Under current laws euthanasia amounts to homicide by the physician, and suicide by the patient, which even the consent of the patient cannot legalize.

Section 309 acts as a block to the protagonists of euthanasia.

In India, attempts to legalize euthanasia began ten years ago, with a bill in the Indian Parliament introduced by M.C. Dega. The bill simply said that "if doctors confirmed that anyone was seriously ill, they could give him a lethal injection if his relatives agreed" ([14], pp. 157-161). This bill provoked strong opposition and it was buried. Recently, however, a "Physician Immunity Bill" was introduced in the Maharashtra Legislative Council by S. Varde, a private member. The bill sought to empower persons to make advance directives regarding treatment decisions if and when they are diagnosed as having a terminal illness. In reading various sections of the bill, one becomes alarmingly aware that, in its present form, it could become an instrument of death by providing full scope for misuse and abuse in the hands of unscrupulous persons. The wording of the bill permits any doctor to certify a case as terminal even if the doctor is not competent or qualified to handle that particular illness. The bill disregards the fact that every patient has a right to withdraw from the treatment if he so wishes, and that no one can compel a doctor to act against his conscience or code of medical ethics to willfully take the life of his patient.

The most disturbing section of the bill is that which makes a competent patient's declaration binding if it is not revoked within 30 days of having been made. It is frightening to this author to think what may happen to a patient who, having made the declaration, then becomes crippled by illness or an accident and loses his power to communicate, but is still of sound mind and has changed his views about the use of heroic measures and does not wish to die. That patient will be unable to change his declaration and will be totally at the mercy of his relatives and doctors because of the binding nature of the clause. Fortunately, the bill has not been passed.

L.H. Hiranandani sees this bill as having been inspired by "the human transplant industry" which is keen on legalizing euthanasia anywhere in the developing world to easier organ transplantation ([14], pp. 157-161). Recently, for example, a dynamic young cardiac specialist issued an appeal to the Prime Minister to legalize euthanasia. It was suggested that with the passing of such a bill, donor hearts could be obtained from the dying patients for heart transplants. However, vehement opposition was raised to his suggestion, with the argument that "the hearts of all the poor patients dying in public hospitals will be removed and made available for transplantation in private hospitals to save rich man's lives. The lives of the poor will be sacrificed" ([14], p. 157).

It must be admitted that passive euthanasia is being practiced in India, as elsewhere in the world, without it being legalized anywhere. If a patient is terminally ill with an incurable disease and cannot be saved, the doctors, motivated by compassion, do not always resort to heroic measures just to prolong a vegetative existence. In fact, if a patient is dying and is in severe pain, drugs to relieve the pain are often given in increasing doses to overcome tolerance to them even if they happen to hasten death. Still, most Indians would agree with the conclusions reached by the consultation workshop on suicide that "suicide is a wrong or criminal act. Human life being of infinite value and a fundamental good, it should undoubtedly be promoted, protected and preserved" ([33], p. 115). This conviction was seen to underlie the concept of the common good of society, which is the basis of all societal laws and restrictions. In this respect, after deep reflection, it was stated that the decriminalization of suicide in India would not promote the common good of society. Given the present environment of an ideological vacuum, collapsing values, increasing corruption, and ever multiplying "bride-burning" and "dowry deaths" it was unanimously reiterated that "the criminality of suicide should be unequivocally maintained".

X. ETHICAL CONCERNS IN AIDS

In India, the first evidence of human immuno-deficiency virus (HIV) infection was detected among ten prostitutes in Madras in 1986. The first AIDS patient was diagnosed in 1987 in Bombay. The patient had undergone coronary bypass surgery in the United States and had received blood there. The second case was a hemophiliac who had been transfused with factor VIII obtained from abroad. Following this, the Government of India stepped up sero-surveillance among groups practicing high risk behavior, such as patients with sexually transmitted diseases, blood donors and prostitutes. As of December 1, 1991, 1,206,055 persons engaging in high risk behavior have been screened and 6,319 have tested positive ([18], pp. 13-21). Of these, the largest number (52 percent) are from heterosexual promiscuous persons (3,005), followed by intravenous drug users (1,352), blood donors (926), recipients of blood transfusions/blood products (127), relatives of HIV patients (51), antenatal women (24) and persons undergoing peritoneal dialysis (21). One hundred and two clinical cases of AIDS were diagnosed, of which 13 are cases involving foreigners. Almost all the states/union territories in India have reported HIV-infected cases. The largest number of reported cases are from the states of Maharashtra, Tamil Nadu, and Manipur. The actual number of cases, however, is far greater than the number reported. It is estimated that cases of HIV infection varied from 0.25 to 1.0 million in 1991. Estimates of AIDS cases range from 2,700-9,000, according to the model used for projection. Maharashtra and Tamil Nadu resemble the pattern of sub-Saharan Africa, where the principal mode of transmission is through heterosexual contact. In contrast to this picture, the principal mode of transmission in Manipur is through the use of infected needles used by intravenous drug users. This pattern appears similar to that in Southern Europe and Thailand.

The government is giving top priority to controlling HIV infection in India ([18], pp. 22-24). Sixty-two centers for sero-surveillance have been set up. These perform AIDS testing using the ELISA technique. In seven centers, facilities for Western Blot have been set up to confirm the diagnosis. Sentinel surveillance is also done to identify cases that are missed by the sero-surveillance centers. Twenty-nine zonal HIV-testing centers for blood transfusions have been arranged in the four metropolitan cities. In the second phase, 37 more centers will be established in the state capitals and cities with a population greater than 500,000. Human resource development is being carried out to train the necessary manpower. Thirteen centers are planned for clinical management of cases, of which four are already in place in Delhi, Bombay, Madras, and Calcutta.

An AIDS Prevention Bill (1989) was introduced in Rajya Sabha (the upper house of Parliament) on August 18, 1989. The bill contained several provisions that infringed on the individual rights and liberties of people who practice high risk behavior (such as prostitutes), and people with HIV/AID. It equated

AIDS with smallpox, required AIDS to be made a notifiable disease, and called for the arrest of persons having AIDS and their confinement. However, once experts explained the true nature of AIDS infection and its modes of transmission to the members of Parliament, the bill was dropped.

A controversial measure initiated by the Government of Maharashtra called for establishing a home to isolate seropositive prostitutes from the profession to prevent their spreading the infection. However, experts believe that this measure is unlikely to help in the control of the AIDS. Another plan to rescue 1,000 prostitutes from Bombay and return them to their families in Tamil Nadu went awry ([29], p. 63). In a carefully planned operation, the Madras police, in cooperation with a voluntary organization, rounded up 983 females, including children, and more than 200 eunuchs from dingy dens in Bombay, who were sent by a special train, aptly named the Mukti (freedom) Express, to Madras. The women were initially kept in an open-air jail. However, before returning them to their families, the government medical professionals subjected them to tests, and almost 60 percent tested positive for AIDS. Consequently, they were placed in remand homes, because officials feared that AIDS would spread if they were set free. Thus, the scare of AIDS provided a sad ending to what might have been a social revolution.

A workshop was organized in Bombay to discuss the ethical concerns posed by AIDS by a multidisciplinary group of national experts ([34], pp. 1-24). Although the view was expressed that "a physician (medical doctor) may in general choose whom to serve, except in an emergency", the participants were largely of the opinion that the physician's primary obligation was to treat the sick without discrimination. Thus, "a physician may not ethically refuse to treat a patient solely because he/she is seropositive for AIDS". However, this is precisely what is happening in practice, as many doctors are refusing to provide treatment to AIDS patients, and many private hospitals and nursing homes discharge patients who test positive for AIDS. The major problem is the unfounded fear doctors have of contracting AIDS if they treat such patients. A major effort in the education of health care professionals is required.

The participants in the workshop were united in their call for adequate counselling of HIV-infected individuals to exercise responsible behavior in preventing the spread of AIDS, to protect their own health, and to alert past and present sexual contacts about their possible infection with the AIDS virus. The view was also expressed that the patient cannot insist that information about seropositivity or AIDS be concealed from the spouse or the sexual partner. Indeed, it was recommended that health care professionals divulge such a diagnosis to those in close contact with the infected individual in the event that the patient refuses to do so.

The workshop participants agreed that informed consent was not required before when undertaking truly anonymous epidemiological testing for AIDS. They also agreed that when patients are admitted voluntarily to hospitals or clinics for investigation and treatment of disease, specific informed consent for

HIV testing is not necessary, but that confidentiality has to be strictly maintained. If, however, the patient is detected to be seropositive for the HIV-virus or found to have AIDS, the group deemed it essential that the patient have access to appropriate counselling and treatment.

With regard to the appropriateness of medical treatment by HIV-positive physicians, the group reached the following conclusion:

> The group a physician who knows that he or she has AIDS infection should not engage in any activity that creates a risk of transmission of the disease to others. Under such circumstances, disclosure of the risk to patient is not enough; patients are entitled to expect that their physicians will not increase their exposure to the risk of contracting AIDS. If no risk exists, disclosure of the physician's medical condition to his or her patients will serve no rational purpose; if a risk does exist, the physician should not engage in that activity.

The workshop group, aware of the international research on therapeutic agents in the battle against the AIDS virus and the tremendous expenditures involved, asked the Government of India not to insist on complying with all the necessary conditions for the recognition of a "new drug" as required under the Drugs and Cosmetics Act and Rules (1950), because full compliance would slow the availability of "new drugs" which are acceptable internationally to AIDS patients. Workshop participants also appealed fervently to the Government of India, to various State Governments, and to non-governmental organizations to establish hospices or centers where seriously and terminally ill patients with AIDS could be cared for with human dignity.

If health care professionals are ethically obliged to treat the sick even at risk to themselves, it is natural to assume that patients also have obligations to caregivers. The most important responsibility of patients is to be truthful in disclosing the intimate details of their personal histories. Such an attitude will enable the health professionals to initiate and adhere to guidelines for preventing the spread of HIV-infection to the community and to themselves. However, educating patients about their responsibility may not be easy. There are some cases on record where professional blood donors, after being tested positive for AIDS, did not disclose their status, but moved to smaller towns to sell blood, thus abdicating their responsibility to society and leaving it to the blood banks to detect their positive status! This is presently an impossible task because 616 of the 1,018 blood banks in the country are operating without licenses. The government plans to amend the rules shortly so that no blood bank can function in the country without a proper license.

All India Institute of Medical Sciences
Genetic Unit, Department of Pediatrics
New Delhi, INDIA

TABLE 1

Reasons for Having a Precise Definition of Death

Medical: (a) to discontinue treatment;
(b) to optimally utilize scarce hospital facilities including intensive care units and life support system;
(c) to avoid adverse psychological effects on health care workers looking after terminally ill patients who have suffered "brain stem" death;
(d) to provide legal protection for doctors and nurses to enable them to withdraw life support systems from patients who have suffered "brain stem" death.

Legal: For purposes of inheritance.

Social: For minimizing emotional trauma and financial hardship to the family members of the deceased.

Religious: For performing religious rites.

TABLE 2

Urban / Rural Differences In Health Care Indicators

	Urban	Rural
Population (%)	27	73
Infant mortality rate*	62	102
Birth rate per 1000	26.3	33.1
Death rate per 1000	7.7	12
Access to safe water (%)	76	50
Population below absolute poverty line (%)	28	40

*Annual number of deaths of infants under 1 year of age per 1000 live births.

TABLE 3
Male / Female Differences In Health Indicators

	MALE	FEMALE
Adult literacy rate[1]	64	39
Primary school enrollment ratio[2]	113	81
Secondary School enrollment ratio	50	27
Population (%)	51	49
Hospital malnourished (ratio)	1	3
Under five mortality rate[3]		20 X higher among females

[1] Percentage of persons aged 15 years and over who can read and write.
[2] Enrollment ratio is the total number of children (of any age) enrolled in a schooling level expressed as a percentage of the total number of children in the relevant age group for that level.
[3] Annual number of deaths of children under five years of age per 1000 live births. It represents the probability of dying between birth and exactly five years of age.

NOTES

[1] Earlier, a similar practice was in vogue for selling blood, which happened particularly in private blood banks. Now that the Government exercises greater control over the private blood banks, and testing blood donors for hepatitis B and AIDS virus has become mandatory in the metropolitan cities, the practice of selling blood has been curbed, although not completely eradicated.

BIBLIOGRAPHY

1. Badhwar, I.: 1990, "Mandal Commission – Dividing to Rule", *India Today* (September 15), 98-103.
2. Bose, A.: 1991, *Population of India –1991: Census Results and Methodology*, B.R. Publishing Corporation, Delhi, 36-71.
3. Census Commissioner and Registrar-General: 1991, *Census of India - 1991 - Provisional Results*, Ministry of Home Affairs, Government of India.
4. Chengappa, R.: 1990, "The Organs Bazaar", *India Today* (July 31), 60-67.
5. Chengappa, R.: 1990, "The Kidney Colony", *India Today* (July 31), 62-63.
6. *Constitution of India*: 1986, Ministry of Law , Government of India. New Delhi.
7. *Criteria of Death Consultation Workshop*: 1985, FIAMC Biomedical Ethics Centre, Bombay, 1-30.
8. Dhanda, A.: 1991, "Distinct Legal Regime for the Mental Retardation – Criteria – Proposal for a Comprehensible Legislation", *Proceedings of Conference on Legislation for the Mentally Retarded*, Vidya Bharti Foundation, Sankara,Vidya Kendra, Delhi, 1-6.
9. Gandhi, R.K. and Vas, C.J.: 1989, "Summary Report – National Seminar on Determination of Death", National Academy of Medical Sciences, New Delhi, 1-29.
10. Ghosh, S.: 1990, "Girl Child in SAARC Countries", *Indian Journal of Pediatrics* 57, 15-20.
11. Grant, J.P.: 1992, *The State of the World's Children, 1992*, Oxford University Press & United Nation's Children's Fund, 71-89.
12. Hassumani, A.: 1987, "Suicide – Judgment in Bombay", in C.J. Vas and E.J. de Souza (eds.), *Suicide - Report of a Consultation Workshop*. FIAMC Biomedical Ethics Centre, Bombay, 92-94.
13. Hingorani, K.: 1990, "Status of Women and Law", *Bhartiya Janani* (Indian Woman), Mahilla Dekshata Samita, New Delhi, Vol 1, No. 1, 15-17.
14. Hiranandani, L.H.: 1990, "The Protection of Life", in C.J.Vas and E.J. de Souza (eds.), *Issues in Biomedical Ethics*, MacMillan India Ltd, New Delhi, 157-161.
15. Jacob, M.M.: 1992, "Crimes Against Women", *Lok Sabha Question Hour*,

Ministry of Home Affairs (March 18).

16. Kapoor, P.N.: 1991, "Implications of Provisional Results of Population Census of India 1991", *Centre Calling* 1991, Vol. 26, No. 5, 3-6.
17. Kulkarni, S.: 1986, "Sample Survey of Sex Determination Tests and Practices in Greater Bombay", Department of Health, Government of Maharashtra, Bombay, quoted in *Report of Central Committee on Sex Determination, 1989*, Ministry of Health & Family Welfare, Government of India, New Delhi, 99.
18. Lal, S.: 1991, "*National AIDS Control Program – India, Country Scenario 1991*", Directorate General of Health Services, Nirman Bhavan, Government of India, New Delhi, 1-62.
19. Lok Sabha Bill No. 155 of 1991: *The Prenatal Diagnostic Techniques (Regulation and Prevention of Misuse) Bill*, 1991, Lok Sabha Secretariat, 1-19.
20. *Maharashtra Live Donor Kidney Transplantation and Prohibition of Commerce in Kidney Bill*: 1990, Department of Health, Government of Maharashtra, Bombay, 1-26.
21. Menon, R. and Dhillon, A.: 1991, "Suicide – Fatal Attraction", *India Today* (February 28), 134-141.
22. Mukherjee, S.: 1991, "The Girl Child in India", *Indian Journal of Pediatrics* 58, 301-303.
23. Pachauri, P.: 1990, "New Reservation Policy-Apartheid, Indian Style", *India Today* (September 15), 100-103.
24. Phatani, P. : 1987, "Medico-Legal Aspects of Suicide and Euthanasia, in C.J. Vas and E.J de Souza, (eds.), *Suicide –Report of a Consultation Workshop*, FIAMC Biomedical Ethics Centre, Bombay, 87-91.
25. Rajya Sabha Bill No. 11.: 1991, *The National Trust for the Welfare of Persons with Mental Retardation and Cerebral Palsy*, Rajya Sabha Secretariat, Government of India.
26. Ministry of Health & Family Welfare: 1991, *Report of the Committee to Examine a Proposal to Enact Legislation on the Removal of Human Organs and Tissues and their Transplantation for Therapeutic Purposes*, Government of India, 1-52.
27. Rohde, J.E.: 1991, "On Closing the Gender Gap – Increasing the Value of Women in India", *Indian Journal of Pediatrics* 58, 295-299.
28. Shah, P.: 1990, "Amniocentesis – the Crippling of Female Psyche" *Bhartiya Janani* (Indian Woman), Mahilla Dekshata Samita, New Delhi, Vol 1, No.1, 28-29.
29. Shetty, K.: 1990, "Prostitutes, Sad Ending – AIDS Scare Hampers Rescue", *India Today* (June 30), 63.
30. Singh, A.: 1992, *Legislation for Women's Welfare*, Ministry of Human Resource Development, Lok Sabha Question Hour (March 17).
31. Vas, C.J.: 1990, "Brain Death in Children", *Indian Journal of Pediatrics* 57, 735-742.

32. Vas, C.J. and de Souza E.J.: 1990, *Brain Transplantation Ethical Concerns*, FIAMC Biomedical Ethics Centre, Bombay, 1-86.
33. Vas, C.J. and de Souza, E.J.: 1987, *Suicide, Report of a Consultation Workshop*, FIAMC Biomedical Ethics Centre, Bombay, 3-116.
34. Vas, C.J. and de Souza, E.J.: 1991, *Ethical Concerns on AIDS - Report of a Workshop*, FIAMC Biomedical Ethics Centre, Bombay, 1-24.
35. Venkataramani, S.H.: 1986, "Female Infanticide – Born to Die", *India Today* (June 15) 26-33.
36. Verma, I.C.: 1988, "Genetic Causes of Mental Retardation in India", in M. Niermeijer and E. Hicks (eds.), *Genetics and Ethical Considerations in Mental Retardation*, Bishop Bekkers Foundation, International Symposium, Reidel Publishing Co, Amsterdam, 99-106.
37. Verma, I.C. and Singh B.: 1989, "Ethics and Medical Genetics in India", in D.C. Wertz and J.C. Fletcher (eds.), *Ethics and Human Genetics, A Cross-Cultural Perspective*, Springer-Verlag, Berlin, 250-270.
38. Verma, I.C.: 1991, "Ethical Issues Arising in Molecular Genetics in Developing Countries", in R.J. Sram, V. Bulyzhenkov, L. Prilipo, and Y. Christen (eds.), *Ethical Issues of Molecular Genetics in Psychiatry*, Springer-Verlag, Berlin, 134-148.
39. Verma, I.C., Fujiki, N., Marwah R.K., Ahuja Y.R., Malhotra K.C., Parikh A.P. and Sharma S.: 1992, "The Common People's Viewpoint on Handicaps and Heredity", *Proceedings of 2nd International Bioethics Seminar, Fukui*, Japan 1992 (In Press).

R. ANGELES TAN ALORA[1]

BIOETHICS IN SOUTHEAST ASIA: 1989-1991

I. INTRODUCTION

Bioethics is a developing and diverse discipline in Southeast Asia. The practice of health care and the concerns of bioethics are as varied throughout the region as are the nations and cultures of the region. There are a number of recurrent influences and concerns that one finds in bioethics discussions. For example there are recurrent concerns about population growth and the allocation of health care resources, but the manifestations of these concerns can vary greatly from country to country. The concern over population, for example, ranges from encouraging population growth in Brunei to discouraging population increases in the Philippines. Throughout the region one also finds strong influences of religion as part of the local culture. Christianity and Islam, for example, are powerful influences on the issues of bioethics in different cultures.

In this report the focus will be on the Philippines, Indonesia, Taiwan, and Brunei. For a number of reasons it was difficult to gather information for this initial report. Nonetheless, these three nations provide a good representation of the region and a powerful contrast for the issues of bioethics and health care that one finds in Southeast Asia.

II. ABORTION

The constitution of the Philippines specifies that "(the state) shall equally protect the life of the mother and the life of the unborn child from conception" [16]. The Revised Penal Code imposes penalties for abortion in all its forms [12]. In spite of these prohibitions the practice of abortion is wide spread. It is commonly estimated that 750,000 abortions occur every year. Abortion practitioners include doctors, nurses, midwives, traditional birth attendants and patients themselves. When abortion cases are brought to court they are usually dropped since the patients often will not testify for fear of being held liable. Often in many cases, there is the use of influence to have the case dropped.

---

[1] Special thanks to Rev. K. Bertens, Center for Philosophy and Ethics, Atma Jaya University, Jakarta, Indonesia, and Dr. Letty Kuan, Philippine General Hospital and Southeast Asia Center for Bioethics.

B. Andrew Lustig (Sr. Ed.), Bioethics Yearbook: Volume 2, 343-353.

There is active, widespread opposition to abortion throughout the Philippines. In the government, there are pending Senate and House Resolutions, against abortion. One Senate bill initiates an inquiry into the use of abortifacients with a view to regulation [17]. Another Senate bill [18] increases the penalties for physicians, midwives, nurses and other parties accessory to abortion. A House bill [19] also seeks to increase the penalties in order to protect the unborn. The Catholic Church, a dominant force in Philippine social, cultural, and political life has condemned abortion. The Catholic Bishops' Conference of the Philippines has stated clearly that "[D]irectly willed abortion, the use of abortifacients, sterilization and contraception are wrong in themselves. They are not wrong because the Church forbids them, the Church forbids them because they are morally wrong"[3]. The Philippine Obstetrical and Gynecological Society, in addition to proposing stricter and more effective implementation of the law against abortionists, has announced the exclusion from its membership list of those members who are known to be performing abortions [11]. The Pro-Life Philippines operates four pregnancy counselling centers in Manila which offer twenty-four hour free services for anyone involved in a pregnancy or anyone who is disturbed or depressed after an abortion.

There also exists a vocal minority favoring the legalization of abortion. They argue that allowing abortion would prevent women from dying needlessly because competent physicians would be allowed to preform the procedure. At the Sixth International Women and Health Meeting, held in Manila in 1990, Dr. Sylvia Estrada-Claudio stated that "large numbers of these women die because of unsafe abortions"[15].

In Taiwan, after an extended and volatile debate, the Eugenic and Health Law was passed in 1985 and remains in force today. The law allows women to obtain an abortion for any of the following reasons: (1) diagnosed fetal abnormality; (2) hereditary disease, mental disorder, or congenital illness in the pregnant woman, her spouse, or their close relatives; (3) if the pregnancy poses a threat to the woman's life or health; or (4) if the pregnancy results from rape, seduction, or a sexual relationship with an individual the pregnant woman is legally forbidden to marry. Abortion procedures must be performed by doctors licensed by the National Health Administration. Unlicensed doctors who perform abortions will be punished by imprisonment of up to three years as well as fines [9].

By contrast, in nations such as Brunei on the northern coast of the island of Borneo, abortion is not an issue of debate. Brunei is a Muslim monarchy which practices Islam according to the Sunni belief. In this practice of faith, all life is regarded as something sacred from God. Health care is shaped by this belief and all measures are taken to preserve life. Life is sacred and must be preserved from any form of harm from the moment of conception.

III. MATERNAL-FETAL CONFLICT

In the Philippines there is a House bill pending which is aimed at deterring the irresponsibility of the mother perpetrated against her own child. House Bill 30751 defines such irresponsible behavior as prenatal child abuse. In defining abuse, the bill states: "It shall be a crime for a pregnant woman intentionally and knowingly take illegal drugs, excess alcohol, and other addicting substances where the use of such substances shall cause harm to the fetus at anytime from the conception to the birth of the child". Any woman found guilty of such abuse faces imprisonment from six months to six years. If the child survives, it is to be awarded to the nearest kin of the offender. The penalties are lifted if the woman has entered a rehabilitation clinic or program [20].

IV. CONFIDENTIALITY

During the years of martial law under the Marcos administration, a presidential decree [21] was issued which required any attending physician or medical practitioners of any health care establishment to report the physical injuries of patients to the nearest government health authority. The aim of this decree was to deter medical treatment for rebels who suffered injuries in fighting. Indeed, some medical practitioners have been killed following reports of treating injured rebels and failing to report such treatments. One of the effects of the decree, however, was a chilling of medical services offered in the countryside, a stronghold of the rebels, where health care needs are often the greatest due to rural poverty. The decree was not only an intrusion of the state into the physician-patient relationship, it also affected the general practice of medicine. A bill has been introduced to repeal the decree [22].

V. EQUITABLE ACCESS TO HEALTH CARE

In the Philippines large segments of the population fail to receive adequate health care because of poverty. The National Statistical Office reported that 49.5 percent of families are below the poverty line, or minimum household income of P2709. In 1987, the Food and Nutrition Research Institute found 72 percent of households to be below the "food threshold" (*i.e.*, the income level needed to afford the minimum necessary dietary allowance). A current economic crisis exists as a result of the repeated coup attempts, the Persian Gulf war, the increase in oil prices, and repeated natural disasters (earthquake, volcanic eruption and floods). In addition to the poverty there is a general lack of available health services. According to the Department of Health there was a ratio of one health center for every 28,328 people.

Public hospitals are overcrowded and indigent patients have to seek care in private hospitals. A number of legislative initiatives have focused attention on the support of, and access to, emergency room facilities. Many have been refused admission because of their inability to pay a deposit. Also, many patients have not been allowed to go home because hospital bills could not be settled. One bill attempts to correct the practice of requiring deposits before tending to, or admitting, emergency patients. A crucial element of the bill is the creation of a state fund to underwrite the cost of the first 48 hours of an indigent patient's stay [23]. Another bill seeks to establish financial support in a fixed amount (P500) for indigent patients [24]. In defining "indigent" it is clear that the patient must not only lack visible means of support, but also lack any form of outside support (*e.g.*, family). The assumption is that families must first assume the costs of medical care. Yet another bill [25] seeks to establish a medical service program in which the government subsidize participatory private hospitals for patients who cannot be admitted to government hospitals. Another bill [26] seeks to provide an increase in the availability of free wards, which are paid for by government appropriations, in private hospitals by requiring that all public health centers should be open at all times to the public. Specific legislation has been introduced to amend existing laws guaranteeing access to health care institutions by increasing penalties for denying access, particularly to emergency patients. The proposed legislation also creates a fund for compensation for the provision of services [23]. There have also been proposals to increase accessibility throughout the country, particularly in rural areas [27]. This legislation sets out the parameters for medical coverage for all citizens of the Philippines who lack existing medical coverage. Those covered under the proposal would pay premiums as set by the Department of Social Welfare and Development. Central to the proposal is the establishment of the Philippines Medical Care Health Fund which is to be funded by contributions from self-employed and irregular earners. Other funding sources are to include money from a sales tax on cigarettes and a percentage from the Philippine Charity Sweepstakes [27]. There have been efforts to promote health care by upgrading equipment and personnel in emergency room service according to standards set by the Department of Health [28]. The financial support for such improvements is left unspecified.

At present, public health care is financed in part by the support of the provinces, cities and municipalities remitting a particular percentage of their annual general income to the national government which, in turn, operates the national hospitals under the Department of Health. It is charged with the construction of hospitals and their maintenance. Cities and municipalities which operate their own locally funded public hospitals are exempted from this requirement. A crucial problem with this system is that many local governments do not remit their full amount to the central government. Also local governments complain that they have no effective way to monitor the prudent use of local aid [28]. The proposed legislation would give local authorities greater control over funding and administration [29].

Brunei provides a sharp contrast to the Philippines. With a citizen population of 188,800, a national reserve of 26 billion dollars funding for hospitals has never been wanting. There are only four hospitals in Brunei, which are fully funded by the Sultanate. People are given the utmost care to the last minute of life, and if the sick individual needs expert care and management, the King of Brunei sends the sick person anywhere in the world in order to provide the best care for the person. Non-Brunei residents manage their own needs because only the natives of Brunei are being taken care of by the King as far as finances are concerned.

In Taiwan, government-sponsored health insurance is a vital aspect of the country's welfare system. At the present time, however, only 40 percent of Taiwanese residents are covered by government insurance. The government is therefore working to establish a comprehensive system by 1995 to cover its population of 20 million persons [5].

VI. ETHICAL ISSUES POSED BY COST-CONTAINMENT MEASURES

In the Philippines, concern for the escalating cost of medical care has been expressed in both bills aimed at expanding medical services. One area of effort at cost containment has been in providing cheaper drugs, encourage herbal use and development, and legalizing the practice of traditional health care providers. A House resolution sought to establish an inquiry into the price of drugs and pharmaceutic production [30]. Others sought to require physicians to include the approximate costs of the drugs on prescriptions [31]. In the area of herbal medicine, the legislature sought to establish herbal medical centers in every legislative district in the Philippines [32], to create a medicinal plants development board [33] and to institutionalize the herbal medicine program of the government [34]. Each initiative is part of an effort to control cost. Also there is an effort to capitalize on the predisposition of many people to use medicines which are part of their culture. Finally, there was also legislation to establish acupuncture and acupressure centers in Metro Manila, cities, and provincial capitals [35].

There is ongoing controversy regarding the Generics Act of 1988 with medical practitioners claiming it to be an imposition of their liberty to practice [2] and the government claiming there is an alliance between practitioner and pharmaceutical companies. An appendix list which enumerates approximately one hundred drugs which can be prescribed by brand names has now been added.

VII. WITHDRAWING/WITHHOLDING TREATMENT FOR DYING PATIENTS

The right of a terminally ill patient to refuse life sustaining procedures was recognized by the Senate when it specified the right of every person of legal age to make a written directive, along with the requirements for such directives to be validly implemented and the conditions related to its execution [36].

The first motion for authority to terminate or discontinue medical treatment and nursing care was filed by the child and co-guardian of a seventy-two year old incompetent widow. The patient had a subarachnoid hemorrhage, brain surgery and meningitis ten years ago. Since then she has needed tracheostomy and gastronomy tubes. For the last seven years she has not communicated, has had various infections and has been retaining fluid. The motion was opposed by the other children.

The outlined legislation deals with directives governing life sustaining procedures in cases of terminal illness. Although active euthanasia is considered killing, and illegal, a Philippine Senate Resolution is inquiring into it [37]. Often such discussions reflect a confusion between withdrawing treatment and killing. There is hope that a public inquiry will clarify this distinction.

In Brunei the fundamental religious and moral commitment is to the sanctity of life. With the availability of resources one finds a societal and government commitment and capability to giving the utmost care up to the last minute of life. The concept of withdrawing treatment or codes such as Do-Not-Resuscitate orders do not exist in Brunei.

VIII. ORGAN DONATION, SALE, AND TRANSPLANTATION

The Philippine Red Cross is the primary source of free blood. There also exist many private blood banks where blood can be bought. Still, the available blood supply is not sufficient and the quality of blood which is available has been questioned. To address questions of quality the Congress has proposed new bills. One [38] requires blood donors and donees to undergo blood examination prior to blood transfusion. Another bill [39] has proposed establishing blood banks in every municipality in the country. These two bills were combined with another bill [40] which promoted the voluntary donation of blood.

Kidneys for transplantation are even more scarce. There has been hesitation due to religious beliefs, fear of physical harm as well as fear of legal implications. To address these fears and misconceptions there are educational campaigns as well as new congressional bills legalizing organ donation after death. Several bills were consolidated into a general Act authorizing the legacy of donation of the human body, or any part of it, after death for medical, surgical, or scientific purposes [41].

Filipino prisoners began donating organs in 1976 as part of a program to reduce overcrowding on death row without resorting to widespread executions. A death row inmate's case would be reviewed after donation, with most donors avoiding execution and some being freed after a few more years in prison. When capital punishment was banned in 1987 prisoners started asking for money in return for donation [2].

Problems with organ supply increased when foreigners competed with Filipinos as recipients for organs. In 1988, for example, 6,373 foreign visitors came to the Philippines for health reasons including organ transplant [15]. Patients came from Japan on a packaged "kidney tour". A (US) $140,000-150,000 fee included air fare, hotel accommodations, hospital fees, doctors' fees and incidental expenses [13]. Trade in human organs prospered. Offers for transplantable kidneys have ranged from $200 to $20,000.

Although organ selling is not illegal in the Philippines, it is generally viewed as an undesirable act. In 1989, a bill was filed in the Senate to ban the selling of human organs or tissues. The Catholic Church forbids the sale or exchange of organs and sees it as amounting to plundering of the body. Today at the National Kidney Institute, donors are either dead patients or living relatives. In other hospitals, however, living non-relatives are accepted [7].

In Taiwan, nine heart transplants were performed from July 1987 through the first six months of 1990. In an attempt to increase the number of hearts available for transplantation, the Government Justice Ministry revised the regulations on public executions in 1990 for convicts on death row who wish to donate their hearts. These convicts will now be shot in the temple rather than being executed, as in the past, by a bullet fired into the heart from the back. For convicts who do not wish to donate their hearts, executions will proceed in the customary way [8].

IX. OTHER ISSUES

A. *AIDS*

There were 277 reported cases of AIDS as of October, 1991 in the Philippines [4]. Various factors have been identified as responsible for the failure of AIDS Prevention Programs in the Philippines: the illusion that as members of a "Christian" country (more than 80 percent are Roman Catholic while 6 percent belong to Protestant denominations), Philippine citizens are more "moral" than others and do not practice the high risk behavior; the cultural barrier toward open discussion of sexuality; and the Roman Catholic Church's hard line opposition to the use of condoms.

A consolidated bill to prevent the spread of AIDS is awaiting funding [42]. One provision of the bill requires all U.S. servicemen to undergo regular HIV testing and authorizes the use of condoms by all clients of Philippine hospitality girls at U.S. bases and areas of recreation for U.S. servicemen. Other sections of the bill establish a national AIDS Control Commission as part of the Department of Health, as well as a national AIDS research laboratory and control clinic. Bills compensating AIDS victims are pending, and bills aimed at preventing the spread of AIDS and establishing an AIDS Center are under deliberation. The Department of Health is presently involved in an intensive preventive/informative program. Centers have been opened for individuals to obtain information about the disease, counselling, and condoms.

According to the National Health Administration of Taiwan, there were eight reported cases of AIDS between 1985 and 1988. Ten cases were already reported during the first ten months of 1989. Health officials therefore fear an increase in the number of AIDS patients and HIV carriers. In response, Chiang Wan-shuan, President of the National University Hospital, has called upon the government to provide free condoms and to expand programs of sex education in secondary schools and universities [6].

B. *Population*

In 1989 the Philippine population was 63.2 million with 39.3 percent below fifteen years of age. The growth rate was 2.3 percent. The Philippine is among the top five countries in the world in terms of population growth. This is accompanied by a very low per capita income with 2.5 million unemployed [28].

At the start of the Aquino regime, Pro-Life had influenced the Population Commission to shift from population control to population development. With foreign funding (*e.g.*, USAID, UNFPA), the program shifted to the Department of Health. The Philippine Family Planning Program (1990-1994) respects the freedom of conscience of the individuals. It emphasizes that it does not aim to reduce fertility or population growth but only to help married couples space their children. The means provided include natural family planning, IUDs, sterilization and oral contraceptives. The Catholic Church, however, has expressed it objections to contraception and sterilization and its reservations regarding the moral acceptability of certain aspects of the program [3]. The Catholic Bishop's Conference of the Philippines in a pastoral letter stated that all who wished to remain faithful to the Gospel cannot associate themselves with the program. Legislation is trying to reverse the country's populations explosion. A number of bills sought to study the rise in population and its affects [37;43;44;45;46] while others sought to establish more effective policies. One piece of legislation sought to establish a Department of Population Control [47] while another sought to require all secondary schools to include in their curriculum the teaching of family planning [48].

In Brunei, in sharp contrast, family planning is not practiced. In fact the King of Brunei awards a good sum of money to every baby born in a year. Unemployment does not exist in Brunei.

C. *Consent and Medical Intervention*

Issues involving informed consent have provided significant discussion in Indonesia. The Indonesian Minister of Health has set forth regulations regarding consent to medical interventions. The physician must provide complete information except in cases where the physician believes such knowledge may prove detrimental to the health interest of the patient. In the latter cases, with the patient's consent, information should be given to close relatives in the presence of a nurse or other paramedical personnel as witness.

The information should communicate the advantages and disadvantages of the proposed medical intervention (diagnostic or therapeutic). In cases requiring invasive procedures (*e.g.*, surgery), the information must be given by the physician himself who conducts the surgery. If for some reason the physician who is to perform the procedure is unable to obtain the consent of the patient he may designate another physician to obtain the consent. A medical intervention is an act, either diagnostic or therapeutic in nature. An "invasive" intervention is a medical intervention which directly influences the wholeness of bodily tissues. The manner of presenting the information and its contents to the patient must be adjusted according to the level of education and condition of the patient. Those interventions which are high risk must have a written consent signed by the patient or the appropriate proxy. It is the responsibility of the physician to execute the consent process. When the medical intervention is carried out in a hospital or a clinic, the concerned institution is jointly responsible, with the physician, in making sure that the consent to medical intervention is given. The primary sanction against a physician, for failing to get the proper consent, is the retraction of license.

One point of controversy about the regulations is the age set for competency. The regulations set 21 years as the age for which consent can be given. Prior to that, consent must be given by the proper guardians of the individual. However, according to Indonesian law, the minimum age for marriage is 16 years and the minimum age for participating in general elections is 17.

In Brunei, when an individual is confined in a hospital, the individual receives a contract between himself and the hospital. This contract is the form for informed consent. There is only one consent form and the explanation is personalized. The Bruneians strictly observe confidentiality and truth-telling because of their strong Muslim beliefs.

South East Asian Center for Bioethics
Manila, THE PHILIPPINES

BIBLIOGRAPHY

Books and Articles

1. Alora, A., *et al.*: 1991, *Casebook in Bioethics*, Southeast Asian Center for Bioethics, Manila, The Philippines.
2. Calleja, H.: 1990, *An Unpalatable Pill Without Therapeutic Effect*, Medical Society, Manila.
3. Catholic Bishop's Conference, *Guiding Principles on Population Control*, Catholic Conference, Manila, The Philippines.
4. Department of Health: 1991, *National AIDS Directory*.
5. "Health Insurance for All a Necessity": 1989, *China Post*, November 6.

6. "Health Officials Worried by Increasing HIV Cases": 1989, *China Post*, November 7.
7. Mendoza, M.: 1991, National Kidney Institute, Manila, The Philippines, Personal Communication.
8. "Ministry Approves New Execution Method to Allow Transplants": 1990, *China Post*, August 16.
9. "New Law Enables Women to Undergo Abortion When They Have To": 1984, *China Post*, June 30.
10. *Philippine Health Care Factbook*: 1990, Center for Research and Communication, Manila.
11. Philippine Obstetrical and Gynecological Society: 1991, "Statement on Abortion".
12. *Revised Penal Code*, Section 2, Articles 256-259.
13. *The Manila Bulletin*: 1989, "Trade in Organs Worries Officials", June 20.
14. *The Manila Chronicle*: 1989, "Kidney Transplant Kindles Manila-Tokyo Row", July 18.
15. *The Philippine Daily Inquirer*, November 5, 1990. Philippine Constitution, Article II, Section 12.
16. *The Philippine Constitution*, Article II, Section 12.

Cases and Statutes

17. Senate Resolution (SR) 450, *Congress of the Philippines, Senate Resolutions*
18. SR 1190
19. House Bill (HB) 32931, *Congress of The Philippines, House Bills*
20. HB 30751
21. Presidential Decree, #169
22. HB 26367
23. HB 4771
24. HB 33758
25. HB 28688
26. HB 22716
27. HB 31246
28. HB 31277
29. HB 22690
30. House Resolution 17
31. HB 31127
32. HB 10799
33. HB 12910
34. HB 21460
35. HB 1561
36. Senate Bill (SB) 1177
37. SR 634

38. HB 17358
39. HB 15665
40. HB 28971
41. HB 11281, HB 17676, HB 19004
42. HB 827, HB 1010, HB 5587, HB 15776
43. HR 40
44. HR 55
45. HR 1035
46. HR 1322
47. HB 15771
48. HB 15835

REN-ZONG QIU AND DA-JIE JIN

BIOETHICS IN CHINA: 1989-1991

I. INTRODUCTION

The field of bioethics has flourished in China since the new policy of reform and openness. After two legal cases, one on euthanasia, the other on artificial insemination by donor (AID), were widely publicized ([30];[54]), bioethics moved beyond narrow academic circles, and become the focus of public debate [49]. In 1988, the First National Conference on Social, Ethical, and Legal Issues in Euthanasia and the First National Conference on Social, Ethical, and Legal Issues in Reproductive Technology were held, respectively, in Shanghai, July 5-8, and in Yueyang, Hunan Province November 3-5 ([43]; [44]). Unfortunately, the proceedings of these two conferences were not published. However, the two legal cases and two conferences stimulated further wide discussions which extended to the beginning of the 1990s. After the political events of June 4, 1989, and subsequent economic difficulties, academic meetings were discouraged, but discussions were still pursued in academic journals, and the media retained a strong interest in bioethics. The legal case on euthanasia, decided in May 1991, prompted further efforts to prepare a Second National Conference on Euthanasia. However, new topics also received attention: the introduction of market mechanisms into health care, mandatory sterilization of the mentally retarded, and ethical issues in the control of sexually transmitted disease (STD). A National Workshop on Control of STD was held in April 1991, and a National Workshop on Ethical and Legal Issues in Limiting Procreation was held in November. But discussion of the topic of mandatory sterilization of the mentally retarded will be postponed until the 1993 regional volume, because most papers on that topic will be published in 1992.

II. REPRODUCTIVE TECHNOLOGY

The reproductive technologies introduced into China during the last decade include artificial insemination by husband (AIH) and by donor (AID), in vitro fertilization (IVF) and embryo transfer (ET). It is estimated that AIH and AID are practiced in the clinics of 23 provinces or municipalities directly under the central government, and that sperm banks have been established in 11 provinces or municipalities. Since 1982, there have been 21 cases of IVF, and 22 test-tube babies have been born.

B. Andrew Lustig (Sr. Ed.), Bioethics Yearbook: Volume 2, 355-377.

A. *Ethical Justification for the Use of Reproductive Technologies*

Chinese ethicists usually argue for the introduction and development of reproductive technologies in China from a consequentialist viewpoint ([16] p. 197; [1989], p.331; [1989], p.75). They make the following points:

First, reproductive technology can help to solve the problems of infertile couples and promote the happiness of these families. According to incomplete statistics, the rate of infertility in China is 5-10 percent among newly married couples. The influence of traditional culture is profound. In Confucianism, lack of offspring is viewed as a violation of the principle of filial piety, and is attributed to a lack of virtue or to the sins of one's ancestors. Thus, infertile couples are under grave psychological pressure to procreate. Moreover, in traditional culture, infertility is always blamed on wives. This stigmatization has led to their abuse, as well as to family quarrels and divorce. Finally, in traditional society, lack of offspring was a justifiable reason for a husband to divorce his wife, even in cases where infertility might have been due to the husband.

Second, reproductive technologies can help to prevent the birth of defective children. If one spouse suffers from a serious genetic disease and the risk of giving birth to a defective child is very high, AID or IVF can help the couple to have a normal child.

Third, reproductive technologies can be used as a kind of "birth insurance". If a husband is engaged in high-risk work, his sperm can be kept in a sperm bank. If he should die, his wife retains the possibility of giving birth to his child through AIH. Fourth, reproductive technologies can help to promote the policy of "one couple, one child", which means not simply limiting procreation, but when necessary, assisting it.

Finally, reproductive technologies can help medical research. For example, IVF can be used to study and improve the effectiveness of contraceptives, to assess the potency of infertile women's eggs, and to measure the effects of toxic or teratogenic substances upon the conceptus.

B. *Ethical Issues in Reproductive Technology*

Most of the discussion of ethical issues posed by the new reproductive technologies has focused on AID. Several concerns have been raised.

1. *Conflict between AID and traditional values*. Those who adhere to traditional values reject AID on two grounds. First, they emphasize the consanguinity or blood ties between parents and children. Within the Chinese patriarchal system, the Confucian principle of filial piety requires one to extend the lives of one's ancestors. In one survey, 54 percent judged the donor to be the real father of a child conceived through AID, while 24 percent identified the non-donor husband as the father, and 22 percent thought that both husband and donor were fathers ([22], p. 104). Second, traditional opponents emphasize

the Confucian principle of chastity. According to the traditional interpretation, AID is construed as adultery ([16], pp. 198-199; [22], pp. 98-105).

2. *Informed consent.* Many ethicists emphasize that AID should be practiced with the full and informed consent of the couple, and that their consent should be legitimized by completing the application form in which their obligations and rights are clearly specified.

3. *Confidentiality.* Chinese ethicists argue that AID should be practiced with informed consent, and in keeping with respect for the confidentiality of all persons involved. Some ethicists suggest that AID should be quintuple-blinded, *i.e.*, blinded relative to the donor, the recipient, the recipient's husband, the physician, and the child conceived. The donor thus would not know the identity of the recipient, her husband or the child, nor would the recipient, her husband, their child, or the physician know the identity of the donor. These ethicists emphasize that the principles of informed consent and respect for confidentiality also prevent destruction of the family structure and harm to the child conceived through AID; ([72]; [16], pp. 200-204; [22], pp. 100-104).

4. *Commercialization of sperm.* As AID develops, the gap between the supply of and the demand for donated sperm is widening. Voluntary sperm donors are difficult to find, because the typical Chinese male thinks sperm is vital to life, and thus will not risk draining the essence of his life. As a result, available sperm is in short supply. One ethicist, Professor Z-X He, contends that commercializing the supply of sperm will solve the shortage problem [22]. He argues that sperm, unlike organs, can be regenerated, and thus should be commercialized, as is blood. Professor Y Ying disagrees, and argues against the commercialization of sperm. In his judgment, sperm, as a part of the human body, should not be viewed as a commodity. Moreover, commercialization of sperm is likely to lower the quality of available supply, as has happened with the commercialization of blood [73].

5. *AID and biological polygamy.* Some scientists worry that, from a genetic viewpoint, a donor's sperm which fertilizes many woman's eggs constitutes polygamy wherein children are conceived by one father and different mothers, or polyandry, wherein children are conceived by one mother and different father (if sperm samples are mixed). One commentator raises the specter of an increasing number of "inbred spouses", although he appears to exaggerate the problem [12].

C. *Legal Issues in Reproductive Technology*

At the 1988 National Conference on Social, Ethical, and Legal Issues in Reproductive Technology, it was suggested that the social regulation of AID should be strengthened through institutional, administrative, legal, and educational means [44]. Since the conference, many authors have argued for legislation on AID in order to limit its negative consequences. Their suggestions include the following:

1. A licensing system for AID should be established in order to assure that personnel at clinics are medically qualified [64] .

2. Informed consent from infertile couples should be authorized by a notary or by completing an application form which is legally binding. Involuntary AID, *i.e.*, practicing AID on a woman without her informed consent, or by deceiving her that the sperm is that of her husband, should be treated as a criminal matter. Instances of medical workers failing to test donor sperm, or unintentionally using the donor's sperm instead of the husband's, should be viewed as malpractice, with required compensation for the loss [19].

3. The legal status of a child born through AID should be determined. S-Y Jia points out that the legal status of married children, unmarried children, adoptive children, and stepchildren brought up by their stepparents is clearly specified in Chinese Marriage Law [23]. However, the relationship between AID children and their social or biological parents differs from all other arrangements. Jia argues that although a donor has blood ties with the child conceived with his sperm, it does not follow that a legal parent-child relationship is thereby established. If the child's social parents relinquish the child, the donor may then voluntarily assume the child's custody, and should be able to establish legal parenthood without the need for formal adoption. Under conditions of informed consent, the relationship between the AID child and the mother is one of consanguinity, but the relationship between the AID child and the mother's husband is an adoptive one, protected and bound by law. In circumstances where consent is lacking, the husband may refuse to accept the child, or may choose to adopt it. M-G Chen suggests, therefore, that the relationship between the AID child and the husband should be treated as a special type of adoption ([5], pp. 238-239).

III. LIMITING PROCREATION

Since 1918, a nationwide policy of "one couple, one child" has been implemented. As this policy has developed, the most controversial ethical issues are raised by late abortions and by the mandatory sterilization of the mentally retarded.

A. *Late Abortion*

Ethical concerns in late abortion involve questions of harm to the mother and to the viable fetus. In discussing two publicized cases involving late abortions, commentators point out that a late abortion exposes the mother to high risk and grave danger, and recommend that physicians should take pains to avoid performing them ([22], pp. 24-28; pp. 133-134). These writers argue that a viable fetus is a person when it exists outside its mother's body, and that its killing constitutes unjustifiable homicide.

The present situation in China is very unfavorable to the fetus in circumstances of conflict. Several commentators note that in conflicted cases, the presumed interest of the viable fetus lies on one side, while more dominant interests (of one's country, one's factory and colleagues, or one's family) lie on the other ([48], p. 347). If the mother agrees with the latter, she tips the balance against the fetus. If she insists on giving birth, or if a late abortion would jeopardize her life, the scale could balance, or the interests of the fetus might take precedence. These commentators draw the following conclusions:

> ... late abortion can be justified ethically in China: 1) if the "one couple, one child" policy is justifiable; 2) if the couple and the physician take the social good into account; 3) if the mother expresses her voluntary consent, ... and 4) if a late abortion will entail only a low risk to the mother's health or life ([48], p. 349).

B. *Mandatory Sterilization of the Mentally Retarded*

A regulation prohibiting procreation by severely mentally retarded persons was promulgated by the Seventh Standing Committee of Gansu Provincial People's Congress on November 23, 1988, and implemented on January 1, 1989. The regulation specified that mentally retarded persons forbidden to procreate and includes: (1) those whose retardation is caused by familial genetic factors, inbreeding, or other congenital factors; (2) those whose IQ is below 49; and (3) those with behavioral disabilities in language, memory, orientation and thinking.

These individuals are permitted to marry only after they have been sterilized. Severely retarded females who are pregnant must undergo abortion and then be sterilized ([27], p. 132). Although the regulation is controversial, other provinces are likely to follow the example of Gansu Province and adopt similar regulations. A National Workshop on Ethical and Legal Issues in Limiting Procreation, held in November 1991, focused on mandatory sterilization of the mentally retarded. Because most of the papers will be published in the 1992 volume *Chinese Medical Ethics*, they will be discussed in the next regional yearbook.

IV. EUTHANASIA

Recent discussion of euthanasia in China has intensified in the wake of a highly publicized first legal case ([49];[59];[76]) and the convening of the First National Conference on Social, Ethical, and Legal Issues in Euthanasia [43]. The debate has attracted scholars from a variety of disciplines.

A. *The Concept of Euthanasia*

The late clinical ethicist Dr. Ai and his colleagues defined euthanasia as "a clinical management" in which a drug or other measure is taken to meet the wishes of those patients who are dying, suffer painful, incurable disease, and want to be quiet and comfortable ([2], p. 67). Others emphasize that the patients upon whom euthanasia is to be practiced should suffer from incurable disease, be unavoidably dying, and in extreme suffering ([66], p. 351). Under such conditions, since meaningful life no longer exists for these patients, to prolong their lives is only to prolong their dying.

"Euthanasia" as usually discussed, includes both active and passive forms. Passive euthanasia, *i.e.*, withholding or withdrawing treatment, is a matter of common practice. But active euthanasia remains a controversial topic among physicians, ethicists, jurists and the public ([1]; [66]; p. 354).

B. *Arguments For and Against Euthanasia*

Most proponents appeal the principles of beneficence, respect for persons, and justice in arguing for euthanasia. In addition, some view active euthanasia as evidence of civilization's increasing control over natural processes ([16], pp. 279-281).

Arguments against euthanasia include the following:

(1) Euthanasia is a bourgeois concept, thus incompatible with a socialist morality which sees even the dying, incurable patient as deserving of rescue efforts [80].

(2) Euthanasia is incompatible with traditional Chinese values. The principle of filial piety requires a child to extend his or her parents' lives as far as possible: "Longer life, more good fortune". To extend another's life is to do good and will be rewarded in this life or in the next. To shorten another's life is to do evil and will be punished [52].

(3) Euthanasia is incompatible with the deontological principles of medical workers: "Heal the wounded, rescue the dying" and "Cure the sickness to save the patient". Even if euthanasia is ethically justifiable, it should not be performed by medical workers, because it will damage their image and erode the trust of patients. In addition, surrendering to the presence of incurable disease will impede the development of medical sciences [73].

C. *Is Active Euthanasia Committing a Crime?*

The debate concerning the appropriate legal status of active euthanasia continues unabated, as reflect in the various opinions expressed about the first case of euthanasia recently brought to trial ([32]; [60]; [61]; [81]). The particulars of the case are as follows.

On June 23, 1986 a 59-years old woman patient *X*, who had suffered from cirrhosis for many years, was admitted in a hospital in the city of Hanzhong, Shaaixi province. Her diagnosis was cirrhosis with ascites, coma and liver-kidney syndrome, and serious bedsores. On June 28, during the hospital director's rounds, the patient's son inquired about his mother's prognosis. The director replied that her condition was "hopeless". The son then asked if his mother's death could be hastened in order to lessen her sufferings. Although the director refused that request, the son and *X's* youngest daughter prevailed upon another doctor *P* to help them by prescribing 100 milligrams of chlorpromazine to *X*. *P* ordered a student nurse to inject *X*. She did so, though at half the prescribed dosage. At midnight, *X* was still alive. The son and the youngest daughter asked the doctor on duty *L* to inject the drug again. According to the instructions *P* had left, *L* agreed and the nurse on duty injected the drug. *X* died at 5:00 a.m. on June 29. On July 3, *X's* two other daughters sued doctors *P* and *L* for murder. On July 4, four doctors at the hospital also jointly filed charges against *P* and *L*. The Public Security Bureau of the city arrested *P*, *L*, *X's* son, and her youngest daughter for murder. After investigation the prosecutor instituted proceedings against these four for murder ([54]; [60]; [61]).

In identifying the cause of *X's* death, the Department of Forensic Medicine of Shaanxi Province High Court concluded, in its report of February 22, 1990, that *X's* condition was so serious that there was little likelihood of improvement. Thus, since treatment could only briefly prolong her life, and the administration of chlorpromazine had only deepened her coma, the drug hastened her death but was not its cause. In trying the case, three different opinions were expressed to the Court. These opinions and their underlying rationales were as follows.

(1) The defendants' action constituted a crime of murder; however, its social harm is insignificant, so the defendants can be exempted from criminal sanction. The treatment of this case, it was argued, should be based on extant Chinese criminal law, not on academic discussions of euthanasia; and the act of the defendants meets the following conditions that constitute a willful murder: (a) The object of a willful murder is to violate another person's life. The defendants infringed upon *X's* right to life. (b) The objective aspect of willful murder is that the agent must act to deprive another person of life. The defendants' act is one which deprived *X* of life. (c) The agent of a willful murder is the person who is legally competent. The defendants, being competent, are legally responsible. (d) The subjective aspect of the act of depriving another person of life is that the act must be willful. The defendants shared the motive to deprive *X* of life, so their act was intentional.

(2) The act of the defendants does not constitute a crime, nor meet the conditions to be a crime, since it is not serious and did not cause any significant social harm. Thus they should be pronounced not guilty. The reasons are: (a) Euthanasia brings benefits to the society; moreover, there is no article in Chinese criminal law which stipulates euthanasia as a crime. (b) The person euthanatized by the defendants was a dying patient. The unavoidability of her death was confirmed by forensic doctors. (c) The motive of the defendants was to relieve the patient from suffering, but not to deprive her of life. (d) Chlorpromazine was injected into *X*, although 87.5 milligrams fell within the limit normally permissible (a toxic dose would be more than 100 milligrams for a first injection). The administration of chlorpromazine, therefore, was not the

direct cause of the patient's death, but only deepened the degree of her coma. Thus, the defendants have no legal responsibility for *X's* death.

(3) In substance, the case was not one of willful murder, though it appeared so. The defendants' act can be judged a crime of killing by euthanasia. The reasons are: (a) Their act illegally deprived *X* of a longer life, and hastened her death; thus their act caused social harm to a degree sufficient to render it a crime. (b) Although there is no article on the crime of killing by euthanasia in Chinese criminal law, it is most similar in nature to willful murder and so it can be said to violate the criminal law. Thus, the defendants should be given criminal sanction. (c) The objective aspect of willful murder includes: first, that there is an act to deprive another person of life; and second, that the act of depriving another person of life is illegal. Killing by euthanasia is similar to willful murder, although differences remain between them. In willful murder, the act directly causes the victim's death, whereas in this case, *X's* death is nearer to natural death, since the main cause of *X's* death is her disease, not the injected drug. In particular, the subjective motive of killing by euthanasia is merciful, for the purpose of relieving the patient's suffering, while in the case of willful murder the agent normally imposes suffering upon the victim.

The lawyer for the defendants argued that euthanasia is in the best interest of the patient if it is limited by three criteria: (a) there is a cause of death; (b) unbearable pain and/or sufferings are present; and (c) it is not against the patient's will [81]. He further contended that there is no willful killing in euthanasia, and that X's case contained no essential subjective element to make the defendants' act a crime. Because there is a prior cause of death, euthanasia should and can be distinguished from homicide, assisted killing, and assisted suicide. And in this case, the defendants did not practice euthanasia, because the dose of chlorpromazine they used was within the limit normally permissible. At 10:00 a.m., on May 17, 1991, the City Court of Hanzhong ruled on the case as follows:

> The defendant Wang Mingcheng, with his mother Xia Suwen being critically and incurably ill, repeatedly requested the physician-in-charge Pu Liansheng to inject a drug into his mother to let her die painlessly. His act obviously is a willful act to deprive his mother of life; however, his act is not significantly serious, its harm is not very grave, [and] so it does not constitute a crime. The defendant Pu Liansheng, under the repeated requests of Wang Mingcheng, prescribed a death-accelerating drug to Xia Suwen, [and] this order played a certain role in hastening Xia's death. However, the dose injected was within the limits of normal use, and did not directly cause Xia's death. His act is also a willful act to deprive a citizen of life, but this act is not significantly serious, its harm is not very grave, and so it does not constitute a crime. According to the 10th and 11th Articles of Chinese Criminal Law, the defendants Wang Mingcheng and Pu Liansheng are ruled not guilty [62].

The prosecutor, and the City Prosecutor of Hanzhong, expressed disagreement with the ruling, and are pursuing further litigation [55]. As this litigation proceeds, a second case of active euthanasia has been reported although no one has been accused [34].

D. *Persistent Vegetative State*

At the First National Conference on Social, Ethical, and Legal Issues in Euthanasia, held in 1988, an initiative was declared, consisting of two parts. Part One recommended initiating a discussion of the concept of death, and the criteria for diagnosing death, and urged drafting guidelines for brain death criteria. Part Two recommended that the wishes of the terminally ill should be implemented by the voluntary use of advance directives. Although no definition of brain death has yet been published, most Chinese medical scientists, ethicists and jurists appear ready to accept brain death criteria. Some questions remain, however, concerning the status of a patient in a persistent vegetative state (PVS), including the following: (1) What is the ontological status of PVS as a condition? (2) Should the patient in PVS be considered living or dead? (3) Is it ethically permissible to discontinue all medical interventions with such a patient? Such issues are pressing, and in need of further serious study.

E. *Euthanasia for Severely Defective Newborns*

Surveys on euthanasia for severely defective newborns were made in some cities ([35]; [58]). In the first survey, 360 respondents, including medical workers, factory workers, private businessmen, judicial workers, students and housewives, were asked whether ending the life of a severely defective newborn (SDN) is ethical. Twenty-five percent judged ending the life of such an infant to be unethical, because the infant has a right to life, and because ending its life is an act of homicide. Forty-three percent concluded that it would be permissible to end the life of a severely disabled newborn who would die shortly, regardless of treatment. Thirty-two percent expressed support for ending the lives of all SDNs, in order to relieve their suffering and that of their families, and to benefit society [35].

The second survey was administered to 2,619 persons. The findings were as follows: (a) In the case of a newborn with a defect that will seriously impede its growth and development, but which can be partly corrected at a later time, 23 percent favored ending its life, but 77 percent favored aggressive therapy. (b) In the case of a newborn with a defect that will seriously impede its growth and development, lead to disability and mental retardation, and which cannot be corrected, 60 percent favored ending its life, while 40 percent recommended aggressive therapy. (c) In the case of a newborn with a defect who will die in a relatively short time, 65 percent favored ending its life, while 35 percent recommended a wait-and-see attitude. (d) In response to the question of whether or not the mercy killing of a severely disabled newborn is ethical, 79.6 percent said yes, while 20.4 percent said no. (e) When queried about whether legislation on the euthanasia of severely disabled newborns is necessary, 69 percent said yes, while 31 percent said no [58].

Some geneticists and ethicists have recently argued that ending the life of a newborn with Down syndrome is ethically permissible ([22], pp. 32-36). These authors conclude that withdrawing treatment from a baby with Down syndrome, because it removes the psychological and financial burden from the family, is in the family's best interest, and because it improves the quality of children born, is also in society's interest. The authors argue that the infant lacks the status of a person in the social sense, and that allowing it to die is better than letting it live with suffering. At the same time, the authors express their support for protecting the status of children and adults with Down syndrome.

F. *Legislation for Euthanasia*

At the third and fourth sessions of the National People's Congress and National Political Consultative Committee, held in May 1990 and May 1991, 36 representatives proposed drafting a law on euthanasia that should include the following points:

(1) Access to euthanasia should be restricted to patients who are terminally ill and who suffer from intractable pain, as confirmed by doctors.

(2) Euthanasia should be practiced only on the basis of patient's own sincere and clearly expressed wishes.

(3) The means of euthanasia should be painless and comfortable.

(4) Medically sound and legally sanctioned procedures of euthanasia are needed.

Many Chinese scholars have called for legislation on euthanasia. One group of commentators suggested that the following restrictions should apply to legalized euthanasia:

(1) The patient must be terminally ill, with death imminent and pain intractable.

(2) The patient's or her/his family's sincere request must be procedurally confirmed.

(3) The request must be approved by an expert committee ([31], p. 108). Another analyst has proposed that a legal procedure for euthanasia should include an application by the patient, consent by the patient's family, ratification by ad hoc expert committee of the hospital, and oversight by the court [59].

Other ethicists, however, have called for further discussion of several concerns raised by active euthanasia, including identifying the actual cause of death, determining what constitute acceptable intentions for decisionmakers, and establishing safeguards to prevent abuse. Because of the need for further discussion, these ethicists prefer to examine the legitimacy of requests for euthanasia on a case-by-case basis.

V. ALLOCATION OF HEALTH RESOURCES AND HIGH TECHNOLOGY MEDICINE

A. *Principles of Allocation of Health Resources*

In China, where only 3 percent of national expenditures are allocated to health care, major questions are raised concerning the proper allocation of limited resources among the different departments of health care (*e.g.*, [20]). Some scholars argue for a distributive principle of the greatest good for the greatest number. They stress any of four principles of priority: (1) giving priority to ethnic minorities and those who are poor; (2) giving priority to rural areas; (3) giving priority to small-scale rather than costly technology; or (4) giving priority to preventive medicine.

B. *Principles of Applying High Technology Medicine*

R-C Peng, Editor-in-Chief of the Chinese journal *Medicine and Philosophy*, recommends limiting the development of high-technology medicine, and spending more resources on "appropriate technology", defined as that technology appropriate to the level of social, economic and cultural development of the local area [42]. Peng emphasizes that appropriate technology should be determined according to cost-benefit analysis.

The rapid introduction of high-technology medicine caused a series of problems. In the decade of the 1980s, 300 computerized tomography (CT) scanners were introduced into China. This equipment was concentrated in a few cities; for example, 150 of the scanners were located in Beijing, Tianjin, Shanghai, and Guangzhou. Beijing alone has 54 CT scanners and 8 magnetic resonance imagers (MRIs). By contrast, vast geographical areas have no medical technology [25].

One commentator has set forth several principles to govern the micro-allocation of high technology medical resources [28]. These principles involve several different social criteria for patient selection: (1) a retrospective principle which gives priority to those who have contributed to society, *e.g.*, veterans, scientists, model workers, etc; (2) a prospective principle which gives priority to those who are likely to make contributions to society in the future; (3) a principle of selection according to family role which gives priority to the family member most responsible for finances, to the single child over the child with siblings, and so forth; and (4) a principle of "scientific" value, which gives priority to patients with rare diseases over other patients.

C. *Organ Transplantation*

The practice of kidney transplantation began in the 1960s, but received publicity only in 1974. The 1970s saw a nation-wide upsurge in the transplantation of kidneys, livers, hearts, and lungs. Since that time, transplants have declined, because of both a shortage of donated organs and the lack of widespread acceptance of "brain death" criteria in China. According to statistics through 1988, there were 122 clinics nationwide that performed kidney transplants. In 1989, kidney transplants were performed on 1,049 patients in 42 clinics. At some hospitals, the one-year graft survival rate reached 80-90 percent, compared to 56 percent in 1985. The major problems facing Chinese doctors in organ transplantation are difficulties in organ procurement, lack of coordination in the nationwide procurement, preservation and distribution of organs, and problems in developing more cost-effective immuno-suppressant drugs. To date, only a few analysts have discussed the ethical aspects of organ transplantation and distribution (*e.g.*, [14]; [22]; [64].

D-L Wei discussed the use of tissue from aborted fetuses in organ transplantation. He claims that such tissue is often used in organ transplantation and recommends legislation to regulate its use. Fetal tissue used in organ transplantation, he argues, should be limited to tissue from abortuses which are no more than five months gestation, or which suffer severe anomalies, such as anencephaly. In addition, procurement of donated fetal tissue should require the informed consent of the parents, and should be reviewed by a hospital committee. Finally, he recommends prohibiting the commercialization of fetal tissue [64].

D. *Gene Therapy and Genome Mapping*

Some authors have discussed ethical issues in genetic diagnosis, gene therapy, genetic engineering and genome mapping ([13]; [15]; [24]; [84]). In genetic diagnosis, issues include whether the fetus with a genetic defect has the same right to life as a normal fetus, and whether parents have the right to decide to abort a fetus with a genetic defect. In gene therapy, issues include the propriety of risk assessment, the principles of justice invoked in patient selection, and the requirements of informed consent and patient confidentiality. Gene therapy, especially when it involves the germ line, may pose excessive risks, given inadequate safeguards at the present time. Genetic engineering may raise issues of justice in distribution of its benefits, as well as environmental concerns and the specter of hybrid forms. Genome mapping, in turn, raises issues of privacy and confidentiality.

VI. CONTROLLING THE SPREAD OF SEXUALLY TRANSMITTED DISEASE (STD)

Sexually transmitted disease, after two decades of dormancy, is now running rampant throughout China. By July 1989, there were 204,077 reported cases of STD. The actual number of cases are estimated to be about five times that number, *i.e.*, about one million cases. There have been 615 reported cases of

AIDS or HIV infection, most of them in Ruili county near the notorious Golden Triangle. Most patients with AIDS or HIV infection are intravenous drug users, but urban patients have been infected through sexual contact. To combat the spread of STD, a law on the prevention and control of infectious disease was promulgated on February 21, 1989 by the Sixth Session of the Standing Committee of National People's Congress [86]. In April 1991, a National Conference on Comprehensive Control of STD was held in Shenzhen [4].

A. *Is STD the Punishment of Misbehavior?*

Many Chinese scholars have attributed STD to sexual indulgence, or to the general deterioration of sexual morality ([6]; [51]; [67]). Whatever the case, the claim that STD is the product of sexual deviance blurs the fact that the etiological factors in STD are micro-organisms. Still, the effect of these microorganisms on the body is facilitated by human behaviors which may be deemed deviant, perverted, immoral, or even illegal. It is this conception of STD as punishment that may lead to discrimination against STD patients.

Many scholars have blamed the spread of STD on the increasing influence in China of Western attitudes about "sexual liberation" (*e.g.*, [65]). Other analysts, while not excluding outside influences, have emphasized more basic internal changes which have taken place in Chinese attitudes in the wake of the reform and openness policy ([8]; [63]; [82]).

One study surveyed sexual attitudes in two groups. One group was composed of prostitutes; the other was a control group of teachers. The proportion of positive answers among each group to certain questions is as follows:

	Subject (%)	Control (%)
Is the purpose of sex reproduction?	38.1	21.5
Is the purpose of sex pleasure?	37.8	54.3
Is the purpose of sex to express love?	55.8	51.1
Is the purpose of sex financial gain?	34.3	0
Is the traditional concept of chastity out-dated?	57.1	15.7
Is premarital sex immoral?	9.3	37.1
Is extramarital sex necessarily immoral?	60.5	24.3
Is premarital sex behavior common?	88.9	49.2
Have you had premarital sex?	80.3	16.2
Is extramarital sex common?	77.6	13.0

The authors conclude that although premarital sex has occurred in only a sector of the population, it is a major factor in the spread of STD. Moreover, because this part of the population is growing, the incidence of STD doubles or triples annually.

W-M Wen points out that basic trends in the transition from a closed, immobile, and egalitarian society to an open, mobile, efficiency-oriented society have led to sexual promiscuity, which in turn has promoted the spread of STD. These trends include a rapid growth in the transient population of the country, a widening of the gap between rich and poor, greater concern about individual rights and interests, greater sensitivity to the flaws in traditional marriages, and increase in sexual stimuli in society [65]. Other analysts have conclude the changed social environment makes the goal of eradicating STD inappropriate and unpracticable (*e.g.*, [67]).

B. *Discrimination against Patients with STD and Protection of their Rights*

Discrimination against high risk groups has the effect of driving them underground, thus depriving them of the chance to be educated about changing their behavior. Some doctors have treated STD patients as criminals, or found amusement in betraying their confidence [71]. In some cases, STD patients have been fired when information about their condition was leaked [53]. In such an atmosphere, STD patients are usually reluctant to visit doctors in public hospitals, and instead visit quacks, often exacerbating their disease. Alternatively, they may use false names and addresses when presenting at the hospital, making follow-up care difficult. Many writers, therefore, have called for legal protection of the privacy of STD patients ([69]; [71]; [78]). However, at the present time, the law on the prevention and control of infectious diseases contains no provision protecting the rights of infected patients.

The issues of mandatory testing and treatment have also been discussed ([71]; [78]). Commentators have concluded that it is unnecessary and impossible for all STD patients to be mandatorily treated, or for all citizens to be mandatorily tested.

C. *IV Drug Use*

In China, more than 80 percent of persons with HIV infection are intravenous (IV) users of illegal drugs, although sexual transmission may soon become the main channel of HIV infection, as has already happened in Thailand. The sharing of needles and syringes appears to be the principal mode of transmission among IV-drug users, who in turn may infect non-drug users through sexual intercourse. Since the time of Opium War in the 1840s, drug use has remained a social problem. Since 1949, the government has enforced stringent prohibitions of opium planting, processing, transporting, sale, and use. The sources of the drug were eradicated, and drug users were enrolled in the local centers for treatment. At the time, virtually all Chinese, including drug users themselves, accepted and supported this approach, which was very successful in eradicating drug use. Now, however, drug use, especially IV injection of heroin, is again on the rise, mainly in border regions. The government still enforces a policy of legal prohibition, and a law prohibiting the planting, sale, and use of drugs was promulgated in 1991. Thus far, there seems to be no objection among the population.

D. *Prostitution*

In 1949, the ancient practice of prostitution was legally prohibited. In the following decades, it was virtually eradicated. Now, however, prostitution is again the upswing, in the wake of new policies of economic and cultural openness. Presently the number of prostitutes is estimated to be several hundred thousand, although that figure may be low. Today, prostitutes are generally found in urban areas, with foreign businessmen and tourists as their primary clients [8]. Prostitutes from rural areas often escape from harsh and brutalizing backgrounds. Some prostitutes from rural areas, however, practice prostitution with the consent of their husbands to improve their financial situation ([65]; [70]).

A few analysts recommend harsher sanctions against prostitution, with the prostitutes themselves as the main target (*e.g.*, [70]). The majority of scholars, however, argue against such measures on several grounds. First, sanctions against prostitutes deal only with the supply side, while the large demand for prostitution comes from a transient male population. It appears unfair to punish only the women involved. Second, legal measures are unlikely to control such private aspects of social life. Finally, police forces are limited, and cannot be expected to arrest such large numbers of prostitutes and their clients. For all these reasons, these scholars conclude that it is impossible to eradicate prostitution in today's social climate (*e.g.*, [67]).

A new law prohibiting prostitution and its patronization was promulgated in 1991, but its practical feasibility remains unclear.

E. *Homosexuality*

No cases of AIDS/HIV through homosexual transmission have yet been reported, although other STDs have been. Homosexuality in China has a long history. In the official books authorized by the Imperial Courts, there are records of some emperors engaging in homosexual behavior. Since the hegemony of Confucianism, however, homosexuality has been deemed immoral. There is no article prohibiting homosexuality in extant Chinese criminal law, although some regulations adopted by the Ministry of Security prohibit homosexual acts. In practice, however, only public acts of homosexuality have led to arrest. Despite moral condemnation, sexual behavior has been seen as of little interest to the law unless it imposes high risks of lethal disease to others. In an era of AIDS, however, societal interests are obvious. Thus, on the one hand, public officials cannot ignore sexual practices that constitute a threat to all, and public educational programs are needed to encourage safer sex. On the other hand, the legal prohibition and/or criminalization of homosexual acts cannot be ethically justified because it violates individual privacy. Instead, efforts to reduce discrimination against homosexuals are more likely to encourage homosexual and bisexual males to disclose the names of their partners and enhance discussion of ways to reduce risks. At the same time, homosexuals and bisexuals should be willing to alter their life-styles in order to prevent the further spread of AIDs.

VII. REFORM OF THE HEALTH CARE SYSTEM

Since the founding of the People's Republic in 1949, the Chinese health care system has consisted of four main elements: labor insurance, public medical service, free preventive inoculation, and cooperative medical service. Under this system, workers in state-owned enterprises and institutions enjoy free health care.

Today, however, free health care for state employees is being modified. Increasing costs are prompting a trial reform of the 40-year-old system. Health care costs for state-owned enterprises and institutions, including schools, research institutes, and government organizations have increased more than six-fold in 12 years, from 3.18 billion *yuan* ($600 million) in 1981 to a total of 22 billion *yuan* ($4.2 billion) in 1990. By contrast, the gross national product only doubled during that time. Free health care was introduced in 1951, when only 4 million people were entitled to receive benefits, compared to 130 million persons today. In addition, the introduction of costly and sophisticated medical equipment and the increasing use of medical services, in part due to an aging population, have driven up health care costs dramatically. The central government, therefore, is currently implementing reforms in four cities: Dandong in Liaoning Province, Siping and Changchun in Jilin Province, and Chongqing in Sichuan Province. Under the trial reform, a ceiling will be set for medical subsidies to be paid by the state, and employers and employees will be required to bear a portion of their own medical costs, *e.g.*, 10-20 percent of the costs of diagnosis, treatment, hospitalization, and other services they receive [68].

A. *Equity in the Health Care System*

R-Z Qiu describes the inequitable consequences of implementing public health care in China [46]. Under the public system, demand always exceeds supply. Every hospital is swollen with patients who complain that the time for queuing is too long and the time for diagnosis is too short. However, high-ranking officials enjoy favored treatment. They do not need to queue for diagnosis in clinics, are admitted to first-class wards, and have access to expensive examinations and treatments, as well as to imported drugs. In addition, the public medical system provides few incentives for physicians, nurses, and other medical workers. The public system may also tend to erode the beneficiaries' sense of responsibility for their own health. For example, 61 percent of Chinese males smoke, and 7 percent of females, for a total of 385 million smokers in China.

B. *Market Mechanisms and Health Care*

Since the policy of reform and openness, a market economy has begun to be introduced into Chinese society. The question of whether market mechanisms should be introduced into health care has provoked significant discussion in recent years. Three perspectives have been developed by various analysts.

One group of analysts conclude that market mechanisms should not be introduced into health care (*e.g.*, [11]). Their arguments in support of this conclusion are as follows:

(1) The aim of health care is to save the patient's life, and to promote the health of the populace. Neither goal can be reduced to monetary terms.

(2) Health care primarily seeks the social benefits, *e.g.*, the cure and prevention of disease and the promotion of people's health, rather than economic benefits.

(3) In health care, doctors and patients have an equal status, whereas in the market, sellers and buyers are equal.

(4) The introduction of market mechanisms requires hospitals to implement a system of leasing contract responsibilities, whose purpose is to increase hospital income. The practical experience of implementing this system has resulted in accelerated medical costs, harm to patients' interests, and the deterioration of medical morality. Health care is public welfare; thus a commercial economy is incompatible with medical morality.

Another group of analysts urge that market mechanisms be introduced comprehensively into health care ([18]; [36]; [56]). Their arguments in support of this conclusion are as follows:

(1) At the preliminary stage of socialist development, the services provided by the health care system have the nature of a commodity with specific characteristics, and medical labor which aims at improving health can be represented in the form of money.

(2) The nature and degree of health services depend on the level of socioeconomic development. Health services have two purposes: a social purpose, *viz.*, to dispel disease and promote health; and an economic purpose, *viz.*, to enable health care workers to improve their lives. Thus, the economic benefits of health care work should not be seen as opposed to their social benefits.

(3) Market mechanisms cannot be introduced into health care piecemeal, because of the nature of market mechanisms. The nature of health care as a commodity makes the comprehensive introduction of market mechanisms unavoidable. Various negative tendencies, such as harm to patients' interests

or skyrocketing medical costs, are not the necessary effects of introducing market mechanisms into health care.

A third group of analysts conclude that market mechanisms can be partially introduced into health care ([3]; [41]; [42]; [79]). The arguments made by these scholars vary. S-Z Zhang holds that the medical services provided by state-owned hospitals are components of welfare, rather than commodities, while medical services provided by private clinics, foreign-owned, or collective hospitals are commodities. Because of the latter, market mechanisms in health care must play the regulative role [79]. G-Y Ai emphasizes that the state of the health care system reflects its socioeconomic level. Thus, market mechanisms can be introduced into a second-tier of services, but cannot be introduced into primary care [1]. F-G Peng recommends a two-tiered approach. Public health and basic health care should be treated similarly to compulsory education and national defense, whereas certain types of high technology medicine are appropriate objects of market mechanisms [42]. R-Z Qiu emphasizes that both proponents and opponents of a market in health care have neglected the unique characteristic of medical care, *i.e.*, the fiduciary relationship between physicians and patients.

VIII. HEALTH MORALITY

A Symposium on Health Morality was held in Shanghai, March 7-10, 1990. The following issues were discussed:

(1) *The concept of health morality*. Some participants suggested that "health morality" is a set of ethical guidelines which will protect individual physical and mental health, regulate the relationships among human beings, and direct relationships between human beings and the natural and social environment.

(2) *The relationships between health morality and medical morality*. On the one hand, health morality is part of medical morality. On the other hand, health morality exceeds medical morality, because the responsibility of health is borne by the whole society, not only by the department of health care [38].

(3) *The basic problems of health morality*. Some maintained that the basic problem of health morality is optimizing human living environment (*e.g.*, [40]). Others argued that the basic theme of health morality should be "health for all, all for health", because this motto embodies the aims of health morality and the responsibility of individuals to promote health (*e.g.*, [26]).

(4) *Health morality and health education*. Some participants emphasized that health education for all members of society is the indispensable basis of health morality, and that health morality will become an essential element of a civilized country ([9]; [83]).

IX. A PROFESSIONAL ETHICAL CODE

In December 1988, the Ministry of Health promulgated an ethical code for medical personnel. This is the first professional ethical code for medical personnel in the history of People's Republic of China. It consists of seven articles:

(1) Rescue the dying and heal the wounded; carry out socialist humanitarianism; always keep the patient's interest in mind; treat the disease and relieve the suffering of the patient by every possible means.

(2) Respect the patient's personality and rights; treat patients as equals without discrimination on the basis of nationality, sex, position, social status, or financial means.

(3) Serve the patient conscientiously; deport oneself in a dignified manner; speak to the patient in a refined manner; be amiable; care for patient with compassion, concern, and solicitude.

(4) Be honest in performing one's duties; conscientiously observe disciplines and laws; do not seek one's own selfish interests through medical practice.

(5) Respect the patient's confidentiality; do not say or do anything to harm the patient's condition.

(6) Learn from one another and respect each other; act properly in one's relationships with colleagues and co-workers.

(7) Be rigorous and dependable in work; be vigorous in spirit and eager to make progress; be committed to continued professional growth; continuously renew one's knowledge and increase one's technical competence [37].

Programs in Bioethics Unit for Philosophy of Science and Medicine
Institute of Philosophy
Department of Social Sciences
Chinese Academy of Social Sciences
Peking Union Medical College Beijing
Beijing CHINA

BIBLIOGRAPHY

Books and Articles

1. Ai, G-Y: 1989, "Discussions on Euthanasia", *Science* Number 162-65.
2. Ai, G-Y *et al.* (eds.): 1989, *Clinical Medicine and Science*, Chinese Science and Technology Press, Beijing.
3. Ai, G-Y: 1990, "Remarks on Whether Market Mechanisms Can Be Introduced into Health Care", *Medicine and Philosophy* 5, 33-34.
4. CAST (Chinese Association for Science and Technology) (ed.): 1991, *Comprehensive Control of STD*, Chinese Science & Technology Press, Beijing.
5. Chen, M-G *et al.* (eds.): 1989, *An Introduction to Medical Law*, Chinese Science Technology Press, Beijing.
6. Chen, Q-Q: 1991, "Incidence of STD and its Countermeasures in Guangzhou", in [4], 740-743.

7. Chen, Y-Y: 1991, "Situation Grim, Task Arduous", in [4], 558-561.
8. Chu, Z-R: 1991, "On Social Countermeasures of STD Spread", in [4], 508-513.
9. Cui, H: 1990, "Health for All: The Ethical Principle of Health Care", *Medicine and Philosophy* 4, 30-32.
10. Cui, X-B: 1991, "Problems of Health Resources Allocation in an Aging Society", *Medicine and Philosophy* 1, 16-18.
11. Du, Z-Z: 1989, "Can Market Mechanisms Be Introduced into Health Care?", *Medicine and Philosophy* 5, 1-4.
12. Fang, F-D: 1989, "Biological Father and Social Father", *Medicine and Philosophy* 3, 9-11.
13. Fang, F-D: 1990, "Ethical Issues in Genetic Engineering", *Medicine and Philosophy* 2, 20-22.
14. Fang, F-D: 1991, "Organ Transplantation and the Concept of Death", *Medicine and Philosophy* 4, 21-22.
15. Fang, F-D: 1991, "Ethical Issues in Human Genome Mapping Project", *Medicine and Philosophy* 10, 22-23.
16. Feng, Z-Y *et al.*: 1990, *Modern Medical Ethics*, Yellow River Press, Jinan.
17. Gai, Y-X: 1989, "A Report on Euthanasia", *Beijing Daily* (March 11), China.
18. Gui, B-W *et al.*: 1989, "Market Mechanisms Should Be Completely Introduced into Health Care", *Medicine and Philosophy* 10, 37-40.
19. Guo, J-Z: 1990, "Social Issues Facing Artificial Insemination and Their Control", *Medicine and Philosophy* 4, 17-19.
20. Guo, X-Q: 1990, "Ethical Principles in the Allocation and Use of Health Resources", *Medicine and Philosophy* 7, 32-34.
21. He, Z-X: 1989, "On Commercialization of Artificial Insemination", *Medicine and Philosophy* 6, 28-29.
22. He, Z-X *et al.*: 1989, *Disputed Cases*, Guangxi Science and Technology Press, Nanning.
23. Jia, S-Y: 1989, "On the Legal Status of AI Babies", in Bureau for Policy and Regulation, Health Ministry (ed.): *Proceedings of First National Symposium on Theory of Health Law*, Supplement of *Chinese Journal of Medical Administration*.
24. Jin, K-L: 1990, "Ethical Issues in Gene Therapy", *Medicine and Philosophy* 2, 23-25.
25. Ke, B-7 *et al.*: 1990, "Selectively and Progressively Developing and Using High Tech Medicine", *Medicine and Philosophy* 11, 10-13.
26. Lang, N-L: 1991, "Remarks on Basic Problem of Health Morality", *Medicine and Philosophy* 2, 55-56.
27. Lei, Z-X *et al.* (eds.): 1991, *Handbook for Family Planning*, Lanzhou, Gansu.
28. Li, C-J: 1990, "Medico-Ethical Issues in the Application of High Technology", *Medicine and Philosophy* 11, 32-34.

29. Li, B-F *et al.*: 1989, *Nursing Ethics*, Science Press, Beijing.
30. Life Weekly: 1986, "A Case on AID" (May 10) China.
31. Lin, F-X: 1989, "Euthanasia and Its Legislation", in [23].
32. Lin, Y-G: 1990, "Euthanasia and Its Legislation", *Law Science* 4, 26-31.
33. Liu, D-S *et al.*: 1990, "Some Thoughts on Health Morality", *Medicine* and Philosophy 4, 27-29.
34. Liu, H: 1991, "She left in Peace and Happiness", *Beijing Daily* (February 2), China.
35. Liu, S-J *et al.*: 1990, "A Survey on Euthanasia of Severely Defective Newborns in Beijing", *Medicine and Philosophy* 10, 29-31.
36. Ma, Q-X: 1989, "On Whether Market Mechanisms Can Be Introduced into Health Care", *Medicine and Philosophy* 10, 41-43.
37. Ministry of Health (MH): 1989, "Ethical Code for Medical Personnel", *Chinese Hospital Management* 3, 5.
38. Minutes of Symposium on Health Morality, 1990, *Medicine and Philosophy* 6, 25-26.
39. Ni, S-Y: 1991, "A Survey of 140 Cases of Homosexuality in Shanghai", in [4], 602-605.
40. Peng, F-G *et al.*: 1990, "Optimizing Human Living Environment: A Basic Problem of Health Morality", *Medicine and Philosophy* 4, 33-35.
41. Peng, R-C: 1990, "Address at the Closing Session of the Fifth National Conference on Philosophy of Medicine", *Medicine and Philosophy* 11, 4-6.
42. Peng, R-C: 1991, "Problems Concerning Development Strategy of Health Care in China (An Outline)", *Studies in Dialectics of Nature* 1, 54-57.
43. Qiu, R-Z: 1988, "A Report on the First National Conference on Social, Ethical and Legal Issues in Euthanasia", *Studies in Dialectics of Nature* 6, 61-63.
44. Qiu, R-Z: 1988, "Social, Ethical, and Legal Issues in Reproductive Technology"; and "The Concluding Report at the First National Conference on Social, Ethical, and Legal Issues in Reproductive Technology" (November 3-5) (unpublished).
45. Qiu, R-Z: 1989, "Play God's Role", *Science and Technology Daily* (January 21), China.
46. Qiu, R-Z: 1989, "Equity and Public Health Care in China", *Journal of Medicine and Philosophy* 14, 283-287.
47. Qiu, R-Z: 1989, "AID Confronts the Law in China", *Hastings Center Report* 6, 3.
48. Qiu, R-Z *et al.*: 1989, "Can Late Abortion Be Ethically Justified?", *Journal of Medicine and Philosophy* 14, 343-350.
49. Qiu, R-Z: 1991, "Morality in Flux: Medical Ethics in China", *Kennedy Institute of Ethics Journal* 1, 16-27.
50. Qiu, R-Z: 1991, "Commodity, Market and Health Care", *Medicine and Philosophy* 6, 38-40.
51. Shao, C-G *et al.*: 1991, "Prevalence of STD in China", in [4], 1-7.

52. Shen, M-X: 1989, "Euthanasia and Cultural Background", *Chinese Medical Ethics* 1, 43-47.
53. Shen, R-N: 1991, "Keep Patients' Secret in STD Clinic and Epidemiological Study", in [4], 147-149.
54. Song, W-L: 1987, "Euthanasia and Murder", *Democracy and Law* 8, 37-38.
55. Wan, J-A: 1991, "The First Legal Case of Euthanasia in China", *Orient Times* (October 22), China.
56. Wan, Y-Z: 1990, "Necessity and Peculiarity of Introducing Market Mechanisms into Health Care", *Medicine and Philosophy* 3, 31-35.
57. Wan, Z-X: 1990, "Interview on AI", *Democracy and Law* 11, 12-13.
58. Wang, B. *et al.*: 1990, "A Survey on Euthanasia of Severely Defective Newborns and Its Analysis", *Medicine and Philosophy* 10, 31-37.
59. Wang, F-Z: 1991, "Some Thoughts on Legislative Procedure for Euthanasia", *Medicine and Philosophy* 1, 45.
60. Wang, H-L: 1990, "Judgments on the First Case of Euthanasia in China and Their Justification", *Chinese Medicine Ethics* 5, 53-55; 6, 57-61.
61. Wang, H-L: 1990, "The First Legal Case of Euthanasia in China", *People Judiciary* 9, 38-40.
62. Wang, H-L: 1991, "The Ruling of the First Euthanasia Case", *Chinese Medical Bulletin* 3, 27.
63. Wang, J-R: 1990, "Thinking Caused AI", *Democracy and Law* 11, 14.
64. Wei, D-L: 1990, "Juristic Analysis of Issues in Transplanted Organ Donation", *Medicine and Philosophy* 1, 18-20.
65. Wen, W-M: 1991, "Sociological Analysis of STD Prevalence in Yunnan Province", in [4], 498-502.
66. Wu, X-Z *et al.* (eds.): 1990, *Modern Clinical Medical Ethics*, Tianjin People Press, Tianjin.
67. Xia, G-M: 1991, "Some Thoughts on STD", in [4], 53-57.
68. Xiao, L: 1991, "High Costs Push State to End Free Health Care", *China Daily* (November 22), China.
69. Xiao, L-B *et al.*: 1991, "Explore the Counter Measures of STD in Shanghai", in [4], 667-670.
70. Xu, W-H: 1991, "Some Issues in Control of STD and Amplification of Necessary Regulations", in [4], 71-73.
71. Yin, T-Y: 1991, "Strengthen the Prevention and Treatment of STD: Keep Medical Confidentiality", in [4], 144-147.
72. Ying, X-D: 1989, "On Ethical Issues in Artificial Insemination", *Chinese Medicine Ethics* 2, 8-10.
73. Ying, Y *et al.*: 1989, "A Survey on Euthanasia", *Chinese Medical Ethics* 7, 48-50.
74. Yu, Z: 1989, "Advances in Philosophy of Medicine", in Xin, B-S (editor in-chief), *Chinese Philosophical Almanac* (1989), Chinese Encyclopedia Press, Beijing, 74-75.
75. Yuan, W-W *et al.*: 1990, "The Change of Goal in the Reform of Health

Care System and Its Sources", *Medicine and Philosophy* 11, 7-9.
76. Zentrum fur medizinische Ethik, Ruhr-Universitat Bochum: 1989, *Case Studies for Bioethical Diagnosis*, Marz.
77. Zentrum fur medizinische Ethik, Ruhr-Universitat Bochum: 1989, *New Case Studies for Bioethical Diagnosis*, Marz .
78. Zhang, J: 1991, "AIDS: Social and Ethical Issues", *Journal of Dialectics of Nature* 2, 26-35.
79. Zhang, S-Z: 1990, "View Health Care and Market Mechanism from the Relationship Between Medicine and Economic Basis", *Medicine and Philosophy* 2, 42-44.
80. Zhang, Z-N: 1989, "Euthanasia: A Materialist Perspective", *Chinese Medical Ethics* 2, 11-13.
81. Zhang, Z-N: 1990, "My Defense for Defendants in the Euthanasia Case", *Reference of Law Science* 4, 38-40.
82. Zhao, Q *et al.*: 1991, "Sociological Perspective of STD Spread", in [4], 491-498.
83. Zhu, X-F: 1990, "Studies in Health Morality: Present Situation and Prospect", *Medicine and Philosophy* 6, 26-28.
84. Zhuang, M-H: 1991, "Biological and Ethical Evaluation of Human Genome Mapping Project", *Medicine and Philosophy* 10, 20-21.

Cases and Statutes

85. *C v. Her Husband*, Shanghai (1987-1989).
86. Law of the People's Republic of China on the Prevention and Control of Infectious Diseases, Promulgated by the Standing Committee of the Seventh National People's Congress, February 21, 1989.
87. Motions of Euthanasia at the Third and Fourth Session of National People's Congress and National Consultative Committee of People's Republic of China, 1990-1991.
88. Regulation on Prohibiting the Reproduction by the Mentally Retarded, Promulgated by the Standing Committee of Gansu Provincial People's Congress, November 23, 1988, implemented on January 1, 1990 .
89. *Wang, Wang v. Pu, Wang*, Hanzhong City, Shaanxi Province (1986-1991).

KAZUMASA HOSHINO

BIOETHICS IN JAPAN: 1989-1991

I. INTRODUCTION

During 1989-1991 in Japan, the Japanese government and professional associations took action on two issues: organ transplants from donors declared dead according to brain-death criteria, and the government's decision to shorten the gestational age limit for legally permissible induced abortions from less than 24 weeks gestational age to less than 22 weeks. Another topic during the past two years, the legitimacy of using liver tissue from live donors, provoked significant discussion but no official action by the government or statements from professional groups. Overall, increasing numbers of people have become aware of the importance of such bioethical concerns as informed consent to medical treatment and experimentation, appropriate criteria for defining death, and standards of decision-making for incompetent patients.

II. INFORMED CONSENT TO MEDICAL DECISION-MAKING

The ethical and legal issues surrounding informed consent came to the fore of Japanese discussion only after the socio-cultural changes wrought by World War II. As Rikuo Ninomiya clarifies in his useful discussion, the doctor's role in traditional Japanese culture was not regulated by ethical codes or legal norms, but by an ethos of virtuous duty toward his patients. Thus, if patients were poor, the physician was not allowed to require a fee for his services; moreover, in this highly interpersonal relationship, medical malpractice was "unthinkable" ([6], p. 926).

The post-War years formed the watershed in Japanese medical ethics because several general trends significantly altered the traditional ethics of doctor-patient relationship. According to Ninomiya, four major factors were involved: (1) a lessening of trust, in large measure the result of "the growth of a health insurance system that has depersonalized the relationship"; (2) a corresponding sensitization of patients to their "rights" in the medical relationship; (3) the indeterminacy of traditional norms in resolving problems ushered in by new medical technologies and procedures; and (4) in conjunction with larger rights claims, the awareness of patients that malpractice suits are available as a recourse to medical malevents ([6], p. 926). Those larger rights claims included the "rights to health" of all persons guaranteed in Japan's

B. Andrew Lustig (Sr. Ed.), Bioethics Yearbook: Volume 2, 379-387.

Constitution, and coverage of all Japanese nationals by the Health and Social Security Insurance established in 1961.

Under the Japanese Criminal Code, physicians are expected to act "so as not to cause possible disorders or adverse effects in their patients" ([6], p. 927). In practice, this requires them to act according to the standard of a reasonable and prudent physician. Moreover, in principle, physicians are required to obtain informed consent from their patients. In keeping with general standards of disclosure, doctors should adequately inform competent patients of the likely benefits and risks associated with treatment options, including the option of non-treatment. However, in Japan, greater discretion is accorded physicians in explaining various treatment options; for example, "the decision to use a curative method that is under medical dispute is left to the discretion of the physician, as long as the method is approved as one of the approaches" ([6], p. 927). In practice, therefore, Japanese physicians appear to retain far more discretion in determining the content of information disclosed than their Western counterparts.

In addition, as Rihito Kimura notes, the traditional ethos, although altered by recent trends, still lingers, so that "even today patients are not expected to raise questions concerning their own bodies or health when they are ill" ([5], p. 457). Despite the general post-War trends, therefore, the Japanese social ethos still permits medical paternalism to go largely unquestioned.

As a result of these characteristic ways of thinking and decision-making of the native Japanese, non-Japanese speaking people living outside of Japan not infrequently have difficulty understanding Japanese national sentiment and the traditional Japanese doctor-patient relationship. In general, a native Japanese is friendly and behaves favorably toward persons inside his or her kinship and social groups, including family members, close relatives, intimate friends, colleagues, "superior" persons in schools, officials in companies or other organizations, and acquaintances with whom he or she has meaningful relations. The same native Japanese will often discriminate against strangers and people outside his/her own groups, behaving coldly or even harshly until those strangers are introduced properly, usually by someone within his or her own group. Most native Japanese are extremely aware of the opinions that others within their social groups hold about them. In Japanese society, insisting upon one's own opinion is usually disdained. Ordinarily, native Japanese do not have the courage of their own opinions and seldom try to speak up. Instead, they tend to value harmonious relations and adjust themselves to the general opinions formed within the group of people to which they belong or to whom they are closely related. Consequently, they are not accustomed to autonomous decision-making in daily life.

This tendency among the native Japanese makes it difficult for Japanese patients to make autonomous decisions about their own medical care. Usually, Japanese individuals tend to view a request for medical care as asking a favor of the physician. As a result, they try to behave as good and obedient patients, without the same concern for their "rights" characteristic of approaches in many other countries.

III. ABORTION

According to J.M. Finnis, Japan's legislation on abortion comports with a model wherein "all abortions are legally permitted if performed by medically qualified persons," with the two objectives of respecting the women's rights over her own body and ensuring competent medical practice ([1], p. 27). Under the 1948 Eugenic Protection Law, abortions had required the approval of district Eugenic Committees based on medical or psycho-medical grounds. However, a 1952 amendment to that law effectively liberalized abortion in Japan by "removing control over individual physicians' decisions". In Finnis's judgment, that liberalization was aimed primarily at controlling population growth, and secondarily, at reducing women's recourse to abortion at the hands of unqualified practitioners ([1], p. 29).

Since the time of that amendment, unrestricted abortions have been allowed for the first two trimesters, until the point when the fetus becomes viable, and even after that time when maternal health is threatened. However, the Ministry of Health and Welfare of Japan announced on March 20, 1990 a ruling by the Vice Minister that the legally permissible gestational period for therapeutic abortion in Japan be shortened from 24 weeks to a period of less than 22 weeks. That new requirement became effective on January 1, 1991.

IV. THE DEFINITION OF DEATH AND ORGAN TRANSPLANTATION

The Japanese Diet has passed a new bill to establish a provisional Commission for the Study of Brain Death and Organ Transplantation. The bill, which was submitted to the Diet on December 20, 1988, was implemented on February 1, 1990 as Law Number 70 by the official advisory committee for the Prime Minister and will become void on February 1, 1992 unless an extension is given. The Commission is chaired by Dr. Michio Nagai, formerly Minister of Education, Science, and Culture, who is assisted by Acting Chairperson Dr. Wataru Mori, formerly president of the University of Tokyo. There are 13 other members.

In light of the currently unsettled situation in Japan with regard to organ transplantation from donors declared brain-dead, the Commission was charged to analyze information and opinions about brain death and transplantation and to make policy recommendations on these matters to the Prime Minister. Upon receipt of the final report from the Commission, the Prime Minister must honor its conclusions.

The first Commission meeting was convened on March 28, 1990 and Prime Minister Kaifu made the opening address. The second meeting was held on

April 26, 1990 and the following topics were selected for consideration:

- the definition of brain death
- the medical and biological definitions of brain death
- ethical and cultural aspects of brain death
- medical difficulties posed by brain death
- legal problems concerning brain death
- the current status of organ transplantation
- the social and ethical problems posed by organ transplantation.

At the third meeting of the Commission held on June 21, 1990, the invited speaker was Dr. K. Takeuchi, who lectured on the topic of brain death. Dr. Takeuchi, as the chief of the study group formed by the Ministry of Health and Welfare in December, 1985, had been responsible for establishing the medical definition of brain death. At this meeting, four councilors were added to the committee, and a fifth councilor was added at the fourth meeting on July 25, 1990. Between the meetings, several groups of Commission members visited various medical institutions to gather expert opinions about brain death. One group visited the Emergency Center at Nippon Medical School in Tokyo on May 16, 1990. A second group visited the Emergency Center at Kitasato University Hospital. A third group visited the Emergency Center and the Intensive Care Units at the Kobe Municipal Central Hospital on August 1, 1990. From July 12-21, 1990, a fourth group visited Seattle and Boston in the United States.

In the fall of 1990, four more Commission meetings were held on September 20, October 31, and November 14 and 28. Invited speakers gave lectures on various subjects. From September 30-October 11, 1990, another group of Commission members visited London and Cambridge in Great Britain, Copenhagen in Denmark, and Cologne and Bonn in Germany. From October 16-23, 1990, a third group visited Brisbane and Sydney in Australia and Bangkok in Thailand. On November 21, 1990, the Commission held its first public hearing in Nagoya. Six members were present and heard the opinions of eight preselected local citizens and two other participants. On December 6-7, 1990, the ninth meeting of the Commission was held. In early 1991, its tenth through thirteenth meetings were held, and its second public hearing was conducted on February 14, 1991 in Fukuoka, Kyushu.

Thus far, the Commission has discussed the following topics:

(1) *What is brain death?* Is it whole brain death, brain stem death, or higher brain death? Is it functional death or brain tissue death?

(2) *Is it possible to determine brain death reliably?* Have there been any cases of recovery from brain-dead conditions after brain death was diagnosed by physicians? Have there been any cases of patients who resumed spontaneous breathing after brain death was diagnosed by physicians? What should be presupposed as exceptions when the diagnosis of brain death is made? Is it necessary to perform such supplemental procedures as a brain circulation test,

or others, when the diagnosis of brain death is made? How long should a patient be observed to determine brain death? Is it necessary to standardize the procedures for diagnosis of brain death in Japan?

(3) *Is brain death considered the death of an individual (biomedical death)?* What is the death of an individual? Is brain death equivalent to death diagnosed by procedures based on the cessation of spontaneous respiration, circulation, and pupillary reflex? Given the present sentiments of Japanese society, is it possible to say that brain death is the death of an individual?

(4) *Is it legally acceptable to determine that brain death is the death of a person?* Is it necessary to obtain social consensus on this matter? Is it necessary to implement brain death as the standard for measuring the death of a person? What deference should be shown to the prior wishes of the patient or the present wishes of family members vis-á-vis such a judgment?

(5) *Should there be conceptual or functional distinctions drawn between the desirability of establishing criteria for brain death and society's need to increase the supply of organs and tissues available for transplantation?*

(6) *What are the bioethical aspects of brain death?* What are the problems left unaddressed in the historical case of heart transplantation performed by Dr Wada in 1968? What medical education should be provided to ensure the competence of medical professionals in determining death according to established standards? How can a follow-up system be developed to monitor professional practice?

In the spring of 1991, the fourteenth to seventeenth Commission meetings were held on March 13 and 29 and April 16 and 24, and the third public hearing was conducted on April 12 in Sapporo, Hokkaido. On May 10, at the eighteenth Commission meeting, there was heated discussion about the content of a draft interim report. After the meeting, Chairperson Nagai announced at a press conference the outline of the draft report. At the nineteenth meeting on May 30, 1991, the draft interim report was again discussed. On June 14, the twentieth Commission meeting was held and the interim report was released to the general public.

The interim report consists of a main text and a supplement. The latter contains a minority report prepared by two Commission members and two councilors. The main text contains the following opinions of the majority of Commission members:

(1) It is possible medically to determine brain death.
(2) Brain death may medically be considered to be the death of a person.
(3) The prior competently expressed wishes of a patient determined to be brain-dead must be respected.
(4) As long as there are patients who cannot survive without organ transplants and persons who are willing to donate their organs for transplantation, organ transplants should be performed to save those patients in need of organs.
(5) A consensus on the appropriateness of brain death criteria and the

legitimacy of organ transplantation appears to have emerged in Japanese society at large.

(6) A serious reconsideration of the Wada case, the first heart transplant performed in Japan more than 20 years ago, underscores the need for ethical standards to be maintained by medical professionals.

On the other hand, the minority group expressed the following opinions:

(1) There must be clear-cut evidence of the wishes of a donor of organs or tissues.
(2) There must be fair and conscientious diagnosis of brain death using the most reliable procedures currently available to determine brain death.
(3) There must be confirmation of the informed consent of both organ donor and recipient.
(4) Hospitals responsible for either donation or transplantation must have established an independent review committee and a system by which patients' self-determination will be respected, not only in the cases of organ transplantation but in all medical cases. All patients treated in the hospitals should in principle be guaranteed free access to and copying of their own medical records.

During the following half year, further Commission meetings were held. In addition, the fourth, fifth, and sixth public hearings were held on August 9 in Tokyo, on September 20 in Osaka, and on October 17 in Hiroshima, respectively. On December 10, 1991, the Commission discussed the contents of the final report, due for submission to the Prime Minister in January, 1992, the draft was publicly released on the same day. The major points of the draft are as follows:

(1) The Commission could not reach a unanimous set of conclusions on the issues it considered.
(2) A large majority of the committee concluded that brain death may be considered the death of a person.
(3) Procedures for determining brain death should incorporate those recommended in the "Guideline and Standard for the Determination of Brain Death" (the so-called Takeuchi Standard)[7].
(4) The prior wishes of the potential donor must be respected, taking precedence over those expressed by any member of the family.
(5) All potential recipients must be given a fair chance to receive an organ transplant.
(6) The prohibition of buying and selling organs must be legally clarified.
(7) A uniform nationwide network for organ transplantation must be organized.
(8) An emergency organ transportation system must be established.
(9) Medical institutes for organ transplantation should be licensed.
(10) An investigatory committee similar to an institutional ethics committee must be established as part of the organ transplantation network.
(11) Comprehensive legislation on organ transplantation must be passed and

enforced.

When the Commission submits its final report to the Prime Minister in January, 1992, the government is expected to implement the Commission's recommendations in their entirety.

V. ESTABLISHING A BONE MARROW BANK

The Tokai Bone Marrow Bank, a private voluntary organization, was established in Nagoya City (in the Tokai Part of Japan) in October, 1989. The Ministry of Health and Welfare deemed it necessary to establish an official bone marrow bank and formed a study group in June 1990. This group submitted a report, "A Study of the Bone Marrow Bank System", in November 1990. In January 1991, the Ministry of Health and Welfare of Japan established a "Special Committee on Bone Marrow Transplantation". This committee released an interim report, and recommended in June 1991 that a Foundation for the Promotion of Bone Marrow Transplantation be established. In November 1991, this Foundation was officially established for the following purposes: (1) as an educational center to promote public understanding of bone marrow transplantation; (2) as an institution which could encourage greater numbers of bone marrow donors; (3) as a communication network on bone marrow transplantation for interested individuals; and (4) as a vehicle for establishing an official bone marrow bank.

VI. HEALTH CARE DELIVERY IN JAPAN

Japan's system of health care delivery is superficially similar to that in the United States, but as several commentators have noted, that resemblance is misleading in key respects ([2];[3];[4];[6]). Unlike the situation in the United States, the Japanese government was involved in providing health insurance to manual workers in major companies as early as 1927. Because of the government's involvement in setting rates, fees were kept low, "favoring ambulatory treatment of acute illness and injury" ([4], pp. 105-106). In subsequent expansions of coverage to an ever greater percentage of the Japanese population, that low fee schedule has been retained as the basis for the general pattern of Japanese health care delivery.

According to Naoki Ikegami, "At a cost that is little more than half of what the United States spends for personal health services, Japan provides its entire population with equitable health insurance that guarantees ready access to virtually all medical facilities" ([4], p. 88). In order to achieve this equity, cross-subsidization among different economic groups and a significant level of governmental oversight and regulation are central constituents of Japanese medical care. Although there are multiple payers that finance health care coverage in Japan, every citizen is obliged to belong to one of the "social insurance" plans, which underwrite the vast majority of services. These plans provide a standard set of comprehensive medical benefits, including medications, long-term care, dental care, and some preventive care. Moreover, insurers and providers are not free to negotiate separate schedules on their own. Physicians therefore receive the same fee for any service provided to a patient, even for a patient on public assistance. The government plays a key role in subsidizing plans that insure individuals who lack coverage through their employers. As a result, according to Ikegami, "Japan has one of the most equitable single-tiered health care systems in the world" ([4], p. 92).

In conjunction with the uniform schedule of low uniform fees, there are also three structural factors that help to contain costs in the Japanese system. First, as noted above, the fee schedule favors physicians who focus on primary care. Second, clinic-based doctors have no admitting privileges to hospitals for their patients. Hence, even highly trained specialists tend to emphasize primary care when they enter clinical practice. Third, a smaller percentage of health care costs in Japan are spent on administrative overhead. As Ikegami observes, "Mandatory coverage and the adoption of a single-fee schedule precludes the need for each individual payer and provider to enter into protracted negotiations over payment and services; this also greatly simplifies claims processing" ([4], p. 99).

At the same time, there is clearly a negative aspect to the uniform fee schedule, *viz.*, the same payment to any and all providers incorporates no incentive to maintain or to improve quality of care. Moreover, there are no formal quality assurance programs, and few specialty boards have assumed oversight responsibilities to insure uniform quality of care. In the future, therefore, greater attention to issues of quality assurance will be required.

Department of Anatomy
Faculty of Medicine
Kyoto University
Kyoto, JAPAN

BIBLIOGRAPHY

1. Finnis, J.M.: 1978, "Abortion: Legal Aspects", in Warren T. Reich (ed.), *Encyclopedia of Bioethics*, Volume 1, New York, The Free Press, 26-32.
2. Iglehart, John K.: 1988, "Japan's Medical Care System", *New England Journal of Medicine* 319:12, 807-812.
3. Iglehart, John K.: 1988, "Japan's Medical Care System - Part Two", *New England Journal of Medicine* 319:17, 1166-1172.
4. Ikegami, Naoki: 1991, "Japanese Health Care: Low Cost Through Regulated Fees", *Health Affairs* 10:3, 81-109.
5. Kimura, Rihito: 1989, "Ethics Committees for 'High Tech' Innovations in Japan", *The Journal of Medicine and Philosophy*, 14, 4 (August), 457-464.
6. Ninomiya, Rikuo: 1978, "Contemporary Japan: Medical Ethics and Legal

Medicine", in Warren T. Reich (ed.), *Encyclopedia of Bioethics*, Volume 3, New York, The Free Press, 926-930.

7. Takeuchi, K.: 1985, *Research Report on Brain Death*, Ministry of Health and Welfare, *Nippon Ijishinpo* (in Japanese) 3187: 104-106; *Nippon Ijishinpo* (in Japanese) 3188: 112-114.

BIOETHICS IN AUSTRALIA AND NEW ZEALAND: 1989-1991

MAX CHARLESWORTH

BIOETHICS IN AUSTRALIA

I. THE NATIONAL BIOETHICS CONSULTATIVE COMMITTEE (NBCC)

In 1988 the Australian Federal and States ministers of health and social welfare established a national body to advise them on developments in bioethics. The National Bioethics Consultative Committee (NBCC) was a multidisciplinary committee with representatives from the biosciences, philosophy, law, theology, and the social sciences. During its brief three years of existence (1988-90) it published a number of significant reports which prompted widespread discussion in the Australian community.

The first report of the NBCC was entitled *Record Keeping and Access to Information: Birth Certificates and Birth Records of Offspring Born as A Result of Gamete Donation* [14]. In this report the NBCC advanced a thesis which was to be a constant refrain in its subsequent reports, namely that there are various possible modes of family formation and that as long as the interests of children and parents and donors are protected, the State should not directly intervene. This report made a number of recommendations about the right of children, born through donor insemination (AID) or in vitro fertilization (IVF) using donor gametes, to have access to information about their biological origins in much the same way as such access is now available in most Australian States to adopted children. The NBCC, however, resisted the idea that the position of children born through gamete donation was wholly identical to that of adopted children, where the interests of the child are paramount. The Committee argued that children born through gamete donation are brought into existence by couples who, because of their infertile condition, choose to undergo AID and IVF using donor gametes to form a family. The situation is not, as in adoption, that of finding a suitable family for an already existing child.

From this perspective, then, the parents have a right, like fertile couples who have children in the traditional way, to have their privacy and their decision to form a family respected. They may choose to inform their children of their biological origins but it is not the business of the State to interfere.

B. Andrew Lustig (Sr. Ed.), Bioethics Yearbook: Volume 2, 389-401.

The second NBCC report, *Surrogacy*, discussed the ethical and legal aspects of surrogate motherhood [15]. After reviewing the debate over surrogate motherhood, the NBCC recommended that non-commercial surrogacy arrangements should be allowed, but that regulatory mechanisms should be put in place to ensure that adequate counselling was provided to all the parties concerned and that appropriate procedures for relinquishment of the child were observed. The report notes that IVF-assisted surrogacy had introduced a new dimension into the discussion, because it enabled a couple to contribute their own gametes for the formation in vitro of an embryo which was, genetically speaking, "theirs", to be transferred subsequently into another woman who had consented to bear the child for them. Again, the NBCC report places great emphasis on the principle of autonomy; *viz.*, the right of individuals in a liberal democratic society to form families in the manner they choose. With regard to women serving as surrogate mothers, the NBCC report invokes the same right of women to control their own bodies that feminists have used with regard to contraception and abortion. If a woman may choose to have an abortion and not to bring a child into existence, the report concludes that she should be able to choose to bear a child for another woman and so bring a child into existence.

Before the NBCC report on surrogacy was completed, a celebrated case of surrogate motherhood had already been widely discussed in the Australian community. A woman in the State of Victoria, who was unable to bear a child because of a prior hysterectomy, was nevertheless able to contribute her ovum, which was then fertilized in vitro by the sperm of a known donor (the woman's husband was infertile). The resultant embryo was transferred to the womb of her sister, who had consented to act as a surrogate mother. The sister had three children of her own, but because her husband had undergone a vasectomy, so long as she remained married to him she was technically in an "infertile situation". (This was an important circumstance since under the Victorian *Infertility (Medical Procedures) Act of 1984*, an IVF embryo cannot be transferred into the womb of a fertile woman.) The embryo transfer was successful and the child was duly born and given into the care of its genetic mother and her husband. Under Victorian law, a child is legally deemed to be the child of the woman who gestates it, so the child in this case had to be formally adopted by the genetic mother.

This case was important in shaping people's perceptions about surrogate motherhood. It was evident that all the parties involved were clear about what they were doing, that all had altruistic motives, that no one was being exploited, and that since the child was genetically the eventual mother's offspring, it was not simply being "sold" or exchanged. Thus, most of the fears that some have had about surrogate motherhood were simply not realized in this case [6].

The NBCC report on surrogacy was criticized by church groups, radical feminist organizations, social welfare bureaucrats, and adoption groups; again, its main recommendations were not accepted by the Australian ministers of health and social welfare, although there is evidence that a majority of people in the Australian community are not opposed to surrogacy arrangements in certain circumstances. Despite the criticism, the NBCC report is one of the best-argued discussions of surrogacy available to inform the general public on the issues raised by the practice.

A third NBCC report, *Access to Reproductive Technology*, discussed criteria for allowing access to publicly funded reproductive technology services [17]. At present, very different criteria are used in Australia. In Victoria, for example, only infertile couples who are legally married have access to IVF programs, while in some other states couples in stable *de facto* unions are accepted. (Homosexual couples are not accepted for donor insemination or for IVF with donor gametes in any states.) Again, particular clinics also have their own criteria. For example, some units discourage couples from non-English-speaking backgrounds, because it is unclear that such couples will be able to comprehend fully the often complex information and instructions given to women in IVF programs. The NBCC report notes that some of these criteria contravene the provisions of the anti-discrimination legislation adopted in several Australian States ([17], Appendix 2, "Access to Reproductive Technology Programs"). In general, the report recommended that infertility be recognized as a serious disability and that the alleviation of its effects by the various forms of reproductive technology be supported by public health resources in the same way as other medical treatments for infertility. In addition, it recommended that each reproductive technology service: (a) make explicit and public its criteria for admission to, or exclusion from, reproductive technology programs and ensure that they are made available to prospective patients; and (b) ensure that persons who are excluded by reference to the criteria be advised of the reasons for their exclusion.

The final work of the NBCC was to produce a set of materials on the subject of embryo experimentation. The committee commissioned a background paper and a bibliography with commentary on the embryo experimentation debate [3]. It also commissioned a poll (based on a similar poll in the United Kingdom) on the attitudes of the Australian community to embryo experimentation and published a survey of the views of Australian research scientists in the field [5].

The public opinion poll found extraordinarily high community acceptance of medical research involving embryo experimentation. Eighty-one percent of respondents agreed that IVF should be available for infertile couples; seventy-five percent approved of embryo research directed at alleviating problems during pregnancy; sixty-eight percent approved of embryo research aimed at the detection of genetic malfunctioning in the embryo; seventy-five percent supported embryo research to enable infertile couples to have children; fifty-eight percent agreed that couples should be able to donate surplus or untransferred embryos for research purposes.

The survey of research scientists revealed a largely pessimistic outlook on the part of Australian scientists about future general research in embryology and in IVF techniques. Until recently, Australian scientists in these areas had been in the forefront of research. The scientists surveyed, however, believed that they would increasingly have to rely upon the work of scientists in other countries.

In addition to the above reports, the NBCC produced a document entitled *Reproductive Technology Counselling* [16]. The Committee also reported annually on new developments in reproductive technology such as DNA research, the new abortifacient RU486, selective fetal reduction, and mass genetic screening programs.

Early in 1991, the NBCC was merged with the Medical Research Ethics Committee, a committee of the National Health and Medical Research Council, to form a new body, The Australian Health Ethics Committee. The latter is described more fully below.

II. NATIONAL HEALTH AND MEDICAL RESEARCH COUNCIL (NH&MRC) AND THE MEDICAL RESEARCH ETHICS COMMITTEE (MREC)

The MREC was, until its termination in 1991, the main ethical committee of the NH&MRC, which is responsible for supporting medical research in Australia. (In 1991-92 the NH&MRC will distribute more than $100 million in research funds.) In the past, the NH&MRC has published ethical guidelines in a number of areas. For example, in 1990 it produced a document, *Discussion Paper on Ethics and Resource Allocation in Health Care* [18]. This brief but useful survey discusses the ethical principles involved in the just distribution and efficient rationing of health care services, a subject of considerable interest in Australia at the present time. A Federal Government inquiry called "National Health Strategy" is currently investigating options for reforming the Australian health system to make it more efficient and cost-effective and more responsive to people's health needs. Thus far it has issued eight background papers and two issues papers but, except for Paper Number 7 on equity in health care, it has not directly concerned itself with specifically ethical issues [11]. The NH&MRC also published in 1991 a paper entitled *General Guidelines for Medical Practitioners on Providing Information to Patients* [4]. This was in response to a significant report, *Informed Decisions about Medical Procedures* produced in 1989 by the Australian, Victorian and New South Wales (NSW) Law Reform Commissions [8]. The latter is discussed further below.

In 1991 the Medical Research Ethics Committee issued a statement, *Guidelines for the Use of Genetic Registers in Medical Research* [13]. These guidelines were drawn up in response to developments in DNA research and technology, particularly those related to genetic diseases in children, and with reference to the stringent safeguards of privacy set forth in the Commonwealth

Privacy Act of 1988. The document also includes a valuable "Statement on Scientific Practice" ([13], Appendix B.).

III. THE AUSTRALIAN HEALTH ETHICS COMMITTEE (AHEC)

As noted above, early in 1991 the NBCC and MREC were merged to form a new body, the Australian Health Ethics Committee, which is now a principal committee of the reconstituted NH&MRC. The membership of the new body is multidisciplinary, and includes individuals with expertise in law, philosophy, medical research, clinical practice, public health research, religion, consumer affairs, and the regulation of the medical profession. AHEC's brief is to (a) inquire into, advise and recommend on ethical, legal and social matters which arise in relation to public heath, health care practice and health and medical research involving humans; (b) develop guidelines to assist in suitable ethical conduct in the health field and to meet the requirements of the Privacy Act 1988; (c) promote community debate, and consult with individuals, community organizations, the health professions and governments on health ethical issues; (d) monitor the workings of institutional ethics committees and advise on these; and (e) monitor international developments in relation to health issues and serve as liaison with relevant international organizations and individuals.

The first inquiry undertaken by AHEC will concern the ethical implications of the allocation of health care resources in Australia and, as already noted, will, to some extent, discuss the ethical assumptions of the Federal Government's National Health Strategy enquiry.

Both the NH&MRC and the AHEC are committed to community consultation. Most of their reports are first published in draft form, and public comment is solicited before a final version is published. Such community consultation on ethical issues is difficult because of the diverse and often polarized community responses to such issues as embryo experimentation and surrogate motherhood. The AHEC has recently issued an informative background discussion paper on this matter: *Consultation: An Appraisal of Community Perspectives* [2]. On a subject which has often engendered a good deal of populist romanticism, this paper provides a practical and realistic assessment of community consultation about bioethical issues.

A. *Institutional Ethics Committees*

An important aspect of the work of AHEC is its responsibility for supporting and monitoring the work of more than 140 institutional ethics committees (IECs) throughout Australia, established under the aegis of the NH&MRC. The IEC's, in effect, regulate ethical standards in medical research institutions. Although there has been some criticism of them for being too compliant toward the interests of medical scientists [12], they have been largely effective bodies. However, the IEC's now face new problems; first, because of the implementation of a new clinical trials system which places considerable responsibility upon them; and second, because of the requirements of the Privacy Act of 1988, which impose new accountability requirements upon such committees.

IV. STATE LEGISLATION ON BIOETHICAL ISSUES

At the state level, there are now three pieces of legislation in existence which regulate IVF and allied research.

A. *The Victorian Infertility (Medical Procedures) Act (1984/1987)*

In 1984, the Victorian State Government enacted pioneering legislation, the Infertility (Medical Procedures) Act, which was subsequently amended in 1987. Under the legislation, a Standing Review and Advisory Committee on Infertility (SRACI) was established to interpret and administer the Act. This committee is a statutory body with extensive powers, in that it does not merely advise the Minister of Health but itself decides whether or not a given piece of research comports with the Act or not. The membership of the committee is multidisciplinary and includes a lawyer, a philosopher, two medical scientists, two religious representatives, a social worker and a community representative. Its chairperson is Professor Louis Waller.

In 1989 a majority of members of SRACI took the view that the Act permits destructive experimentation on post-syngamy embryos formed by IVF procedures which have not been implanted in a woman. (Under an amendment of 1987 the Act permits experimentation on embryos formed expressly for that purpose up until the point of syngamy at about 20 hours of development). Such experimentation requires the consent of the couple who have contributed the gametes for the formation of the embryo, and must be in accordance with rigorously monitored conditions laid down by the committee.

This decision of SRACI raised considerable controversy in the State of Victoria. The then Minister of Health requested the committee to review the whole issue of "post syngamy" embryo experimentation and to make recommendations for possible amendments to the Act. After eighteen months of deliberation, the SRACI has recently published its report in three parts: Part I: Background and Legislation; Part II: New Developments, Policies and Procedures and Community Concerns; and Part III: Recommendations for Amendment of the Infertility (Medical Procedures) Act. This latter part includes a revised "plain English" version of the Act, which will be considered by the Victorian Parliament in 1992 [24].

In general, the SRACI has reaffirmed its majority view that destructive embryo experimentation may be permitted, under certain conditions, on up to ten untransferred embryos until the fourteenth day of embryonic development. It recommends that couples in a stable *de facto* union should be allowed access to reproductive technology programs. It also recommends an enlargement of the SRACI, in order to make it more broadly representative.

The SRACI report is an invaluable source of information on the implementation of the Act, as well as on the subject of embryo experimentation in general.

B. *The South Australian Reproductive Technology Act (1988)*

The South Australian *Reproductive Technology Act* establishes an 11-member Council on Reproductive Technology to monitor IVF practices and research and to advise the Minister of Health on matters of reproductive technology. The Act specifies provisions to guide the Council in formulating a code of ethical practice in IVF; for example, it prohibits embryo flushing, the freezing of embryos for more than ten years, and research that "may be detrimental to an embryo". Finally, it provides for licensing IVF procedures and research

In a recent statement, the chairperson of the Council, Professor L.W. Cox, has set forth the conditions governing access by infertile couples to reproductive technology. In effect, the South Australian Council views access to reproductive technology similarly to adoption and, invoking the child's interests as paramount, restricts access on much the same grounds as those which regulate adoption. Thus, it recommends that treatment be denied: (a) to couples suffering from any illness or disability which could interfere with their capacity to care for a child; (b) to couples one of whom has been convicted of a offense; (c) to couples convicted of child abuse; or (d) to couples who have had children removed from their care by community welfare authorities ([17], p. 21). Others have questioned the use of such non-medical or social criteria to exclude couples from access to reproductive technology, since they are not used with fertile couples having children.

C. *The Western Australian Human Reproductive Technology Act (1991)*

The Western Australian *Human Reproductive Technology Act* sets up a Reproductive Technology Council to regulate the practice of reproductive technology. However, the Council is prohibited from approving any kind of destructive embryo experimentation or diagnostic testing on embryos. It is worthwhile to cite the preamble to the Act.

A. In enacting this legislation, Parliament is seeking to give help and encouragement to those eligible couples who are unable to conceive children naturally or whose children may be affected by a genetic disease.
B. Parliament considers that the primary purpose and only justification for the creation of a human egg in the process of fertilization or embryo in vitro is to so assist these couples to have children, and this legislation should respect the life created by this process by giving an egg in the process of fertilization or an embryo all reasonable opportunities for implanting.
C. Although Parliament recognizes that research has enabled the development of current procedures and that certain non harmful research and diagnostic procedures upon an egg in

the process of fertilization or an embryo may be licit, it does not approve the creation of a human egg in the process of fertilization or embryo for a purpose other than the implantation in the body of a woman.

D. Parliament considers the freezing and storage of a human egg in the process of fertilization or an embryo to be acceptable:
(i) only as a step in the process of implanting; and (ii) only in extraordinary circumstances once the freezing and storage of eggs can be carried out successfully [21].

Surrogate motherhood and access to identifying information for children born from donor gametes are subjects for further legislation still being developed.

The Western Australian, *Human Embryology Act* has been vigorously opposed by medical scientists and by members of infertile patients groups for its draconian stance on embryo experimentation. They point out that if such experimentation is prohibited in Western Australia, IVF clinicians will have to rely upon the scientific research being conducted in other Australian states and other countries where such experimentation is allowed.

V. LAW REFORM COMMISSIONS

Most of the State Law Reform Commissions in Australia have standing references on bioethical issues and have played a notable part in the public discussion of those issues. As noted above, an important joint report by the Australian Law Reform Commission, the Law Reform Commission of Victoria and the New South Wales Law Reform Commission, *Informed Decisions About Medical Procedures*, was published in 1989 [8]. This document clearly affirms the right of the patient to determine the mode of his or her medical treatment and the corresponding obligation of doctors to provide sufficient information for patients to make informed decisions about treatment. The report makes four recommendations concerning the legal implications of informed decision-making. First, the common law standard of reasonable care, which now applies to the provision of information to patients about a proposed treatment or medical procedure, should not be replaced by a statutory standard. Second, guidelines for the provision of information concerning a proposed treatment or procedure should be formulated by the National Health and Medical Research Council. Third, legislation should be enacted requiring that, in an action for professional negligence, the guidelines will be admissible in evidence and the courts will consider them in deciding whether a doctor has acted reasonably in relation to the provision of information. Fourth, the relevant State *Medical Practitioners Acts* should each be amended to provide specifically that professional misconduct includes a failure to provide adequate information to a patient concerning a proposed treatment or medical procedure.

After extensive public discussion, the State of Victoria *Medical Treatment Act* of 1988 established the right (already recognized in common law) of a competent patient to refuse medical treatment in certain circumstances. Under this Act, a certificate is required from the doctor (and a witness) that an adult patient who is informed and competent wishes to refuse medical treatment for a particular illness. The certificate is not an open-ended "living will". If the doctor can produce this certificate, he or she cannot be charged with negligence in refusing treatment. However, as one commentator pointed out, "the certificates are not a defense for a doctor acting with criminal intent to murder or to aid or abet suicide by neglecting to provide reasonable care". Nor are "the certificates ...to be used to refuse reasonable means of delivery of nutrition and hydration" ([25], p. 16). Another piece of legislation, the *Medical Treatment (Enduring Power of Attorney) Amendment Act* (1989), enables a patient to nominate another person to make decisions about medical treatment for the former should he or she become incompetent. In addition, the Law Reform Commission of Western Australia has published a *Report on Medical Treatment for the Dying* (1991) which largely follows the Victorian legislation [10]. In New South Wales, however, a recent attempt to introduce similar legislation was defeated.

In 1989, the Law Reform Commission of Victoria published a document entitled *Genetic Manipulation* [7]. On the one hand, the report recommends that specific legislation should be enacted to control experimental releases of recombinant organisms into the environment. On the other hand, the report generally does not favor special legislation to control genetic manipulations since most genetic manipulation work is safe and only a small proportion of such work involves significant or unknown risk. The latter should be reported to the appropriate authorities in the same way as is done with other potentially hazardous work. Apart from this, genetic manipulation should be regulated in the same way as other scientific work and over-regulation should be avoided. Gene therapy on human patients should continue to be regulated by NH&MRC guidelines and monitored by institutional ethics committees [7]. The report marks a significant move away from legislative solutions to the problems raised by the new biotechnology.

As noted before, privacy legislation has focused attention on the confidentiality of medical records. A number of reports have also addressed this issue, including the *Report on Confidentiality of Medical Records and Medical Research* by the Law Reform Commission of Western Australia [9]. The Commission's states its views in these terms:

> The Commission acknowledges that the law of confidentiality is one of the most important mechanisms evolved by the courts for the protection of individual privacy. The full potential of that law to protect personal information has only recently been recognized. In the medical context, the obligation of confidentiality underpins the bond of trust between health professional and patient which is so important for medical care. The bond is essential not only from the individual patient's point of view but also from that of the general community which has an interest in the health of each of its members. Nevertheless, the Commission agrees with the views of health bodies both in Australia and overseas that the benefits to be derived from medical research justify disclosure of name-identified records without patient consent for that purpose, providing strict safeguards are maintained....To avoid any misconception, the Commission emphasizes that its

proposal is simply intended to free the record keeper from the legal duty of confidence which would otherwise apply. It would not create a legal duty to disclose, even if all the safeguards...were complied with. Under the legislation the record-keeper would remain as free as before to refuse access if he or she thought fit ([9]; pp. 5-6.).

VI. *THE BABY M CASE*

A recent coroner's inquiry in Melbourne, Victoria into the withdrawal of medical treatment by physicians at the Royal Children's Hospital from a child ("Baby M") with severe spina bifida provoked a public controversy. Although the parents of the child agreed with the physicians that further treatment was pointless, and should be withdrawn, and that sedatives should be given to the child, members of the Right to Life Association intervened on the grounds that if the baby were sedated, she would in effect die from starvation. The baby did die, but at a subsequent coroner's inquiry, the parents and physicians were exonerated from any wrong-doing and the members of the Right to Life Association were severely criticized. However, because treatment decisions for disabled newborns pose legal difficulties for both parents and physicians, guidelines are clearly necessary.

The Law Reform Commission of Victoria has proposed establishing a committee to prepare such guidelines for circumstances in cases similar to the Baby M case. The concerns raised by such cases include the following: (1) the need of the whole medical team including nurses, to consult about decisions; (2) the need to consult fully with parents or next of kin, both initially and as the case is reviewed; (3) the factors to be considered in choosing "conservative management" rather than invasive surgery; (4) the circumstances in which independent medical consultants should be asked for advice; (5) the principle that it is lawful to give whatever medication is necessary to alleviate pain, even if that has the incidental effect of shortening life; (6) the type of pain relief that is appropriate in these cases, or a method of obtaining advice about the appropriate pain relief in a specific case; (7) the matters to be noted by doctors and nurses concerning a patient's condition from time to time, especially the pain and distress that would justify the administration of increasing doses of analgesics or sedatives; (8) the timing of when a death is reported to the Coroner under the Coroner's Act; and (9) what review of procedures should be considered after a coroner's inquiry [23].

The Proceedings of the National Conference on Neonatal Intensive Care, sponsored by the New South Wales Department of Health in 1989 also discussed the care of disabled newborn children. This important collection of papers analyzes the medical, economic, ethical and legal issues raised by neonatal intensive care. Of particular interest is the keynote address by Mr Justice Michael Kirby, who has played a major part in bioethical discussion in Australia [19].

The New South Wales Department of Health has also sponsored a conference on *The Ethics of Allocating Health Resources*, and a report is available [20].

VII. OTHER BODIES

A number of other bodies and organizations contribute to the ongoing bioethical debate in Australia. One of the most influential is the Monash University Centre for Human Bioethics; directed by Professor Peter Singer and Dr. Helga Kuhse, who are also the editors of the quarterly journal *Bioethics* (Blackwells, Oxford). Singer and Kuhse have recently edited *Embryo Experimentation*, which contains essays by the multidisciplinary group of scholars – scientists, philosophers, and lawyers – associated with the Monash Centre [22]. The volume, one of the best studies to date, provides up to date information on the status of embryo experimentation in Australia.

Roman Catholic views on bioethical issues are well represented by the St Vincent's Bioethics Committee in Melbourne; the South Australian Southern Cross Bioethics Institute, which produces the monthly *Bioethics Research Notes*; the Western Australian L.J.Goody Centre for Bioethics, which issues a regular newsletter for Catholic doctors; and the Queensland Provincial Bioethics Centre, attached to the Mater Misericordiae Hospital in Brisbane. The Australian Catholic bishops also have a permanent secretariat, under the direction of N. Tonti-Filippini, to advise them on bioethical issues. In general, Australian Catholic views on bioethical issues tend to be conservative. Few Australian Catholic thinkers in this area seem akin to liberal Catholic moral theologians such as Richard McCormick or Charles Curran in the United States. Other church-affiliated bodies, such as the Kingswood Centre for Applied Ethics in Perth, Western Australia and the Anglican Social Responsibilities Commission, also contribute to the public debate on bioethics.

In addition, feminist scholars (for example, Dr. Robyn Rowland and Dr. Renate Klein) associated with the Feminist International Network of Resistance to Reproductive and Genetic Engineering (FINRRAGE) have played a part in bioethical discussion in Australia. One of the notable aspects of opposition to the NBCC's report on surrogate motherhood was the coalition between FINRRAGE and several more conservative Roman Catholic bioethicists.

In 1991 the Australian Bioethics Association was formed and the first annual conference, with the theme "Bioethics and the Wider Community", was held. The proceedings of the conference give an excellent idea of the breadth and vitality of bioethical discussion in Australia at the present time [1].

Department of Philosophy
Deakin University
Victoria, AUSTRALIA

NOTES

[1] For a fuller discussion of the Report's implications, see p. 415, [4].

[2] Health Boards have overall responsibility for health care in their region, and a very high proportion of secondary and specialist medical care is provided by this state funded system.

[3] It should be noted that there are no separate clinical ethics committees in New Zealand.

BIBLIOGRAPHY

1. Australian Bioethics Association: 1991, *Proceedings of the First Annual Conference:Bioethics and the Wider Community*, Christine Martin (ed.) Mercy Maternity Hospital, Melbourne, Victoria.
2. Australian Health Ethics Committee (AHEC): 1991, *Consultation: An Appraisal of Community Perspectives*. National Health and Medical Research Council, Canberra, A.C.T.
3. Dawson, K.: 1991, *Human Embryo Experimentation: A Background Paper and Select Bibliography*.
4. Dawson, K.: 1991, *General Guidelines for Medical Practitioners on Providing Information to Patients*.
5. Dawson, K.: 1991, "Human Embryo Research: A Survey of Scientists' and Clinicians' Opinions". The Roy Morgan Research Centre, in *Survey on Human Embryo Research*. (The above three publications were prepared by the National Bioethics Consultative Committee but published under the auspices of the National Health and Medical Research Council.)
6. Kirkman, M and L.: 1989, *My Sister's Child*, Penguin Books, Ringwood, Victoria.
7. Law Reform Commission of Victoria, Australian Law Reform Commission, New South Wales Law Reform Commission: 1989, *Genetic Manipulation*, Melbourne, Victoria.
8. Law Reform Commission of Victoria, Australian Law Reform Commission, New South Wales Law Reform Commission: 1989, *Informed Decisions About Medical Procedures*, Melbourne, Victoria.
9. Law Reform Commission of Western Australia: 1990, *Report on Confidentiality of Medical Records and Medical Research*. Perth, W.A.
10. Law Reform Commission of Western Australia: 1991, *Report on Medical Treatment for the Dying*.
11. McClelland, A.: 1991, *In Fair Health? Equity and the Health System*, National Health Strategy Unit, Department of Health, Canberra, A.C.T.
12. McNeill, P.M. *et al.*: 1990, "Reviewing the Reviewers: A Survey of Institutional Ethics Committees in Australia", *The Medical Journal of*

Australia, 152, 289-296.

13. Medical Research Ethics Committee: 1991, *Guidelines for the Use of Genetic Registers in Medical Research*, National Health and Medical Research Council Canberra, A.C.T.
14. National Bioethics Consultative Committee: 1989, *Record Keeping and Access to Information: Birth Certificates and Birth Records of Offspring Born as a Result of Gamete Donation.*
15. National Bioethics Consultative Committee: 1991, *Surrogacy.*
16. National Bioethics Consultative Committee: 1991, *Reproductive Technology Counselling.*
17. National Bioethics Consultative Committee: 1991, *Access to Reproductive Technology*
18. National Health and Medical Research Council: 1990 *Discussion Paper on Ethics and Resource Allocation in Health Care.*
19. Neutze, J.M.: 1989, "A Standard for Ethical Committees", letter to the Editor, *New Zealand Medical Journal*, 102, 111.
20. New South Wales Department of Health: 1989, *Proceedings of the National Consensus Conference on Neonatal Intensive Care*, Sydney, New South Wales
21. Parliament of Western Australia: 1991, *Human Reproductive Technology Act*. Perth,W.A.
22. Singer, P, *et al.*: 1990, *Embryo Experimentation*, Cambridge University Press, Melbourne.
23. Skene, L.: 1991, *The Baby M Inquest: Treating Children with Severe Spina Bifida*, Law Reform Commission of Victoria, Melbourne, Victoria.
24. Standing Review and Advisory Committee on Infertility (SRACI): 1990-1. *Report to the Minister for Health on Matters Related to the Review of Post-Syngamy Experimentation*; Part I: Background and Legislation; Part II: New Developments, Policies and Procedures, and Community Concerns; Part III: Recommendations for Amendment of the Infertility (Medical Procedures) Act 1984, Department of Health, Melbourne. Victoria.
25. Tonti-Filippini, N.: 1990, "The Doctor-Patient Relation: Legislative and Professional Changes", in H. Caton (ed.) *Trends in Biomedical Regulation*, Butterworths, Sydney.

ALASTAIR V. CAMPBELL

BIOETHICS IN NEW ZEALAND

In the period under review, developments in bioethics in New Zealand have been dominated by the aftermath of a Judicial Enquiry into events taking place in the National Women's Hospital, Auckland. Popularly known as the Cartwright Report (after Dame Silvia Cartwright, the judge conducting the enquiry), this document, published in 1988, set the agenda for a series of legislative initiatives and national guidelines in the fields of clinical research and clinical treatment and brought about major changes in the teaching of ethics in medical education. In addition to implementation of the Cartwright recommendations, there has been increasing attention to other issues, notably the policy of the medical profession in relation to HIV/AIDS and of the government in relation to smoking and health. Some discussion has also begun regarding the ethical implications of genetic manipulation and developments in biotechnology.

I. THE CARTWRIGHT REPORT

The *Report of the Cervical Cancer Enquiry* [hereafter referred to as The Cartwright Report] is a complex document, which deals in part with the details of allegations against some medical staff at the National Women's Hospital in Auckland regarding failures to treat women with cervical cancer adequately and the enlisting of these women in a research project without their consent [7]. The Report is of wide interest beyond its findings on these specific allegations because of its general recommendations regarding the changes required to prevent a recurrence of the problems uncovered.[1] The specific allegations focused on the management of carcinoma in situ (CIS) by a senior academic clinician at the hospital, Associate Professor Herbert Green. Professor Green believed that CIS, a symptomless condition consisting of a lesion on the lining of the uterus or other areas of the genital tract, was not necessarily a precursor of invasive cervical cancer. This was a view at variance with most expert opinion at the time (the early 1960s). Normal practice was to detect CIS by observation or by cervical smear and then to eradicate the lesion through the use of a cone biopsy technique. Professor Green believed that this degree of intervention was unnecessary and proposed to demonstrate this through a combination of "conservative treatment" and regular observation. Thus, he was, in effect, conducting a clinical research project with his patients (without their knowledge or consent), and he continued to research his thesis, despite

B. Andrew Lustig (Sr. Ed.), Bioethics Yearbook: Volume 2, 403–416.

numerous expressions of concern from some of his colleagues, for approximately twenty years. The report gave findings on nine separate terms of reference in relation to these events. The most serious findings related to failures in treatment and failure to gain informed consent. With regard to the former, the judge found as follows:

> The outcome of treatment for the majority of women has been adequate, although a significant number were not managed by generally accepted standards over a period of years. For a minority of women, their management resulted in persisting disease, the development of invasive cancer and, in some cases, death ([7], p. 210).

With regard to the research involved, the judge was gravely concerned both about the failure of the researcher or of any other responsible person to take proper notice of the serious expressions of concern about the project being voiced by medical colleagues. She also found that there was a major failure to obtain informed consent, both in respect of the unusual approach to treatment and in respect of the attempt to prove a hypothesis about the nature of CIS. The report allocated responsibility widely for these failures:

> The fact that the women did not know that they were in a trial, were not informed that their treatment was not conventional, and received little detail of the nature of their condition were grave omissions. The responsibility for these omissions extends to all those who, having approved the trial, knew or ought to have known of its mounting consequences and design faults and allowed it to continue ([7], p. 61).

The effects of these major findings of medical failure were far reaching and continue to influence present attitudes about medicine in New Zealand at the present time. Although articles in professional journals had raised questions about Dr. Green's treatment policy and research, it was a long report in the popular journal, *Metro*, which drew the matter to the public's attention and provoked the decision to hold a judicial enquiry. The enquiry itself and the subsequent Report were given extensive coverage by the media, and the issues of adequate consent to treatment and research and proper monitoring of the medical profession and the research community remain matters of intense public concern and interest. The medical profession has also paid serious attention to the findings, particularly in relation to consent procedures and responsibilities for the oversight of medical research. Disciplinary procedures were instituted against Dr. Green by the Medical Council (the statutory licensing body for medical practitioners in New Zealand), but were dropped because of his incapacity to plead on grounds of ill-health. Similar procedures were then instituted against the head of Professor Green's academic department, and he was found guilty of "disgraceful conduct in a professional respect". The Medical Council subsequently produced a report on informed consent (see below), and it is currently preparing one on the responsibilities of heads of department in relation to research.

The Cartwright Report produced a number of recommendations which were accepted by the Minister of Health and assigned for implementation to a task force within the Department of Health. These recommendations included the establishment of a national cervical cancer screening program, a follow-up of patients involved in Professor Green's research project, the establishment of a Patient Advocacy Service and an Office of Health Commissioner to deal with patients' rights and patients' complaints, the development of national guidelines for review of research and treatment protocols, and changes to medical education to improve knowledge of ethical guidelines for treatment and research and to improve communication skills. Implementation of these recommendations has proceeded at a variable pace. Some are well advanced (*e.g.*, the follow up of patients and the establishment of a national screening program); others have been the subject of various reports and parliamentary bills, but are not yet fully implemented.

II. CARTWRIGHT IMPLEMENTATION

A. *Consent to Treatment*

A major concern of Judge Cartwright was the failure of medical staff to inform patients adequately and the omission of valid consent. Three subsequent reports have sought to address this issue: a New Zealand Health Council Working Party Report on informed consent (1989), a Report of the New Zealand Medical Council (1990), and a Report of a Department of Health Working Party on informed consent (1991).

1. *New Zealand Health Council Report*

The first report, entitled: "Informed Consent, A Discussion Paper and Draft Standard for Patient Care Services", was prepared by a working party established by the New Zealand Health Council with the specific purpose of providing a discussion document for the use of hospital and area health boards [19]. The intention was that discussion would be promoted among "staff, communities, Iwi [*i.e.*, Maori tribal] authorities, and consumer groups" ([19], p. 5). The report made it clear that its concerns stemmed in part from the Cartwright Report, but it also made reference to the need to implement the partnership principle of the Treaty of Waitangi. (This is a treaty drawn up in 1840 between the British Crown and the indigenous Maori community which establishes a basic biculturalism in New Zealand. The practical implementation of the principles of the Treaty is the subject of much current debate, in view of the disregard for its requirements in the past.) The basic philosophy of the Report is clearly conveyed in the opening sentences of the Introduction:

The relationship between doctors and patients is changing under pressure from Maori groups, consumers, women's groups, and a better educated and knowledgeable public. The emphasis is moving away from paternalistic and authoritarian medical practice towards health care as a partnership between a client and a health care professional ([19], p. 5).

The report (which was published in both English and Maori) was divided into two parts. Part One gave an outline of the "key issues". A definition of informed consent was provided in terms of four essential elements: (1) information on which to make a decision; (2) comprehension of the information; (3) competence to make a decision; and (4) absence of pressure or coercion.

This part of the report also gave a theoretical grounding for the necessity of informed consent in order to enhance autonomy and to promote trust. It also listed the requirements for the content of consent form, and briefly discussed the legal issues in the context of existing New Zealand statutes. Part Two of the Report was entitled "Draft Standard for Informed Consent in Patient Care Services". In fourteen sections, each with subsections, it laid out draft guidelines for obtaining consent across the whole range of health care interventions, and provided sample admission and consent forms.

This document, despite its relative brevity of 37 pages, is undoubtedly the most comprehensive statement about informed consent in health care to be produced in New Zealand. It did provoke the desired discussion and debate; over 200 submissions were received by the Department of Health, commenting on its provisions and suggesting problems, amendments, or further clarification of issues. In some quarters, the report provoked strong opposition, based on the opinion that it was unnecessarily legalistic and wholly impracticable to apply in many clinical situations. Some felt that the Working Party was insufficiently representative of those involved directly in health care delivery. (Of the Working Party's fifteen members, two were nurses and two were medical practitioners, but all four in administrative posts. Other disciplines represented were health service management, law and philosophy.) An impression of the opposition which the Report provoked among some clinicians can be gained from the following extract from a special article in the *New Zealand Medical Journal*:

.... it is difficult not to see the proposed standard as window dressing, aimed at reassuring the public that its interests are being safeguarded by having doctors policed at every step...... The proposed standard is grossly over-prescriptive and invites the development of a new era of litigation, benefiting no one ([1], p. 350).

2. *New Zealand Medical Council Statement*

A feature of the New Zealand scene which makes debate about informed consent of great significance is that litigation against doctors for failure to gain consent (or for other medical failures) is virtually unknown, and can be instituted only for punitive or exemplary damages, not for simple negligence.

The reason for this absence of litigation is the existence of a national accident compensation scheme which provides "no fault" compensation for all types of accident, including "medical misadventure". An important ruling by the Court of Appeal on a case arising from the Cartwright Report (*Green v. Matheson*) explicitly included failure to inform under "medical misadventure". Thus, no clear legal sanctions exist to ensure that standards for informed consent are properly implemented by health care professionals. For this reason, any statement published by the Medical Council of New Zealand assumes great importance, since it can influence practice by making compliance with consent procedures part of the requirements for good conduct by registered practitioners.

The statement, which was adopted by the Medical Council in 1990, was drafted by a three-member working party consisting of a medical dean, a medical practitioner with a doctorate in philosophy, and the lay member of the Medical Council [17]. A background paper accompanying the statement pointed out that, in view of the legal situation in New Zealand with regard to negligence cases, the Council inevitably has the responsibility to adjudicate on deficiencies in informed consent procedures by medical practitioners and, when required, to take appropriate action through its disciplinary tribunals. The background paper sounded a warning note about impracticality and the creation of undue suspicion of doctors, which were perceived to be weaknesses of the Health Council report:

> The medical profession (and indeed the Council), while welcoming the [Report] as a catalyst for discussion was concerned both at the rather specific details, often judged to be clinically impractical, as well as the air of distrust of doctors which seemed to prevail ([17], p. 8).

Yet the Medical Council statement itself is uncompromising in its identification of the essential elements of informed consent and in its warning regarding disciplinary action if they are not observed. The critical section of the statement relates to grounds for medical misconduct:

> The Medical Council affirms that if it can be shown that a doctor has failed to provide adequate information and thereby has failed to ensure that the patient comprehends, so far as is possible, the factors required to make decisions about medical procedures, such failure could be considered as medical misconduct and could be the subject of disciplinary proceedings ([17], p. 3).

The statement also recognizes that guidelines will be required to make plain to practitioners what is required of them and to provide a standard against which misconduct could be judged. The Council opposed enacting such guidelines in legislative form and suggested instead a number of bodies by which such guidelines could be issued, including the medical colleges, the Medical Association, and the Ethics Committees of Health Boards.

3. *Department of Health Report*

The situation had to some extent been clarified by the new Medical Council policy, but the absence of agreed upon guidelines made this policy somewhat general in character. In the meantime, the New Zealand Health Council had been disbanded. (It was an advisory body to the Minister of Health consisting mainly of the chairpersons of the Area Health Boards.) The task of dealing with the responses to their working party report was delegated to a new working party established by the Department of Health, consisting of representatives of the New Zealand Medical Association, the New Zealand Institute of Health Management, the Medical Council of New Zealand, the Auckland Women's Health Council, and the Maori Nurses Association, plus the professional advisory members of the Department, an academic lawyer and the director of a bioethics center. This group was given the task of sifting through the responses to the Health Council Report and producing a set of guidelines for all health care providers and planners regarding the obtaining of informed consent.

The resulting report was a much shorter document than its predecessor (14 rather than 37 pages of text). Entitled "Principles and Guidelines for Informed Choice and Consent", the report offered three general principles for guidance: respect for autonomy; acceptance of individual and shared responsibility; and allocation of accountability [10]. The title of the report, with its use of "choice" as well as "consent", conveys its overall approach. The emphasis throughout is on the *process* of choice or decision making, of which the formal procedures of consent are only a relatively minor part. The stated purpose of the specific guidelines contained in the document is stated "to help providers achieve high quality and appropriate 'user centered' 'health care'" ([10], p. 8).

If the former draft guidelines were criticized for being too detailed and specific, those contained in the newer document are likely to be regarded as being too general and vague. The fullest detail is to be found in the guidelines for implementing the principle of respect for autonomy. These lay down minimum standards for the information to be conveyed and include, in the addition to details about outcomes, risks and benefits and alternative options, a requirement that the name, status, and relevant experience of the provider of the service be supplied. (This requirement has major implications for teaching hospitals in which much of the direct clinical work is carried out by persons under training. The guidelines require that users of the services be properly informed about supervisory arrangements for those under training.) The report is also quite specific about the need to take the wishes of children into account, rather than relying on proxy consent only, and about the need for a user-appropriate advocacy service. Recognition of cultural differences is also stressed.

The later sections of the report, dealing with responsibility and accountability, are very brief (one and two pages respectively). Although guidelines are given, they are very broad. The prime responsibility of the provider in ensuring informed choice is emphasized, but the user's responsibility to enable a collaborative relationship is also mentioned. Under "accountability", the report requires written consent in most circumstances and opposes notions of implied or non-specific consent. Finally, the importance of documenting of decisions is emphasized, so that there will be evidence that the process of informed choice has been correctly followed. This documentation should be available to users to check and copy if they wish.

4. *Conclusion*

The need for proper national standards for information and consent, which was identified by the Cartwright Report, has been extensively addressed over the past two years, but as yet no comprehensive and authoritative standards have been set. It may be that this is an impossible task, in light of the diversity and range of health care interventions and of the widespread distrust within the professional groups of over-legislation in this area. Nevertheless, some standards are required, at the lowest level to provide a benchmark for professional discipline, and, at the highest, to provide a practical model for genuinely collaborative health care relationships. Clearly the definitive document has not yet been produced.

B. *Oversight of Treatment and Research*

In October 1988, a draft document was produced by the Department of Health for discussion by chairpersons of Area Health Boards prior to its implementation as a national standard for all hospital and health board ethics committees established to review research and treatment protocols [9]. Following consultation with Boards and with the (then) Medical Research Council of New Zealand, the document (usually referred to simply as the "National Standard") became the accepted source of authority for the constitution and functioning of the fourteen ethics committees operating in New Zealand under the Health Board structure.[2] The status of the National Standard, however, was never entirely clear. The document has been under continual discussion since its first draft, and has drawn some criticism from the research community ([18];[22]). A revised standard is due to be published in 1992 (and will be described in the next edition of the yearbook), but it will function only as advice or guidelines from the Department of Health, without any statutory force.

However, a significant change came about with the passage of the Health Research Council Act in 1990. This legislation replaced the New Zealand Medical Research Council with The New Zealand Health Research Council (HRC). The Act shifted the emphasis in government funded research towards issues in community health, although it made plain that a large amount of funding should still be allocated to biomedicine. In addition, the Act established an Ethics Committee as one of the standing committees of the HRC, giving it a range of important functions, including the provision and review of ethical guidelines for HRC-funded research and power to assess other ethics committees in terms of their competence to review research proposals submitted to the HRC ([13], Section 25). Since the Ethics Committee of the HRC has chosen to adopt the National Standard as its benchmark, at least for the time being, this has had the effect of giving the Standard statutory authority in a very significant area of research activity in New Zealand.

Much of the National Standard follows guidelines already well established internationally by the Declaration of Helsinki and its subsequent revisions, but the standard is also distinctive in a number of respects. First, it lays down detailed prescriptions for informed consent, with particular emphasis on the quantity and quality of information to be provided. Second, it insists on cultural and gender balance on committees, requires that at least half the membership must be lay (*i.e.*, non-health professional) and that the chairperson must be lay. Third, it encourages committees to assess treatment protocols as well as research protocols and to consider matters of wider ethical concern in relation to health care.[3] The emphasis on treatment protocols reflects the concern raised by the Cartwight Report that novel or idiosyncratic treatment regimens can be as harmful as poorly controlled research projects. These three features (particularly the emphasis on significant rather then token lay membership) have produced a distinctive approach to research ethics in New Zealand.

C. *Refusal of Treatment, Advocacy and Complaints*

The enabling of patients to take more control of their own treatment was advanced by two further legislative moves in the wake of Cartwright. The New Zealand Bill of Rights, which became law in August 1990, affirmed two civil and political rights in relation to experimentation and medical treatment: (1) "Every person has the right not to be subjected to medical or scientific experimentation without that person's consent."; and (2) "Everyone has the right to refuse to undergo any medical treatment" ([2], Parts II, X, and XI). In September 1990, a Bill was introduced to Parliament, entitled the "Health Commissioner Bill". The purposes of the Bill are well described by its own Explanatory Note:

> This Bill provides for the appointment of a Health Commissioner who will investigate complaints against persons or bodies that provide health care, and who will have general functions in relation to the protection of the rights of health consumers. The Bill also provides for the establishment of a Health Consumer Advocacy Service, and for the promulgation of a Code of Health Consumers' Rights [12].

These far-reaching proposals were explicitly designed to meet the concerns of the Cartwright Report regarding the inadequacy of the current complaint-systems in New Zealand, and to provide, through the Advocacy Service, a "patient's voice" in health care. However, not long after the introduction of the Bill, a new government was elected, and, although a parliamentary committee has received submissions on the Bill, it remains unclear which, if any, of the original proposals will ever be enacted.

D. *Ethics in Medical Education*

The Cartwright Report specifically recommended that medical education should have greater emphasis on both ethics and communication skills ([7], p. 216). In fact, as the Report itself noted, a notable start had already been made by the Otago Medical School by its introduction of a formal curriculum in ethics [3]. There was rapid development in the period 1988-1990, with the establishment of some ethics courses in all the New Zealand Clinical Medical Schools [16], and with the foundation at Otago University of a Bioethics Research Centre [5], staffed by a Professor of Biomedical Ethics (the author of this entry) and a Senior Lecturer in Medical Ethics (Dr. Grant Gillett). The expansion continues as the Bioethics Research Centre increases its complement of research fellows and further posts are established for the teaching of medical ethics.

E. *AIDS, Confidentiality, and Testing for HIV Status*

Although the incidence of AIDS appears to be relatively low in New Zealand, the medical profession has been concerned about the risks of transmission of the disease both to sexual partners of AIDS or HIV-positive patients and to the health professionals involved in their care. Following widespread debate and consultation, the New Zealand Medical Association (NZMA) adopted Policy Statements on both confidentiality and HIV testing. In a policy adopted by its National Assembly in September 1990, the NZMA advised all medical practitioners that they should take the following steps if they considered a breach of medical confidentiality was necessary to protect a third party:

1. Take all reasonable steps to educate, counsel and support the HIV positive person to discuss his or her HIV status with sexual and intravenous drug sharing partner/s.
2. If that person then refuses to discuss their HIV status with their sexual partner/s and there is a clear risk to an acknowledged sexual partner/s, the medical practitioner should discuss with a senior colleague, or the Central Ethical Committee if necessary, whether confidentiality should be maintained.
3. The matter should be discussed with the practitioner's medical protection or defense advisor.
4. Having reached a decision, the practitioner should then consult with the HIV-positive person, advising him if it is the practitioner's intention to disclose the information to the third party and to present [him or her] with a written confirmation of this.
5. A final opportunity should be given to the patient to change his stance and inform the third party of [his] condition.

6. If the patient again refuses to respond, the practitioner should notify the third party of the risk. This would involve the opportunity for a consultation and to initiate steps to provide the third party with appropriate counselling and medical advice [20].

A more complex protocol was adopted by the National Assembly of the NZMA in April 1991 with regard to testing of patients and practitioners for HIV status. The essence of this policy statement is that testing without consent is impermissible except in emergency situations (not defined), but that a doctor involved in invasive procedures has the right to withdraw from a case if a patient "at risk of HIV" persistently refuses permission for testing. A duty is also imposed on doctors to monitor their own HIV status "in any appropriate circumstances", and to seek appropriate advice from a specialist in the event of a positive test, but nothing is stated regarding any possible duty to inform patients of a practitioner's HIV positive status [21].

F. *Anti-Smoking Legislation*

Significant changes in legislation designed to reduce mortality and morbidity caused by smoking occurred in the period under review. In 1986 the Toxic Substances Board had recommended the elimination of tobacco advertising and promotion. The Government did not accept that recommendation, but continued voluntary agreements with the tobacco industry. These agreements restricted some types of advertising, but not the amount of money spent.

However, by July 1988, the discouragement of smoking had become a stated government policy. The Toxic Substances Board's concerns continued. In September 1988, it re-established the Tobacco Subcommittee and, in May 1989, released its report, *Health or Tobacco: An End to Tobacco Advertising and Promotion* [24]. This report surveyed the data concerning tobacco promotion and its effects, and the promotion policies and tobacco consumption trends in 33 countries. The authors were convinced that there was enough evidence to warrant the total elimination of tobacco advertising and sponsorship in New Zealand, and recommended this policy to the government. The report emphasized that society's right to protect the future health of its young people must prevail over the freedom of the tobacco industry to continue to promote its products.

The Tobacco Industry commissioned its own *Independent Scientific Review of the Toxic Substances Board Report* [23]. This review described the Report as "full of untenable assumptions, misleading data, faulty statistical methods, errors and contradictions" ([23], p. iv.). The review also argued that to restrict advertising is to put democracy at risk: "..in any democratic nation it is axiomatic that the freedom of commercial speech is critical to freedom of choice and the autonomy of the individual. If the state controls the flow of commercial speech...political and economic democracy are no longer viable" ([23], p. iv).

At the time the Toxic Substances Board Report was released, the Coalition against Tobacco Advertising and Promotion was formed. Founding members included The NZMA, The Cancer Society, The Heart Foundation, The Public Health Association, The Asthma Foundation, and the New Zealand Nurses Association. This coalition conducted an intensive, high-profile campaign aimed at influencing public opinion and lobbying the government.

The Government accepted the recommendations of the Toxic Substances Board, and passed the Smoke-free Environment Act of 1990. This act creates smoke-free indoor environments in public, places, workplaces, and institutions; provides for complaints and specifies fines for which employers, operators, bodies corporate and individuals will be liable. It also places severe limits on advertising, promotion and sponsorship by tobacco companies, and establishes a Health Sponsorship Council to promote health and healthy lifestyles and to provide sponsorship for individuals and organizations involved in sporting, artistic, and cultural activities.

There has been some pressure from tobacco companies and sporting bodies to repeal parts of the Act or to soften the legislation concerning sports sponsorship. By June 1991, Action on Smoking and Health (ASH) and other groups were lobbying the government to prevent such changes, and thus far the legislation remains unchanged.

G. *Genetic Engineering*

Since agriculture and livestock breeding are major parts of the New Zealand economy, the concept of genetic engineering as it applies to plant development and to stock production is already well known to a significant section of the public. However, a survey published in 1990 revealed that there was also substantial public concern about possible future developments, particularly as these might be applied to animals or humans [8]. Fifty-eight percent of respondents regarded manipulation of genetic material in humans as unacceptable, and 44 percent had similar objections to genetic manipulation of animals. Two recent reports have been prepared to improve the awareness of the scientific community and the general public about what is currently happening in New Zealand and to explore some of the ethical issues raised by genetics. *Genetic Engineering: A Perspective on Current Issues* was produced as a background paper for the staff of the New Zealand Department of Scientific and Industrial Research, but was also aimed at a wider readership [15]. The paper summarizes current applications of genetic engineering worldwide in industry (including the pharmaceutical industry), in agriculture, and in animal breeding. Ethical issues are discussed, including the possibilities of applications to humans, and an extensive review of safety issues is provided. The report is broadly supportive of further development in plant and animal genetic manipulation, but perceives areas of uncertainty and recommends fuller public debate, possibly leading to more comprehensive regulation of research and further controls over the field release of genetically modified organisms.

The second report, *Genetic Engineering in New Zealand: Science, Ethics and Public Policy*, was commissioned by the Centre for Resource Management of Lincoln University, a research and teaching organization supported by the Ministry for the Environment [14]. (The report, however, does not necessarily reflect Ministry policy.) This report draws similar conclusions regarding the need for more public consultation and better understanding of the issues prior to any legislative change. It surveys the medical implications in greater detail and draws attention to some of the ethical hazards arising from the human genome project and from the possible development of germline therapy in humans.

In view of the commissioning bodies for these two reports, it is hardly surprising that only minimal attention is paid to those areas of biotechnology (largely researched outside New Zealand) which directly affect medical therapy and human health. However, a gap currently exists in New Zealand legislation and policy guidelines in relation to human genetics, and appropriate advice should be sought before the policy makers are overtaken by events. Detailed reports on these issues are required in order to stimulate the necessary controls and/or legislative changes.

H. *Epilogue*

The period covered by this report necessarily excludes a very recent development in bioethics in New Zealand. In July 1991, the New Zealand Health Department published a document proposing radical changes in the whole structure of the New Zealand Health Service [11]. There is little doubt that this report will provoke a major ethical debate in the two-year period now commencing, and readers may expect a full discussion of the problems of justice in health care delivery in New Zealand in the next edition of the yearbook.

Bioethics Research Centre
University of Otago
Dunedin, NEW ZEALAND

NOTES

1 For a fuller discussion of the Report's implications, see [4].

2 Health Boards have overall responsibility for health care in their region, and a very high proportion of secondary and specialist medical care is provided by this state funded system.

3 It should be noted that there are no separate clinical ethics committees in New Zealand.

BIBLIOGRAPHY

1. Agnew, T.M. *et al.*: 1990, "Informed Consent: Discussion Paper by the New Zealand Health Council's Working Party", *New Zealand Medical Journal* 25 July, 348-350.
2. Bill of Rights: 1990, No. 109, New Zealand Government, Wellington.
3. Campbell, A.V.: 1987, "Teaching Medical Ethics: Reflections from New Zealand", *Journal of Medical Ethics*, 13, 137-138.
4. Campbell, A.V.: 1989, "An 'Unfortunate Experiment'", *Bioethics* 3, 59-66.
5. Campbell, A.V.: 1991, "Ethics after Cartwright", *New Zealand Medical Journal*, 104, 36-37.
6. Cancer Society of New Zealand Inc: 1990, *Professional Bulletin Report* 1.
7. Committee of Inquiry into Allegations Concerning the Treatment of Cervical Cancer at National Women's Hospital and into other Related Matters: 1988, *The Report of the Cervical Cancer Inquiry*, Government Printing Office, Auckland.
8. Couchman, P.K. and Fink-Jensen, K.: 1990, "Public Attitudes to Genetic Engineering in New Zealand", *DSIR Crop Research Report Number 138*, DSIR, Christchurch, NZ.
9. Department of Health: 1988, *Standards for Hospital and Health Board Ethics Committees Established to Review Research and Treatment Protocols*, New Zealand Department of Health, Wellington (unpublished document available from Department of Health, P.O. Box 5013, Wellington, New Zealand).
10. Department of Health: 1991, *Principles and Guidelines for Informed Choice and Consent* (May), New Zealand Department of Health, Wellington.
11. Department of Health: 1991, *Your Health and the Public Health*, (July), New Zealand Department of Health, Wellington.
12. Health Commissioner Bill: 1990, New Zealand Government Printing Office, Auckland.
13. Health Research Council Act: 1990, Number 68, New Zealand Government, Wellington.
14. Macer, D., Bezar, H. and Gough, J.: 1991, *Genetic Engineering in New Zealand Science Ethics and Public Policy*, Centre for Resource Management, PO Box 56, Lincoln University, Canterbury, New Zealand.
15. Macer, D.: 1990, *Genetic Engineering: A Perspective on Current Issues*. DSIR Crop Research, Private Bag, Christchurch, New Zealand.
16. Maclaurin, C.: 1989, *The Teaching of Biomedical Ethics* (Proceedings of a Symposium), The Glaxo Foundation for Medical Education,Palmerston North, New Zealand.
17. Medical Council of New Zealand: 1990, *A Statement for the Medical Profession on Information and Consent*, available from Medical Council

of New Zealand, P.O. Box 9249, Wellington, New Zealand.

18. Neutze, J.M.: 1989, "A Standard for Ethical Committees", letter to the Editor, *New Zealand Medical Journal*, 102, 111.
19. New Zealand Health Council: 1989, *Informed Consent: A Discussion Paper and Draft Standard for Patient Care Services*, Department of Health, Wellington.
20. New Zealand Medical Association: 1990, *Policy on HIV Status and Patient Confidentiality*, unpublished, available from NZMA, PO Box 156, Wellington, NZ.
21. New Zealand Medical Association: 1991, *Policy on HIV Testing, Patient Care and Responsibility*, unpublished, available as above.
22. Paul, C.: 1989, "A Standard for Ethical Committees", letter to the Editor, *New Zealand Medical Journal*, 102, 20.
23. The Tobacco Institute of New Zealand: 1989, *Independent Scientific Review of the Toxic Substances Report Summary*.
24. Toxic Substances Board: 1989, *Health or Tobacco. An End to Tobacco Advertising and Promotion*. Government Printing Office, Auckland.

NOTES ON CONTRIBUTORS

Francesc Abel, S.J., M.D., Ph.D., Director, Institut Borja de Bioetica, San Cugat de Valles, Barcelona, Spain.

Robert Arnold, M.D., Division of General Internal Medicine, Department of Medicine, University of Pittsburgh, Pittsburgh, PA, 15261, U.S.A.

Athena Beldecos, University of Pittsburgh, Pittsburgh, PA, 15261, U.S.A.

R.L.P. Berghmans, Associate, Institute for Bioethics, Maastricht, The Netherlands.

Professor Alastair V. Campbell, Professor of Biomedical Ethics, Bioethics Research Centre, University of Otago, Dunedin, New Zealand.

Professor Max Charlesworth, Department of Philosophy, Deakin University, Victoria, Australia.

Maurice A.M. de Wachter, Ph.D., Director, Institute for Bioethics, Maastricht, The Netherlands.

G.M.W.R. de Wert, Associate, Institute for Bioethics, Maastricht, The Netherlands.

Gustavo Pis Diez, Department of Medical Humanities, Universidad Nacional de La Plata, La Plata, Argentina.

Martyn Evans, B.A., Ph.D., Fellow in Philosophy and Health Care, Centre for the Study of Philosophy and Health Care, University College Swansea, Singleton Park, Swansea, West Glamorgan, SE Wales, UK.

Professor Anne Fagot-Largeault, Department of Philosophy, Universite de Paris-X, F-92001 Nanterre, France.

David Greaves, M.B., M.A., M. Litt., Fellow in Philosophy and Health Care, Centre for the Study of Philosophy and Health Care, University College Swansea, Singleton Park, Swansea, West Glam.

Kazumasa Hoshino, M.D., D.Med.Sc., Department of Anatomy, Faculty of Medicine, Kyoto University, Kyoto, Japan.

Professor Da-Jie Jin, c/o Institute of Philosophy, The Chinese Academy of Social Sciences, Beijing, People's Republic of China.

Reidar K. Lie, M.D., Ph.D., Professor of Medical Ethics, Center for Medical Ethics, University of Oslo, Oslo, Norway.

Jose Alberto Mainetti, M.D., Ph.D., Professor of Medical Humanities, Universidad Nacional de La Plata, La Plata, Argentina.

Michael Manolakis, University of Tennessee, Knoxville, TN, U.S.A.

Derek Morgan, B.A., Senior Fellow in Health Care Law, Centre for the Study of Philosophy and Health Care, University College Swansea, Singleton Park, Swansea, West Glamorgan, SE Wales, UK.

Dr. Richard Nicholson, Editor, *Bulletin of Medical Ethics*, 31 Corsica Street, London, England.

Ma Pilar Nunez, M.D., Institut Borja de Bioetica, San Cugat de Valles, Barcelona, Spain.

Lisa Parker, Ph.D., Department of Human Genetics, University of Pittsburgh, Pittsburgh, PA, 15261, U.S.A.

Jens Erik Paulsen, Associate, Center for Medical Ethics, University of Oslo, Oslo, Norway.

Neil Pickering, B.A., M.A., M.Phil., Centre for the Study of Philosophy and Health Care, University College Swansea, Singleton Park, Swansea, West Glamorgan, SE Wales, UK.

I. Ravenschlag, Associate, Institute for Bioethics, Maastricht, The Netherlands.

Professor Ren-Zong Qiu, Director, Program in Bioethics, Institute of Philosophy, The Chinese Academy of Social Sciences, Beijing, People's Republic of China.

Hans-Martin Sass, Ph.D., Professor of Philosophy, Director, Zentrum fur medizinische Ethik, Ruhr-Universitat Bochum, F.R.G.; Senior Research Fellow, Kennedy Institute of Ethics, Georgetown University, Washington, D.C., U.S.A.

A.K. Simons-Comecher, Associate, Institute for Bioethics, Maastricht, The Netherlands.

Angeles Tan Alora, M.D., Executive Director, South East Asian Center for Bioethics, Manila, Philippines.

Juan Carlos Tealdi, M.D., Faculty of Medicine, Universidad Maimonides, Argentina.

Henk A.M.J. Ten Have, M.D., Ph.D., Professor of Medical Ethics, Catholic University of Nijmegen, Nijmegen, The Netherlands.

R.H.J. Ter Meulen, Ph.D., Institute for Bioethics, Maastricht, The Netherlands.

Hugh Upton, B.A., M.Phil., Fellow in Philosophy and Health Care, Centre for the Study of Philosophy and Health Care, University College Swansea, Singleton Park, Swansea, West Glamorgan, SE Wales, UK.

Ishwar C. Verma, F.R.C.P., F.A.M.S., F.A.A.P., Professor of Pediatrics, All India Institute of Medical Sciences, Genetics Unit, Department of Pediatrics, Old Operation Theatre Building, Ansari Nagar, New Delhi, India.

Jos V.M. Welie, M.A., M.M.S., J.D., Assistant Professor of Medical Ethics, Catholic University of Nijmegen, Nijmegen, The Netherlands.

Lissa Wettick, University of Pittsburgh, Pittsburgh, PA 15261, U.S.A.

John R. Williams, Ph.D., Director of Ethics and Legal Affairs, Canadian Medical Association, Ottawa, Ontario, Canada; formerly Principal Research Associate, Center for Bioethics, Clinical Research Institute of Montreal, Quebec, Canada.

H.A.E. Zwart, Associate, Institute for Bioethics, Maastricht, The Netherlands.

INDEX OF TOPICS

INDEX OF AUTHORS, CASES, DOCUMENTS, AND LEGISLATION DISCUSSED